The Art and Practice of Statistics

The Art and Practice of Statistics

Lisa F. Smith
*University of Otago,
Te Whare Wānanga o Otāgo*

Zandra S. Gratz
Kean University

Suzanne G. Bousquet
Kean University

WADSWORTH
CENGAGE Learning

Australia • Brazil • Japan • Korea • Mexico • Singapore • Spain • United Kingdom • United States

The Art and Practice of Statistics
Smith, Gratz, and Bousquet

Psychology Editor: Eric Evans

Development Editor: Ian Lague

Assistant Editor: Gina Kessler

Editorial Assistant: Rebecca Rosenberg

Technology Project Manager: Amy Cohen

Marketing Manager: Michelle Williams

Marketing Assistant: Melanie Cregger

Marketing Communications Manager:
Linda Yip

Project Manager, Editorial Production:
Tanya Nigh

Creative Director: Rob Hugel

Art Director: Vernon Boes

Print Buyer: Judy Inouye

Permissions Editor: Bob Kauser

Production Service: Pre-Press PMG

Text Designer: Lisa Henry

Illustrations: Pre-Press PMG

Cover Designer: Eric Handel

Cover Image: Mike Kemp, Getty Images

Compositor: Pre-Press PMG

For product information and technology assistance, contact us at
Cengage Learning Academic Resource Center, 1-800-423-0563

For permission to use material from this text or product,
submit all requests online at **cengage.com/permissions.**
Further permissions questions can be e-mailed to
permissionrequest@cengage.com.

Library of Congress Control Number: 2008921229

ISBN-13: 978-0-495-09708-2

ISBN-10: 0-495-09708-X

Wadsworth
10 Davis Drive
Belmont, CA 94002–3098
USA

Cengage Learning is a leading provider of customized learning solutions with office locations around the globe, including Singapore, the United Kingdom, Australia, Mexico, Brazil, and Japan. Locate your local office at **international.cengage.com/region.**

Cengage Learning products are represented in Canada by Nelson Education, Ltd.

For your course and learning solutions, visit **academic.cengage.com.**

Purchase any of our products at your local college store or at our preferred online store **www.ichapters.com.**

Printed in Canada
1 2 3 4 5 6 7 12 11 10 09 08

For Our Families

Brief Contents

Contents

Lisa F. Smith received her Ed.D. in 1993 in the Educational Statistics and Measurement Program of the Educational Psychology Department at Rutgers University. She is Professor of Education—Research and Development at the University of Otago on the South Island of New Zealand. Previously, she taught at Kean University with Zandra and Suzanne. Lisa delights in teaching statistics and in including her students in her research projects. She was the recipient of the 1999 Emerging Researcher Award sponsored by the New Jersey Psychological Association and has been the recipient of teaching awards in both hemispheres. Lisa's research focuses on affective factors that influence test performance, both on standardized tests and in the classroom. In addition, she conducts research on the psychology of aesthetics and is coeditor of the APA journal *Psychology of Aesthetics, Creativity, and the Arts*. She has coauthored one book; authored or coauthored over 60 articles, reviews, monographs, book chapters, grants, and conference proceedings; presented more than 50 papers at national and international conferences; and given numerous invited presentations in settings as varied as the Metropolitan Museum of Art and the Russian and Chinese Ministries of Education. Lisa is married to her colleague in statistical crime and favorite travel partner, Jeffrey K. Smith, and has one daughter, Kaitlin, who learned about probability from a young age by predicting the likelihood of getting her favorite flavor of vitamin from knowing how many of each color were in the container.

Zandra S. Gratz received her Ph.D. in educational psychology from Pennsylvania State University in 1976. After graduate school, she worked in areas related to applied research, measurement, and statistics for the Delaware Department of Public Instruction, in the Stamford Connecticut public school system, and at SUNY–Farmingdale. She joined the faculty of Kean University in 1990, where she is currently a professor of psychology. Zandra has worked as an independent evaluation consultant for over 30 years; in this capacity, her projects have ranged from family literacy programs to efforts to retrain the defense industry workforce. She has over 50 publications and presentations, most dealing with aspects of program evaluation and several related to the teaching of psychology. She is married and has three children (Adam, Pauline, and Lori). Zandra has always identified with the student who has difficulty with math. In 11th grade, a teacher told her not to take any more math classes. Following this less-than-sage advice, she managed to earn a bachelor-of-arts degree without taking any math courses. Although in graduate school she found this a regrettable decision, she developed a love for statistics. This, graduate school, and motherhood have fueled her interest and efforts to support the hesitant or fearful statistics student.

Suzanne G. Bousquet received her Ph.D. in cognitive psychology at Rutgers University in 1982. In 1984, she joined the faculty at Kean University (then Kean College of New Jersey), where she is currently professor and chairperson of the Department of Psychology. At Kean University, she teaches courses in cognition and perception, but her favorite course to teach is undergraduate statistics. Suzanne has received several awards for her teaching, including the 2006 Distinguished Teacher Award, sponsored by the New Jersey Psychological Association; the Teacher of the Year in 1997, awarded by the Alumni Association of Kean College; and the Outstanding Teacher Award in 1993, awarded by the Kean College chapter of Alpha Sigma Lambda, the National Honor Society for Part-Time Students. Suzanne's published research includes studies of visual illusions and the perception of time, and she has collaborated on projects with Lisa on the psychology of aesthetics and with Lisa and Zandra on issues of academic integrity. Suzanne is married to Richard A. Bousquet; they have one daughter, Danielle. Over the years, Suzanne has cultivated an appreciation for the many forms of spirituality and has found "mindfulness"—being present in each moment—to be relevant to the challenges of life, including the teaching of statistics to anxious students.

Foreword: To the Student

The Art and Practice of Statistics is designed to gently nurture you as you learn introductory statistics. The aim is to give you a basic understanding of the important applications of statistical concepts in the behavioral sciences, while using minimal mathematical computations. We rely on the computer to generate statistics, and we focus on statistical reasoning and the presentation of results. We offer only what are known as "definitional formulas" to encourage your conceptual understanding of the material.

In *The Art and Practice of Statistics*, you will learn both the basic statistical procedures that form the foundation of the discipline and the latest statistical methods. Rather than treat the computer as supplementary, we use it as an integral part of the content of this textbook. You will learn how to compute basic statistics and perform statistical analyses with the Statistical Package for the Social Sciences (SPSS) computer software. We provide carefully constructed sets of sample data with detailed instructions for analysis and interpretation of output. We think you will find that this approach gives you insight into statistical reasoning and theoretical principles with far greater success than you might attain by doing complex mathematical calculations.

The examples we use to illustrate statistical concepts reflect current research and practical applications within the social sciences. We emphasize the **practical** nature of this approach in the title: *The Art and Practice of Statistics*. In addition, we translate results from the examples into a professional presentation form, using the style specified by the American Psychological Association, or APA (2001). This will help you learn to read and write within the social science research community, which includes such disciplines as psychology, sociology, and education.

Although the purpose of the text is seemingly simple, mastery of the practice of statistics (even at an introductory level) can be challenging and may be considered an art. It is in this spirit that we entitled the book *The Art and Practice of Statistics*. Many students of statistics bring deep anxieties to the task. If this is true of you, that anxiety may cloud your thinking and can create obstacles to your success. By offering you the opportunity to reflect and practice, we hope that you will quiet your mind when learning what may appear to be a mysterious and somewhat unapproachable domain.

We encourage you to value practice and to cultivate patience. Our advice to you is "Do not hurry." Do not rush through a reading assignment or a homework problem. Do not hurry through the practice just to be finished quickly, with the mind-set that work in statistics is a chore. Just as one does not rush through practice when learning to play a musical instrument, learning to drive, or learning to dance the tango, one cannot rush through practice in statistics if one wants to derive its greatest benefits. In a similar way, we suggest that you try not to focus solely on obtaining **the** answer; instead, true understanding emerges as the result of the process of practice, which is actually designed to instruct you.

We also suggest that you embrace the minor failures that are sure to come. We have found that one must suffer failures through the efforts of practice before seizing on the concepts and principles of statistics. In beginning the journey toward the mastery of any art form—be it archery, flower arrangement, fencing, or statistics—we recognize that the path will be marked by occasional mistakes and frustrations. These errors are not to be seen as discouraging, but rather are to be accepted as part of the process.

We will remind you of these truths and other paradoxes along the way in the form of reflections to ponder within the text. It is our hope that *The Art and Practice of Statistics* will encourage mindfulness and calm consideration as you master this introduction to statistics.

> **Reflect**
> All journeys begin with a single step.

Organization of Each Chapter

Each chapter begins with a preview of the material presented, in order to help you "**Focus**." The overview includes a general introduction to the central topic and the major terms and concepts that will be covered within the chapter. Only what are known as definitional formulas are presented, to help you understand the particular statistical procedure or analysis. Each chapter is written so that the examples we use help to illustrate the properties of the statistic or statistics presented. It is the experiences you gain from entering sample data into the computer, following step-by-step SPSS instructions, and interpreting the results of each procedure that will promote your understanding and statistical reasoning. You will also see how the results of analyses are transcribed into APA presentation format. In addition, an "**It's Out There...**" feature presents an example from published research for each statistic. Thus, you will experience the entire process, from conceptualization of the problem to the presentation of findings and conclusions. This experience will help you develop the essential skills for critical thinking in statistics.

Within each chapter and where appropriate, we ask you to "**Reflect**" on points that are ponderous, paradoxical, or amusing. In addition, to make certain that you remain fully present in the statistical moment, we encourage you to **"Be Here Now"** from time to time in each chapter. [This concept is loosely based on the spiritual writings of Ram Dass (1971).] Each time we invoke you to "Be Here Now," we will be posing a brief question or problem that is designed to check whether you understand and can apply the concept or principle that is currently under discussion. We suggest that you treat these moments as opportunities to measure your progress in mastering statistics. At the end of each chapter, we recommend that you **"Practice."** The practice problems give you the chance to apply all of the principles and the statistical techniques that you learned in the chapter. By working with these practice problems, you will gain an even deeper understanding of the concepts, a greater feeling of comfort with the technical procedures, and an exposure to a range of applications within the social sciences. This practice should be neither dreaded nor hastily completed. To encourage you to check your work immediately, the solutions to all practice problems are presented in a final section in each chapter, entitled **"Solutions."**

Organization of the Book

The book progresses from basic ideas to more complex statistical procedures. Each chapter is written so that it can stand on its own; this approach allows you to customize your journey through the material. At the end of the book, appendices complete the statistical tool kit. All necessary statistical tables are presented in a separate appendix. Another appendix contains a complete listing of formulas for reference, and there is a glossary of terms.

> **Reflect**
> The map is not the journey.

Introduction

In This Chapter

Why Statistics?

Statistical Approaches
- Experimental Studies
- Correlational Studies
- Descriptive or Observational Studies

Qualitative and Quantitative Data

Important Concepts and Terms
- What Are Statistics?
- Using Statistics in Real Life

Variables

Scales of Measurement

Statistical Notation

FOCUS | This chapter introduces you to some basic concepts and terms that we use in statistics. You'll also learn some symbols that we use, called *statistical notation*. The information presented here lays the foundation for what will follow. Take your time and become comfortable with the material; you'll be glad you did!

Why Statistics?

Statistics help us understand life. There are many ways to understand life: reading poetry, talking to friends and loved ones, listening to great music, studying philosophy or biology, and challenging yourself. Each of these activities broadens us as human beings and helps us grow as individuals. Each has its advantages and drawbacks. Statistics provide another vehicle for understanding life. It, too, has advantages and drawbacks. In this book, we want to show you how statistics can help us understand ourselves. In the process, we will highlight the strengths and weaknesses of statistics so that you can become a sophisticated, but not cynical, consumer of statistics.

Maybe you are reading this introduction with eager anticipation. You've been waiting to take statistics and now here you are, ready to begin. If that's the case, you're not exactly like most of the students we encounter. Instead, let's say you are asking yourself why you are taking a statistics course. You're wondering who decided that you needed to learn statistics to get your degree. You're not a numbers person, you're not fond of math, and you would be happy to just escape intact at the end of the semester. Well, if that's the case, you're just like 98% of the students we've taught in our 60-plus years of teaching. Oops, we slipped in a statistic: 98%. More on that later—and we're not that old: that's 60-plus years of *combined* teaching experience.

Actually, we have statistics all around us. For example, take the 98% mentioned in the previous paragraph. You know that 98% is a lot; in fact, out of 100%, it's almost everyone. The truth is, you already have lots of experience with statistics and you know more than you may realize.

Statistics also provide a way to answer questions that are of interest to us. Life is full of questions. These questions may be very specific ("What time is dinner?"), or they may be quite broad ("What would be the consequences of changing to the metric system in the United States?"). Some such questions may take a lifetime to think about and ultimately may not be answerable in a concrete sense. Asking "What is beauty?" would fall into this category. Some questions can be answered by looking them up or by asking someone. These include questions like which state has the longest coastline or what happened on the last episode of your favorite TV show. But other questions—for example, "Will this vaccine for influenza stop the spread of the disease?" "Which type of psychological therapy will help victims cope with trauma?" and "Are people who wear hats bad drivers?"—have to be researched in a systematic fashion. To answer these questions, we need to perform empirical research and collect data.

Statistics are concerned with questions that are researchable. Research involves collecting information from a number of sources. Statistics requi that we collect data.

> **Reflect**
> Life and statistics are inseparable.
> Understanding statistics will help you understand life.

Be Here Now

Try to come up with two questions for each of the following categories:

- Personal
- Historical
- Factual
- Philosophical
- Market Research

Now decide whether you need to conduct empirical research to answer each of your questions.

Answers

The following are some possible questions:

Category	Question	Requires Research?
Personal	What is your favorite color?	No
Historical	When was the Battle of Hastings?	No
Factual	What is the capital of Maine?	No
Philosophical	What is the meaning of life?	No
Market Research	Which type of coffee bean makes the best latté?	Yes

Statistical Approaches

Researchers may study the impact of interventions under controlled conditions. For example, different people may be given different levels of medication to study the impact of the medication on pain. Or one group of students may be assigned to listen to classical music while studying for an exam, while a second group of students is assigned to study without music. In this instance, the researcher can then examine whether listening to classical music affects the amount of information recalled on the exam. These are examples of **experimental studies.**

Experimental studies are controlled studies.

Other times, we are interested in the relationship between two aspects of life. For example, is there a relationship between gender and the number of speeding tickets people get? Or is intelligence related to athletic ability? Or is your score on a college entrance test related to how well you actually do in college? These are examples of **correlational studies.**

Correlational studies examine relationships.

Experimental studies and correlational studies are the two major approaches to answering questions by using statistics. For both experimental studies and correlational studies, the data that are obtained are analyzed with the use of statistics, and conclusions are drawn on the basis of the statistical results. Of course, statistics are not limited to data about people; however, in this book we often focus on people rather than on animals or types of corn, for example.

Descriptive or observational studies describe observations on a single case or a group.

We are not limited to experimental and correlational studies. Sometimes, we conduct **descriptive** or **observational studies,** in which we describe observations on a single case or a group. This kind of study may seem simple, but don't be deceived: descriptive studies are a very important method of informing us about life, and they also help generate ideas for other research. Without a comparison group, however, we cannot predict relationships or make statements about cause and effect.

The key to understanding research is to recognize that the outcome is not information about statistics, but information about life. We can conceptualize the process as follows:

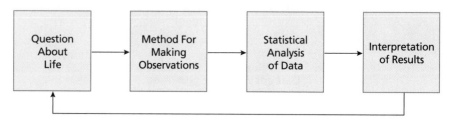

We start with a question about life that interests us. We then develop a method for observing people. We may randomly assign people to groups and treat the groups differently, or we may just ask questions and observe behavior. Once the data have been collected, we analyze the data by means of statistical techniques, and once the data have been analyzed, we are faced with the task of interpreting the results. Our interpretation uses the statistical results as the foundation or starting point, but it is not about the statistical results. It is about using the statistics to better understand some aspect of life. One consistent finding in the history of science is that the answer to one question invariably leads to another question or a series of questions—and that is good.

Reflect
How we answer questions in life is as interesting as the questions themselves.

Qualitative and Quantitative Data

Qualitative data are in-depth verbal descriptions of behaviors, attitudes, or knowledge.

When collecting information to answer questions, we can use a qualitative or a quantitative approach. **Qualitative data** take the form of narratives. Qualitative data tend to be in-depth verbal descriptions of behaviors, attitudes, or knowledge.

Such descriptions could be oral histories told by survivors of the Holocaust, interviews with participants in welfare programs, or descriptions of the social interactions of five-year-olds.

Quantitative data are numerical descriptions of behaviors, attitudes, or knowledge.

Quantitative data consist of numerical information. Quantitative data can take the form of achievement test scores, responses on attitudinal surveys, or measurements of reaction time.

This book focuses on quantitative data, and the remainder of this chapter will introduce you to the basic ideas behind statistics. We will also present some basic statistical notation.

Important Concepts and Terms

Descriptive **statistics** tells us how to organize, analyze, and interpret data.

Sample refers to the group of people observed.

Population refers to everyone you are interested in.

A **statistic** is a mathematical characteristic of a sample.

A **parameter** is a mathematical characteristic of a population.

What Are Statistics?

The term **statistics** refers to the process of collecting, organizing, analyzing, and interpreting data. We typically compute various statistics on the basis of a **sample** of individuals. That sample is part of a larger group: the **population.** The population comprises all the individuals you are ultimately interested in talking about. A **statistic** is a mathematical characteristic of a sample; a **parameter** is a mathematical characteristic of a population. It's pretty easy to remember this distinction if you realize that the words *statistic* and *sample* both begin with the letter "s" and *population* and *parameter* both begin with the letter "p."

Be Here Now

Which of the following represent a sample and which represent a population?

- All people in the United States over the age of 65
- 25 senior citizens selected to participate in a study on vitamin use
- Randomly selected sixth-grade students asked to participate in a testing program
- Everyone who took the SAT exam in 2006

Answers

- Population
- Sample
- Sample
- Population

Generally, we talk about two broad types of statistics: **descriptive statistics** and **inferential statistics**.

Descriptive statistics are used to simplify or organize data so that the data are easier to interpret. With descriptive statistics, we summarize a group of scores in a way that represents or *describes* the group of scores. For example, after you take your first exam in statistics and you get your grade, you will probably want to know what the class average is for the exam. The class average is a type of descriptive statistic. It is a single score that is used to represent all the scores on the exam.

Inferential statistics help us use data that were obtained from a sample and make generalizations about a population. For example, let's go back to everyone's grades on your first statistics exam. Let's say that we wanted to know whether your class section was different from another section of introductory statistics. Inferential statistics would help us answer that question.

This book will introduce you to the various types of descriptive statistics and will then move on to examine inferential statistics. For each type of statistic, we will explain what it means; how it is calculated with SPSS, a statistical software package; how to interpret it; and how it is used in real life.

Descriptive statistics
simplify data, making it easier for us to understand the data.

Inferential statistics
allow us to generalize from a sample to a population.

Using Statistics in Real Life

When was the last time you checked your weight? What is the hottest temperature you remember enduring in the summer? How tall are you? Have you ever had your blood pressure taken? Have you noticed that we are asking questions again?!

Weight, temperature, height, and blood pressure are all examples of numbers that we use in our everyday life.

You have used statistics if you have ever checked movie ratings or referred to a person as a "10" when talking to a friend. You have a grade point average, an interest rate on your credit card, and your best score on your favorite video game. In fact, one of the very first things that happened to you when you were born was that the doctor gave you an "Apgar rating," which is a statistical index of how healthy a newborn is.

Be Careful. Inferences are only as good as the data on which they are based. You must be careful when interpreting statistics, so that your conclusions are appropriate. For example, it would not be useful to calculate the average of all the student ID numbers in your class. Although it is possible to compute this statistic, it would be of little or no value.

Variables

A **variable** is a characteristic or idea about something that can take on different values.

When we conceptualize a characteristic or an idea about something that can take on different values, we have invented a **variable**. Shoe size is a variable, as is visual acuity, the speed of a pitch in baseball, and/or how good we think we are in mathematics. These are concepts that have specific definitions as to what they are and that can take on different values. Psychodynamic therapy is not a variable—it is an approach to working with people with psychological problems. The Democratic Party is not a variable; it is a political organization. However, type of psychological therapy is a variable, as is political party affiliation. These are concepts that are clearly defined and that can take on different values for different entities. Notice that the values do not have to be quantitative; they can be qualitatively different. Variables differ in type: some variables differ just in name, others involve a quantitative difference.

Independent variables are manipulated or controlled, either naturally or by the researcher.

An **independent variable** is a variable that is manipulated or controlled, either naturally or by the person conducting the research. A variable that changes naturally might be age. You may want to compare different age groups, and you can choose to observe different age groups, but you can't "assign" a person to be a particular age. A variable that you can change or manipulate is amount of exercise performed. You might want to look at different age groups and, within each age group, randomly assign half of the sample to perform a certain kind of exercise and the other half to being couch potatoes for purposes of your study. Independent variables describe which groups will be observed.

Dependent variables are the outcomes you measure.

A **dependent variable** is a variable that you measure to answer the question you have raised in your hypothesis. This is the outcome that you are ultimately interested in. It is what comes out of your observations and tells you about the effect of any manipulations done or

> **Reflect**
> Uncertainty is the very condition to impel man to unfold his powers.
> Erich Fromm

relationships observed with any naturally occurring independent variables. If you sorted everyone in your class into groups based on political party affiliations and then asked everyone to rate how well the president is doing, those ratings would be the dependent variable for your survey.

Be Here Now

For each of the following, name the independent and the dependent variable:

1. You randomly divide your class into three groups to see which reward (praise, a movie pass, or extra points) most effectively encourages the completion of homework.

2. You ask participants in a study to evaluate paintings by either Monet or Picasso, using a scale from 1 (low) to 10 (high).

Answers

1. Independent variable: three groups (praise, movie pass, extra points)
   Dependent variable: homework completion rates

2. Independent variable: the two types of paintings (Monet, Picasso)
   Dependent variable: the ratings given

Scales of Measurement

**Scales of measure-
ment** describe the type of
data collected.

Scales of measurement help you determine the type of variables that you are working with and, also, the type of procedures that you can perform on those data. There are four levels of measurement that progress in order: **nominal, ordinal, interval,** and **ratio.** Each of the subsequent levels has all of the properties of the levels that precede it. If you can remember the French word *noir*, meaning *black*, you'll have an acronym (*n*ominal, *o*rdinal, *i*nterval, *r*atio) for the four levels of measurement. You can use this acronym as a memory aid (or mnemonic).

The nominal level is the first level. At the nominal level, you are simply using a number as a name or label. There is no notion of more or less, or better or worse. Gender is a good example of the **nominal level of measurement.** We may label male as 1 and female as 2, or female as 1 and male as 2, or female as 12 and male as 13. This arbitrary labeling of levels of variables is possible because there is no sense of one gender being more or less than the other, only different. Really. Types of juice are another example: we can label or *code* orange juice as 1, grape juice as 2, cranberry juice as 3, and tomato juice as 4. Another term for nominal is *categorical*. Because the nominal level of measurement addresses only a difference in *kind*, data on this level are considered *qualitative* in nature.

The **nominal level of
measurement** uses a
name or label, such as flavor of ice cream.

At the **ordinal level of measurement**, labels can be applied and the data can be put in rank *order*: first, second, third, etc. Letter grades—for instance, A, B, C, D, and F—are a good example of the ordinal level of measurement. Another example is classification in the army: private, sergeant, captain, and general, among other ranks. The ordinal level of measurement builds on the nominal level of measurement to produce a sense of more or less, higher or lower. We can perform mathematical operations on ordinal measurement scales, even though we cannot claim that equal distances between numbers represent equal amounts of the variable being measured.

The **ordinal level of
measurement** uses
rank order, such as movie
ratings.

At the **interval level of measurement**, labels can be applied, the data can be put in rank order, and there are equal distances between any two units. An example of the interval scale of measurement is scores on an IQ test. The distance between a score of 100 and 120 is the same as the distance between a score of 120 and 140. Another example is Fahrenheit temperature: If it is 46 degrees outside, that is 1 degree less than 47 degrees, which is 1 degree less than 48 degrees, and so on. The amounts of heat measured between degrees are identical.

The **interval level of
measurement** involves
equal distances between
any two units.

As with the interval level of measurement, at the **ratio level of measurement** labels can be applied, the data can be put into rank order, and there are equal distances between any two units. But ratio measures have an additional characteristic: an absolute zero, or absolutely nothing. This means that we can make ratio comparisons. (Often, ratio scales involve measuring the attributes of concrete variables, such as the height of a bookcase.) Income is a good example of the ratio level of measurement: a person who earns $40,000 per year makes twice as much as someone who earns $20,000 per year and four times

The **ratio level of
measurement** has an
absolute zero, and com-
parisons of ratios can be
made.

Scale or Level	Description	Example
nominal	names, labels, codes	academic major
ordinal	rank order	class rank
interval	equal distances between any two units	SAT scores
ratio	can have absolute zero and can make ratio comparisons	length of a term paper (zero if you don't write it!)

FIGURE 1 Scales of measurement.

as much as someone who earns \$10,000 per year, and it is also possible to have zero earnings in a year. Another example of the ratio level is the number of children a person has. By contrast, it is not possible to have an IQ score of zero, nor is someone with an IQ of 150 twice as intelligent as someone with an IQ of 75. That is why IQ is an interval-level variable and not a ratio-level variable.

Figure 1 summarizes the scales of measurement.

Be Here Now

Which scale of measurement would each of the following be?

- 32 degrees Fahrenheit
- favorite color
- class rank
- amount of credit card debt

If you answered interval, nominal, ordinal, ratio, then you are here with us now!

Statistical Notation

Statistical notation refers to symbols that help you label your variables or that tell you what to do with your data.

Statistics use symbols that are known as **statistical notation.** A way to think about statistical notation is as a type of shorthand that tells you what to do with your data. It is used for formulas or simply to denote variables.

Statistical notation for individual variables uses letters. If we have only one variable, we use X, and if we have two variables, we use X and Y. Of course, if we know the name of the variable, we use the name. So, X may be your age and Y may be a measure of how much you like text messaging, on a scale from 1 to 10. We aren't going to get more complicated than that in this book, but you can imagine that more variables would require more letters.

When you have data about a sample (meaning only those people in your study) and you want to denote the total number in the sample, you use a lowercase n. When you have data about a population (meaning everyone you are ultimately interested in), you use a capital N. So, if you take a survey of 100 college seniors in your school to see whether they have ever been to an art museum, you would have $n = 100$ in your sample. If you wanted to use those results to generalize (use in a wider sense) to all college seniors in your state, you might have something like $N = 13{,}265$.

Summation notation, or sigma, is the symbol Σ, which tells you to add everything that follows it.

The most fundamental of statistical notations is called **summation notation.** It has a very simple definition: "add all of these things up." The symbol for summation notation is the Greek capital letter sigma, which looks like this: Σ. Here's what it looks like in a simple expression: ΣX.

ΣX means "Take all of the values of the variable X, and add them all up." If the variable X consisted of the numbers 3, 5, 2, 7, and 9, then

$$\Sigma X = 3 + 5 + 2 + 7 + 9 = 26$$

That was pretty simple, right? Let's make it just a little more elaborate:

$$\Sigma(X - 1)$$

This means "Take each value in the variable X, subtract 1 from it, and then add up all of the resulting numbers." So we would have

$$(3 - 1) + (5 - 1) + (2 - 1) + (7 - 1) + (9 - 1) = 21; \text{ therefore,}$$
$$\Sigma(X - 1) = 21$$

This is probably a good time to learn and remember the mnemonic **P**lease **e**xcuse **m**y **d**ear **A**unt **S**ally. "Please excuse my dear Aunt Sally" tells us the correct order of mathematical operations to follow: **p**arentheses, **e**xponents, **m**ultiplication, **d**ivision, **a**ddition, and **s**ubtraction. So, with $\Sigma(X - 1)$, we have to do what is in the parentheses before we can do the addition. That is why we subtract 1 from each number before summing.

Now, here's a trick question: Using our original values for X, what is $\Sigma X - 1$? If you said 25, you are right. Remember, there are no parentheses, and according to "Please excuse my dear Aunt Sally" addition comes before subtraction. So, here is what you would do:

$$\Sigma X - 1 = 3 + 5 + 2 + 7 + 9 = 26, \text{ and then } 26 - 1 = 25$$

It's important to pay attention to what you are being asked to do. As you can see from the two examples, $\Sigma X - 1$ and $\Sigma(X - 1)$ are not the same and do not give the same answers for any given set of numbers! This situation also applies to addition.

Use the following values of X to calculate ΣX, $\Sigma X + 2$, and $\Sigma(X + 2)$: 4, 8, 2, 10

Answers

$$\Sigma X = 4 + 8 + 2 + 10 = 24$$
$$\Sigma X + 2 = 24 + 2 = 26$$
$$\Sigma(X + 2) = (4 + 2) + (8 + 2) + (2 + 2) + (10 + 2) = 32$$

How did you do? Remember, $\Sigma X + 2$ does not equal $\Sigma(X + 2)$

Let's look at another example with statistical notation. For the X values that follow, we want to calculate ΣX^2. It helps to first make a column for X^2 and then multiply each value for X by itself:

X	X²
3	9
1	1
4	16

Now you can add the values in the **X²** column to obtain the answer:

$$\Sigma X^2 = 9 + 1 + 16 = 26$$

By now you can probably guess that ΣX^2 is not equal to $(\Sigma X)^2$. To solve $(\Sigma X)^2$, because you have parentheses, you must first add up all of the X values and then square the resulting number. So,

$$(\Sigma X)^2 = 3 + 1 + 4 = 8 \text{ and } (8)(8) = 64$$

There are a number of mathematical concepts that are commonly used in statistics; each has its own notation. We'll cover these concepts in the chapters ahead.

SUMMARY

This chapter introduced statistical approaches and concepts, statistical terms, and statistical notation, all of which will support your journey through statistics. To assist you on this journey, take some time to work on the practice problems that follow. This will help you become both comfortable and competent with the concepts and terms.

PRACTICE

Concepts and Definitions

1. Identify whether each of the following requires an experimental study or a correlational study:

   a. relationship between age and dexterity in playing video games

    b. determining whether vitamin C can prevent colds

    c. studying whether type of music played affects level of relaxation before a statistics exam

2. Identify each of the following as a descriptive statistic or an inferential statistic:

    a. average height on a college basketball team

    b. flexibility of the members of a swimming team versus flexibility of the members of a gymnastics team

    c. range of SAT math scores in your class

3. You divide a group of people according to age groups and count how many sit-ups each age group can do in 1 minute. Identify the independent variable and the dependent variable.

4. One group of babies listens to classical music in utero during their mothers' pregnancies. Then, when they are 1 month old, they listen to classical music and rap. The amount of time they look in the direction of the classical music, compared with the amount of time they look in the direction of the rap music, is measured. Identify the independent variable and the dependent variable.

5. Identify the scale of measurement that applies to each of the following:

    a. number of keys you carry

    b. academic major

    c. values on a number line

    d. responses to a survey asking how much people like broccoli, on a scale ranging from strongly disagree to strongly agree

    e. favorite type of vacation

    f. running time in the 100-yard dash

Statistical Notation

For problems 6–12, use the following data set:

X	Y
8	5
6	10
2	3
1	6

6. What is ΣX?

7. What is $\Sigma(Y + 4)$?

8. What is $\Sigma(Y - 3)$?

9. What are $\Sigma(X + 3)$ and $\Sigma X + 3$?

10. What are ΣY^2 and $(\Sigma Y)^2$?

11. What is $\Sigma X^2 + (\Sigma X + 1)$?

SOLUTIONS

Concepts and Definitions

1. **a.** Correlational Study
    b. Experimental Study
    c. Experimental Study

2. **a.** Descriptive Statistic
    b. Inferential Statistic
    c. Descriptive Statistic

3. Independent variable: age
    Dependent variable: number of sit-ups

4. Independent variable: type of music
    Dependent variable: time spent gazing

5. **a.** Ratio
    b. Nominal
    c. Interval
    d. Ordinal
    e. Nominal
    f. Ratio

Statistical Notation

To do problems 6–12, it would be helpful to add some columns to the data set and then carefully complete the columns:

X	X + 3	X²	Y	Y²	Y + 4	Y − 3
8	11	64	5	25	9	2
6	9	36	10	100	14	7
2	5	4	3	9	7	0
1	4	1	6	36	10	3

6. $\Sigma X = 8 + 6 + 2 + 1 = 17$

7. $\Sigma Y + 4 = 9 + 14 + 7 + 10 = 40$

8. $\Sigma Y - 3 = 2 + 7 + 0 + 3 = 12$

9. $\Sigma(X + 3) = 11 + 9 + 5 + 4 = 29$
    $\Sigma X + 3 = 17 + 3 = 20$

10. $\Sigma Y^2 = 25 + 100 + 9 + 36 = 170$
    $(\Sigma Y)^2 = 24^2 = 576$

11. $\Sigma X^2 + (\Sigma X + 1) = 105 + (17 + 1) = 105 + 18 = 123$

-2-

Frequency Distributions

In This Chapter

Organizing Data

Application and Interpretation: Creating Frequency Distribution Tables
- Computer Creation of Frequency Distribution Tables
- Frequency Distribution Tables for Grouped Data
- Presenting Frequency Distribution Tables in APA Style

Application and Interpretation: Creating Frequency Distribution Figures
- Histograms and Frequency Polygons
- Bar Graphs and Pie Charts
- Stem-and-Leaf Diagrams and Box Plots

Interpretation: Shapes of Distributions
- Skewness
- Kurtosis

Presenting Figures in APA Style

FOCUS

Frequency distributions summarize and organize data.

Simple distributions present all raw scores

One way of summarizing a set of data into an objective, organized form is by using a **frequency distribution**. We must choose the best organizational scheme, depending upon the complexity of the data and the range of numbers that are included in the data set.

   Simple distributions summarize sets of data that require little additional organization; typically, the data span a relatively narrow range of values or categories, and all raw data are shown.

Grouped frequency distributions summarize sets of data that are distributed across a relatively broad range of values. For example, we would use a grouped frequency distribution to organize a set of SAT or IQ scores.

We can choose to display frequency distributions in the form of a table or a graph. If we present a frequency distribution in graphic form, the **scale of measurement** (nominal, ordinal, interval, or ratio; see Chapter 1) will help determine the choice of graphic.

The graphic forms of frequency distributions described in this chapter include **histograms, frequency polygons, bar graphs,** and **pie charts.**

Grouped frequency distributions summarize across a wide range of raw scores.

Organizing Data

Imagine that your professor asks all of the students in your class about their anxiety related to taking this statistics course. Imagine that all of you are asked to use a number from 1 to 10 to represent your personal level of anxiety, where 1 means that you experience very little anxiety and 10 means that you are extremely fearful. Everyone in the class is asked to report his or her anxiety level out loud in turn. At the end of the exercise, do you think that you could remember everyone's anxiety level? Probably not. Instead of relying on memory regarding a set of data, in statistics we can use **frequency distributions** to summarize the data.

Frequency distributions are used to objectively summarize a set of data. To facilitate the computation of certain statistics, we organize the data from the highest possible value to the lowest possible value (in other words, in descending order), and we recommend that you do the same. Then we count (or let SPSS do the counting for us) to determine the frequency of the scores—or how many scores occur—for each value.

For example, consider the following set of anxiety data from a class:

TABLE 1
Anxiety Scores

Anxiety Level	f
10	1
9	2
8	2
7	5
6	4
5	3
4	0
3	3
2	2
1	2

NOTE: 1 = Very Little Anxiety to 10 = Extremely Fearful.

Anxiety Scores

7	10	2	3	7	3	6	6	9	1	7	5
9	5	7	8	6	1	7	8	2	6	5	3

Because there are quite a few scores here ($n = 24$), it is a challenge to get any sort of organized impression about how anxiety is distributed in this class. We need to reorganize the data into a more usable form that allows for a clear and immediate impression about the class responses.

Application and Interpretation: Creating Frequency Distribution Tables

Let's organize the set of anxiety scores into a **frequency distribution table**. We have done this in Table 1.

The first column of Table 1 presents the possible anxiety scores. We use **X** to symbolize anxiety level, because **X** can stand for any variable name. However, if we were formatting the table to appear in APA style or for presentation, we would use a brief variable name, and not the generic "**X**," to make the interpretation easier for a reader.

The symbol *f* signifies **frequency** in a frequency distribution table.

In the second column, we use lowercase *f* to signify **frequency**. Frequency is the number of students who responded with a particular value of anxiety level. We do not need to spell out the word *frequency*, because *f* is universally used to symbolize frequency, but we can choose to spell the word out if we wish.

All of the values in the frequency column should tally to the original number of scores ($n = 24$ in our example); therefore, the sum of the frequencies is equal to the number of scores in our data set. In statistical notation,

$$\Sigma f = n$$

It is always advisable to check to make sure that all of the scores in the data set are represented in the frequency distribution table. Checking is relatively simple if you calculate Σf.

Now that the scores are organized into a frequency distribution table, it is easy to see precisely how anxiety levels are distributed in the class. There are more students reporting higher levels of anxiety (ratings over 5) than lower levels of anxiety. It is important to point out that a frequency of 0 is an important entry in the frequency distribution. It is just as meaningful to find that no one in the group reported an anxiety level of 4 as it is to say that only one person identified 10 as his anxiety level.

> **Reflect**
>
> *No one (zero frequency) is as important as someone.*

Frequency distribution tables can also be used for variables that are nominal (represented as categories). For example, if we were interested in summarizing the number of people using various types of antidepressant medications, we could use a frequency distribution table in which the types of medications are listed alphabetically, or by manufacturer, or even arbitrarily.

Consider the following example: You survey your classmates as to their political party affiliations. In your class, 8 report that they are Republicans, 10 say they favor the Democratic Party, and 6 classmates indicate that they are independents. As shown in Table 2, you can create a frequency distribution to share this information.

In addition to listing frequencies in frequency distribution tables, we can show **proportions (*p*)** or we can show **percentages. Proportion** refers to how many of the total *n* share the same value or category; proportion is

TABLE 2
Political Party Membership

Political Party	*f*
Democrat	10
Independent	6
Republican	8

Proportion (*p*) is expressed as a decimal. It represents the portion in the sample that shares the same value.

TABLE 3
Anxiety Scores

Anxiety Score	f	p
10	1	.04
9	2	.08
8	2	.08
7	5	.21
6	4	.17
5	3	.13
4	0	.00
3	3	.13
2	2	.08
1	2	.08

NOTE: 1 = Very Little Anxiety to 10 = Extremely Fearful.

Percentage (P) is a proportion multiplied by 100.

Proportions sum to 1. **Percentages** sum to 100.

expressed as a decimal that ranges from 0.00 to 1.00. Proportions are symbolized by a lowercase p. To calculate a proportion for any given value, you need to know the total number in the sample and the frequency of responses (f) for that value. In other words, we use f and n to calculate proportions. The formula looks like this:

$$p = f/n \quad \text{or} \quad p = f/\Sigma f$$

Let's calculate proportions to add to our frequency distribution of anxiety scores. This is done in Table 3.

Remember, the highest value for a proportion is 1.00. Therefore, the larger the number, the greater is the proportion of the class sharing that specific anxiety level. The smaller the number (the lowest value for a proportion is 0.00), the smaller is the proportion of the class sharing that specific level of anxiety. Also, it's important to point out that if we add all of the proportions, the sum should be 1.00. (You may be off a bit because of rounding.)

$$\Sigma p = 1.00$$

By converting raw frequencies into proportions, we create a universal scale that allows proportions to be compared across different sample and population sizes. No matter how small or how large n is, we know that a proportion close to 0.00 reflects a small minority and a proportion close to 1.00 represents a vast majority.

In some circumstances, we also use **percentages**. In this case, we report the percent that shares a value or category of membership. Percentages are simple to calculate: just multiply each proportion by 100. Percentages are symbolized by a capital P. To calculate a percentage, we use the following formula:

$$P = p(100)$$

Let's again consider the class anxiety frequency distribution, this time including percentages. We present the data in Table 4.

TABLE 4
Anxiety Scores

Anxiety Level	f	p	P
10	1	.04	4
9	2	.08	8
8	2	.08	8
7	5	.21	21
6	4	.17	17
5	3	.13	13
4	0	.00	0
3	3	.13	13
2	2	.08	8
1	2	.08	8

NOTE: 1 = Very Little Anxiety to 10 = Extremely Fearful.

The sum of the percentages should be 100% (or close to it). So, to check our calculations, we can do the following formula:

$$\Sigma P = 100$$

Cumulative Percent (C%) is the percent of the sample at or below a given value of X.

We can add one more column to our table: **cumulative percent (C%).**

To compute the cumulative percent, begin in the last row of the frequency table. The cumulative percent for an anxiety score of 1 or less was 8 percent. Next, move up one row. What percent of the class had an anxiety score of 2 or less? As you look at Table 4, you must add the percent of the class that chose 1 (8 percent) to the percent that chose 2 (8 percent): 8 + 8 = 16 percent.

Next, move up one row. What percent of the class chose a 3 or less? You must add the percent that chose 1 (8 %), to the percent that chose 2 (8 %), to the percent that chose 3 (13 %): 8 + 8 + 13 = 29%.

Continue up the table until you reach the top. The cumulative percent for the highest score should be 100; however, depending on how you round the numbers, the actual value of C% for the highest score may range from 98 to 102. We have added a column for cumulative percent for our anxiety data and presented it in Table 5.

As you look at Table 5, notice that the cumulative percent for an anxiety level of 4 is 29. Even though no one used a 4 to describe his or her anxiety, you must consider those who used a 4 *or less.*

Computer Creation of Frequency Distribution Tables

Although creating a frequency distribution table is relatively easy when the data set is small, we can still use SPSS to count the frequencies for us. This is particularly helpful when we have a large set of data.

 **Begin by creating an SPSS file with the data.**

- Click on the "Variable View" tab, define the variable "Anxiety."
- Then, click on the "Data View" tab, and enter the data.

TABLE 5
Anxiety Scores

Anxiety Level	f	p	P	C%
10	1	.04	4	100
9	2	.08	8	96
8	2	.08	8	88
7	5	.21	21	80
6	4	.17	17	59
5	3	.13	13	42
4	0	.00	0	29
3	3	.13	13	29
2	2	.08	8	16
1	2	.08	8	8

NOTE: 1 = Very Little Anxiety to 10 = Extremely Fearful.

Be Here Now

The nutritionist at the university's cafeteria asked a sample of students how many vitamins and herbal supplements they took weekly. He collected the following counts from the student sample:

7, 4, 2, 0, 1, 0, 2, 2, 7, 6, 3, 1, 0, 0, 1

Create a frequency distribution table including columns for raw frequencies, proportion, percentages, and cumulative percentages.

Answer

TABLE 6
Weekly Regimen of Vitamins and Supplements

Intake	f	p	P	C%
7	2	.13	13	101
6	1	.07	7	88
5	0	0	0	81
4	1	.07	7	81
3	1	.07	7	74
2	3	.20	20	67
1	3	.20	20	47
0	4	.27	27	27

- To create a frequency distribution table, you need to use

  Analyze → Descriptive Statistics → Frequencies.

- SPSS will create frequency tables for each of the variables that have been moved from the list of variables on the left to the box labeled "Variable(s)."

Figure 1 presents the SPSS output of results for the anxiety scores presented in Table 1.

In the first section of the output, SPSS presents basic information: **Anxiety** was the name of the variable that we analyzed.

N	**Valid = 24**	We analyzed 24 records.
	Missing = 0	There were no missing data points.

In the second section of the output, the frequency distribution table appears.

The first column in the frequency distribution table presents the anxiety scores that were analyzed. Note that the scores *are not* in the descending order we suggest you use. Note also that the anxiety level of 4 is missing, because no one in the class selected that value.

Frequencies

Statistics

Anxiety

N	Valid	24
	Missing	0

Anxiety

Valid	Frequency	Percent	Valid Percent	Cumulative Percent
1.00	2	8.3	8.3	8.3
2.00	2	8.3	8.3	16.7
3.00	3	12.5	12.5	29.2
5.00	3	12.5	12.5	41.7
6.00	4	16.7	16.7	58.4
7.00	5	20.8	20.8	79.2
8.00	2	8.3	8.3	87.5
9.00	2	8.3	8.3	95.8
10.00	1	4.2	4.2	100.0
Total	24	100.0	100.0	

FIGURE 1 SPSS output: Frequency distribution table for the variable *anxiety level*.

Across the top row of the frequency distribution table are labels that define the statistics calculated by SPSS:

Frequency	The second column reports the frequencies: the number of students who identified each score as representing their anxiety level.
Percent	The third column reports the percent of the class that selected each score.
Valid Percent	The fourth column reports the valid percent. Sometimes a respondent fails to provide a response to a question or item. Valid percent is based on the number of respondents who actually answer the question.
	When there are no missing data, Percent = Valid Percent.
Cumulative Percent	The fifth column presents the percent of respondents who fall at or below each score.

SPSS does a fine job tallying the anxiety levels of the class, organizing the values in ascending order, and then presenting frequencies and percentages. Even though it may be tempting, the frequency distribution table created in this case by SPSS is not something that you should cut and paste and use "as is" in another document. There are several reasons for this. First, the SPSS table is not in APA format. Second, the anxiety level "4" is not listed in the SPSS frequency table, because there were no values of "4" in the data set that was entered. Yet, as we noted previously, the fact that no one selected "4" to represent anxiety is important and needs to be included in the table.

Frequency Distribution Tables for Grouped Data

Grouped frequency distribution tables are used to organize data with large ranges of data. The data are grouped into smaller ranges, or intervals.

Intervals are ranges of values of equal size; taken together, intervals include all of the data in the data set.

Occasionally, we encounter data that span a large range of values. IQ scores, SAT scores, college credits earned, and annual income are all examples of data that may be suited for **grouped frequency distributions**. Whenever there is a need to condense a large range of possible values into a more manageable form, we group the data into **intervals** to make the data easier to comprehend. Intervals are ranges of values. Each interval range is the same size, or width. All of the data in the data set must be accounted for within the ranges used. Notice that, in this context, intervals are ranges of values and do not imply a particular scale of measurement. Let's begin with an example.

In trying to create a profile of students in a statistics class, the professor asks each student to report the number of college credits earned thus far. Each student in the class identifies the total number of college credits that he or she earned out of a possible 120 credits required for graduation. The data are as follows:

College Credits Earned

68	100	72	30	37	80	36	16	94	51	27	65
91	49	17	58	76	61	47	58	22	85	35	3

Although in many instances the SPSS output provides a good approximation to an ideal presentation of the data, in this case the SPSS output is not as helpful. However, we can use the output from SPSS to help us create the grouped frequency distribution table.

➡ **Begin by creating an SPSS file with the data.**

- To create a frequency distribution table, you need to use

    Analyze ➙ *Descriptive Statistics* ➙ *Frequencies.*

- SPSS will create frequency tables for each of the variables that you move to the box labeled "Variable(s)."

Figure 2 presents the SPSS output. In the first section of the output, **Credits** is the variable analyzed.

Frequencies

Statistics

Credits

N	Valid	24
	Missing	0

Credits

Valid	Frequency	Percent	Valid Percent	Cumulative Percent
3.00	1	4.2	4.2	4.2
16.00	1	4.2	4.2	8.3
17.00	1	4.2	4.2	12.5
22.00	1	4.2	4.2	16.7
27.00	1	4.2	4.2	20.8
30.00	1	4.2	4.2	25.0
35.00	1	4.2	4.2	29.2
36.00	1	4.2	4.2	33.3
37.00	1	4.2	4.2	37.5
47.00	1	4.2	4.2	41.7
49.00	1	4.2	4.2	45.8
51.00	1	4.2	4.2	50.0
58.00	2	8.3	8.3	58.3
61.00	1	4.2	4.2	62.5
65.00	1	4.2	4.2	66.7
68.00	1	4.2	4.2	70.8
72.00	1	4.2	4.2	75.0
76.00	1	4.2	4.2	79.2
80.00	1	4.2	4.2	83.3
85.00	1	4.2	4.2	87.5
91.00	1	4.2	4.2	91.7
94.00	1	4.2	4.2	95.8
100.00	1	4.2	4.2	100.0
Total	24	100.0	100.0	

FIGURE 2 SPSS output: Frequency distribution table for college credits.

N   **Valid = 24**   There were twenty four records analyzed.

Missing = 0   There were no missing data points.

In the second section, the frequency distribution table appears.

Across the top row of the frequency distribution table are the labels that define the statistics calculated in SPSS:

Frequency   The second column reports the frequencies: for each number of credits, the number of students who earned that number of credits.

Percent   The third column reports, for each number of credits, the percent of the class which earned that number of credits.

Valid Percent   The fourth column reports the valid percent. Sometimes a respondent fails to answer a question or leaves an item blank. Valid percent is based on the number of respondents who actually answer an item.

When there are no missing data, Percent = Valid Percent.

Cumulative Percent   The fifth column presents, for each number of credits, the percent of respondents who earned that number of credits or less.

TABLE 7

College Credits Earned

Credits Earned	f
0–29	5
30–59	9
60–89	7
90–120	3

SPSS can help us create a better frequency distribution table, because it provides a complete accounting of the data in ascending order; still, the output does not come close to an optimal organizational scheme. Before we begin to impose additional order, we need to think carefully about the best way to group the data. For example, do we wish to group the data the way the university registrar might group students, by using groupings to reflect the general class level (freshman, sophomore, junior, and senior)? Table 7 indicates how the university registrar might organize the data.

As before, we can double-check our accuracy by verifying that $\Sigma X = 24$.

How did we get a frequency of 9 in the "sophomore" category of 30–59 credits earned? In using SPSS to assist us in the tallying, we had to take care to notice that one score, 58, had a frequency of 2.

Notice that the intervals of credits earned do not overlap. That is, we *do not* specify 0–30, 30–60, 60–90, and 90–120, because if we did, we would incorrectly represent any scores falling on a boundary. Then, in our example, the person who had earned 30 college credits would be counted twice because 30 credits would fall into two categories: 0–30 *and* 30–60. Of course, counting an individual twice would be incorrect.

Although using the organizational scheme of the university registrar makes intuitive sense, we could group the data differently. We could group the data into intervals of 10, 25, 50, or whatever makes sense, given the variables in the data set. It is important to note that smaller intervals provide more detailed information about a data set, whereas the use of larger intervals

TABLE 8
College Credits
Earned, Interval of 10

Credits Earned	f
100–109	1
90–99	2
80–89	2
70–79	2
60–69	3
50–59	3
40–49	2
30–39	4
20–29	2
10–19	2
0–9	1

results in a loss of detail. However, larger intervals make the distribution easier to visualize when you have a large range of data.

When the data do not have any meaningful divisions, such as grade level, you can determine the best width for an interval by subtracting the lowest score from the highest score and dividing by 10:

$$\frac{\text{High Score } - \text{ Low Score}}{10}$$

Next, round the result of this computation to a number that people would find easy to count by (for example, 2, 5, or 10).

Following this guideline with our example of college credit data (see Figure 2), we would take the high score of 100 minus the low score of 3 and divide the result by 10. This will result in a frequency distribution table having approximately 10 rows.

$$\frac{100 - 3}{10} = \frac{97}{10} = 9.7$$

Our result, 9.7, rounded to the nearest easily countable number, would yield an interval size of 10, as shown in Table 8.

Notice that the first number in each interval is a multiple of the interval width. This feature makes the groupings easier to interpret.

Grouped frequency distributions are also useful in situations in which the data span a large range of decimal values. As another example, consider cumulative grade point averages (GPAs). They do not span a large overall range (as 0–120 credits earned do), but instead span a large range of decimal values (0.00–4.00). In this case, we would not want to list every possible value for the cumulative GPA (4.00, 3.99, 3.98. 3.97, 3.96, etc.), but instead would organize the data into a grouped frequency distribution.

Be Here Now

A cognitive psychologist was interested in the time it takes a sample of undergraduate students to complete a challenging Sudoku puzzle (a number puzzle that taps logic and reasoning abilities). Twenty students were asked to log the amount of time it took them to solve the puzzle; the data are reported as total minutes:

128, 99, 213, 117, 83, 201, 79, 156, 88, 91,
23, 65, 141, 43, 52, 81, 126, 113, 166, 142.

Create a grouped frequency distribution table for this example.

Answer

First, we choose an interval width. The highest number of minutes taken to solve the puzzle was 213, the lowest 23. Applying the formula

$$\frac{\text{High Score } - \text{ Low Score}}{10}$$

we calculate $(213 - 23)/10 = 190/10 = 19$. We will round to an interval of 20 minutes. Table 9 presents the grouped frequency distribution.

(Continued)

TABLE 9
Puzzle Completion Time

Minutes	f
200–219	2
180–199	0
160–179	1
140–159	3
120–139	2
100–119	2
80–99	5
60–79	2
40–59	2
20–39	1
0–19	0

Presenting Frequency Distribution Tables in APA Style

The *Publication Manual of the American Psychological Association* (2001) presents detailed instructions for presenting data in a table. Notice in each case that the table heading contains the table number and, on a separate line, a brief, but clear, explanatory title is provided in italics. Standard abbreviations and symbols may be used (such as f, p, and P or %), but we could also choose to spell out the words "frequency," "proportion," and "percent." Table headings (such as "Credits Earned" and "Anxiety Level") should be brief, but informative. Headings identify the data that are presented beneath them within the body of the table.

In any APA document, we refer to the table by its number and not its title. For example, if we were referring to Tables 7 and 8, we might write the following:

Table 8 presents the total number of credits earned by members in the class.

Or we might write

The total number of credits earned by members in the class was grouped according to year of undergraduate study (see Table 7).

When referring a reader to a table, we do not repeat everything that is presented in the table. Rather, we highlight the important points of the data contained in the table.

Shin and Koh (2007) examined teachers' beliefs relative to classroom management. They described their sample in their article. As shown in Table 10, most teachers surveyed in both the United States and Korea had more than 16 years of classroom experience.

TABLE 10
Teaching Experience of Teachers Surveyed in the United States and Korea

Years Teaching	United States		Korea	
	N	*P*	*N*	*P*
1 to 6 years	24	22.0	38	23.0
7 to 15 years	25	23.0	69	41.5
16 or more years	60	55.0	59	35.5

Application and Interpretation: Creating Frequency Distribution Figures

In some situations, a graphic representation of a frequency distribution is preferred over the use of a table. In APA format, graphic representations of data are called *figures*. We would not present the same data in both a table and a figure; as with tables, figures should enhance—not duplicate—the presentation of data.

There are several ways in which we can create figures to present frequency distribution data; choices include **histograms, frequency polygons, bar graphs,** and **pie charts.** Deciding which figure to use is a function of both the type of data represented (specifically, we are referring to the scale of measurement; see Chapter 1) and the most informative format for presenting the data.

Histograms and Frequency Polygons

When we have data that are continuous (data capable of representation on the ratio, interval, or ordinal scale of measurement), we have the option of using a histogram or a frequency polygon. In contrast, nominal scales are categorical in nature. The levels for any variable on a nominal scale are qualitatively different (for example, male or female, Denver Broncos fan or Chicago Bears fan or New York Giants fan). To present nominal data, we would select a bar graph or pie chart. The bars on a bar graph do not touch, so they represent the categorical (or qualitatively different) nature of the data. If we have percentages or proportions to report, the data can be presented in a pie chart.

Let's consider again class anxiety level, but let's utilize a frequency distribution different from what we considered previously in the chapter. This new set of class data is presented in Table 11.

TABLE 11
Anxiety Scores

Anxiety Level	*f*
10	2
9	2
8	3
7	0
6	4
5	3
4	5
3	2
2	2
1	1

NOTE: 1 = Very Little Anxiety to 10 = Extremely Fearful.

A **histogram** is a graph that uses bars to represent frequencies; the bars touch.

In a **histogram**, each possible anxiety level value is presented on the x-axis (the horizontal axis, or abscissa). The frequencies are represented on the y-axis (the vertical axis, or ordinate). Bars are used to represent the various frequencies, with the height of each bar reflecting the frequency for the value represented by the bar. The taller the bar, the higher is the frequency that is represented. The bars touch each other because the data are continuous. Also, the bars are drawn centered over the values on the x-axis. Notice in Figure 3 that both the x-axis and y-axis are labeled with the variable names: "Anxiety" and "Frequency."

Notice in Figure 3 that the highest frequency coincides with the anxiety level 4. The histogram gives the reader an immediate impression of the distribution of anxiety scores across the entire range of possible values.

A **frequency polygon** is a graph that uses data points to represent frequencies; the points are connected with lines.

An alternative to the histogram is the **frequency polygon**. The frequency polygon organizes the x- and y-axes identically to the way the histogram organizes them, but instead of bars whose heights denote frequencies, data points are used. Because what we have is a frequency **polygon**, we connect the data points to each other to "enclose" the shape of the distribution. In addition, the end values are connected to the x-axis. In our example (see Figure 4), these end values extending down to the x-axis indicate that there are no values for an anxiety level of –1 (too bad!) and no values for an anxiety level of 11 (thank goodness!).

When you connect the data points in a frequency polygon, you get the same general visual impression of the distribution of data as you do in a histogram. Comparing Figures 3 and 4, notice that the zero frequency for the anxiety level of 7 is apparent in both. In the histogram, there is no bar centered over the anxiety level 7; in the frequency polygon, the connections

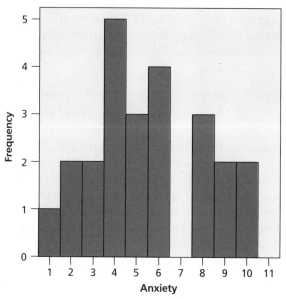

FIGURE 3 Histogram for class Anxiety scores.

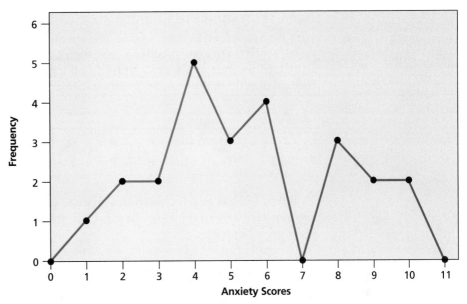

FIGURE 4 Frequency polygon for the variable, class anxiety level.

between data points dip down to 0 on the *x*-axis at the value 7, making it clear that no one in the class selected the anxiety level 7.

In this way, the general shape of the frequency polygon is similar to the general shape of the histogram (although the polygon looks like jagged jack-o'-lantern teeth and the histogram looks like a city skyline). This similarity is illustrated in Figure 5.

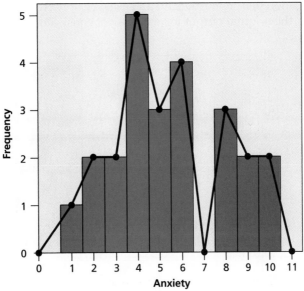

FIGURE 5 Class anxiety: Frequency polygon superimposed on a histogram.

Bar Graphs and Pie Charts

A **bar graph** is a graph that uses separated bars to represent frequencies.

Bar graphs are used to organize and present data that are categorical. The bars do not touch each other. In other words, each category is qualitatively distinct.

Consider data and a bar graph (called a **bar chart** in SPSS) for a hypothetical distribution of gender in our statistics class. The data are as follows:

Gender

M	F	F	F	F	F	M	M	F	F	F	F
F	F	F	M	F	F	F	F	M	F	F	F

From the bar graph shown in Figure 6, you can quickly see that there are more females in the class. In this case, the use of the bar graph creates great visual impact.

A **pie chart** is a graphic representation of the proportion (or percentage) of each category in a set of data.

Instead of using a bar graph, we could choose to use a **pie chart** to represent the gender distribution in the class. Percentages or proportions (not raw frequencies) are the measures reported in pie charts. Like bar graphs, pie charts have great visual impact and are commonly used in newspapers and news magazines because they are easy to interpret.

The pie chart for the gender distribution in our class is depicted in Figure 7. Notice that the whole "pie" is divided into sections (or slices) based upon the percentage or proportion of males and females in the class. From the pie chart, we can easily appreciate the fact that females account for more than 75% of the class.

SPSS can assist us in creating some types of figures. In a few cases, SPSS can create perfect figures; in other circumstances, SPSS produces figures that can be used as drafts to help us envision the data and adapt them to more usable formats. In all cases, figures created in SPSS, unmodified from the original output format, are not consistent with APA style.

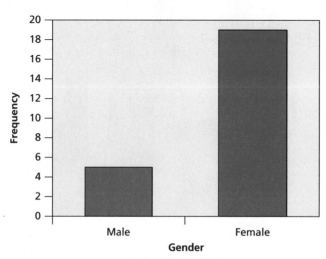

FIGURE 6 Bar graph distribution of gender.

Frequency

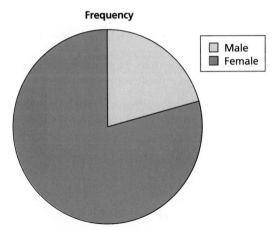

FIGURE 7 Pie chart distribution of gender.

Stem-and-Leaf Diagrams and Box Plots

Stem-and-leaf diagrams (or stem-and-leaf plots) and box plots are methods used to explore data graphically. Stem-and-leaf diagrams are similar to histo-grams of grouped data turned on their side. They are helpful in examining the shape of a distribution. Stem-and-leaf plots are relatively easy to create without the aid of a computer. Consider the following 20 achievement test scores:

Achievement Test Scores

88	58	89	65	79	58	99	100	96	87
78	67	75	90	74	66	76	84	75	69

To create a stem-and-leaf diagram, we first need to create the stem. In this case, the stem is the tens column of the scores of participants; 5 is used for scores in the 50's, 6 for scores in the 60's, 7 for scores in the 70's, and so forth. Next, each score is listed alongside the stem as a single number. For example, a score of 58 would be listed alongside the stem 5 (for 50) as an 8 in the form of a numeric leaf. In our example, there were two scores of 58. Notice in Figure 8 that alongside 5 (the stem) are two 8's (two leaves). Notice also that we have chosen to create our stem-and-leaf plot so that it goes from low to high. We did this so that if you turn the plot on its side, it will look like a histogram.

```
 5 | 8 8
 6 | 5 6 7 9
 7 | 4 5 5 6 8 9
 8 | 4 7 8 9
 9 | 0 6 9
10 | 0
```

FIGURE 8 Stem-and-leaf diagram of achievement test scores.

We created the stem-and-leaf by hand rather than SPSS because that produced by SPSS is not useful for these data (the SPSS default stem value is 100).

Box plots (box-and-whisker diagrams) graphically present the 50th percentile and identify the middle 50 percent of the scores and the highest and lowest scores.

Box plots, also known as **box-and-whisker diagrams**, are another way to present data graphically. Box plots present

- 50% of the scores and below which lie 50% of the scores (also called the 50th percentile rank);
- the range that includes the middle 50 percent of scores; and
- the highest and lowest scores.

We will use SPSS to create a box plot of the college credit data already presented in this chapter. As a reminder, the data are as follows:

College Credits Earned

68	100	72	30	37	80	36	16	94	51	27	65
91	49	17	58	76	61	47	58	22	85	35	3

→ **Begin by creating an SPSS file with the data.**

- After clicking on the "Variable View" tab, define the variable "Credits."
- Then, click on the "Data View" tab and enter the data.
- To create a box plot, you need to use

    Analyze → Descriptive Statistics → Explore.

- Move the variable "Credits" to the box labeled "Dependent List."
- In the section labeled "Display" → check the *Plots* choice.
- Click on the button labeled *Plots* → check *Box Plots, Factor levels together.*

The output has several sections, including a summary of the analysis and the box plot.

Figure 9 presents the summary of the analysis.

Case Processing Summary

	Cases					
	Valid		Missing		Total	
	N	Percent	N	Percent	N	Percent
Credits	24	100.0%	0	.0%	24	100.0%

FIGURE 9 Summary of cases processed.

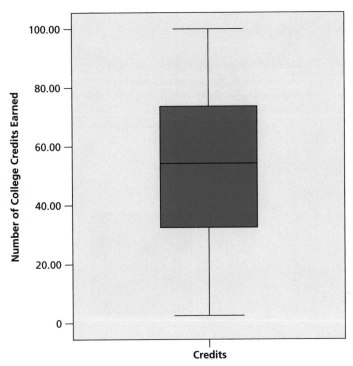

FIGURE 10 Box plot of achievement data.

Credits	The variable analyzed was Credits.
Valid	The number of cases with data for Credits. There were 24 valid cases.
Missing	Missing provides the number of cases for which the credit information was missing. In our example, there were no (0) missing cases.
Total	In total (valid plus missing), there were 24 cases.

Figure 10 shows the box plot. As you examine Figure 10, note that the box in the center marks off an area that includes the middle 50 percent of the scores. The box spans the range from approximately 35 to approximately 72. The horizontal line in the center of the box is at the 50th percentile rank (half of the scores in the entire data set are above this line, half of the scores below). The center line corresponds to a score of 54.5. The box plot also indicates, in the form of short horizontal lines, the lowest score (3) and the highest score (100).

Interpretation: Shapes of Distributions

In some cases, sets of scores are distributed or patterned in such a way that we can refer to their graphic shape. One way that we can describe the shape of a

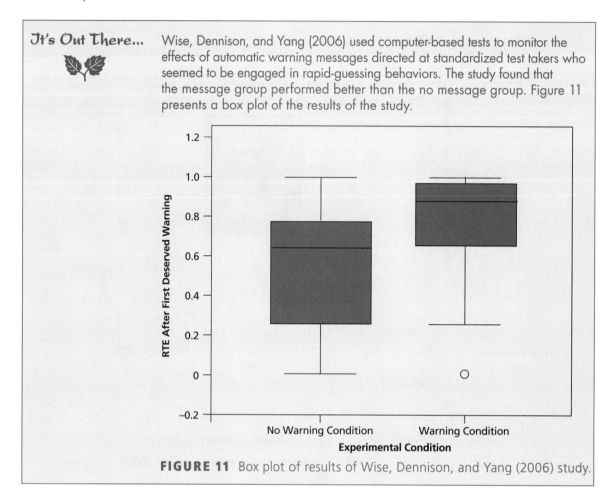

It's Out There... Wise, Dennison, and Yang (2006) used computer-based tests to monitor the effects of automatic warning messages directed at standardized test takers who seemed to be engaged in rapid-guessing behaviors. The study found that the message group performed better than the no message group. Figure 11 presents a box plot of the results of the study.

FIGURE 11 Box plot of results of Wise, Dennison, and Yang (2006) study.

distribution is by observing whether the data are distributed in a symmetrical or an asymmetrical manner. Another way to describe the shape of a distribution is by observing whether the data are distributed in a tall, pointed shape or whether the data are relatively flat and spread out.

Skewness

A **symmetric distribution** forms a mirror image if a vertical line is drawn down the center.

Consider a **symmetric distribution** of student ratings of a professor (see Figure 12). The rating scale ranges from 1 (poor) to 11 (excellent). Looking at Figure 12, we could describe the data by saying that the ratings are symmetric. Note that most of the ratings are "piled up" in the middle range.

Now, what if the data were distributed asymmetrically? To answer this question, consider a different histogram, shown in Figure 13. The data depicted in this histogram would be described as a **negatively skewed** distribution. The data are not symmetric. There are relatively few lower ratings. The majority of the ratings are "piled up" in the direction of the higher values.

A **negatively skewed** distribution is asymmetric; there are relatively few values on the low end of the scale.

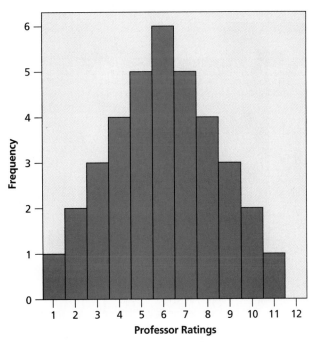

FIGURE 12 Histogram for professor ratings, showing a symmetrical distribution.

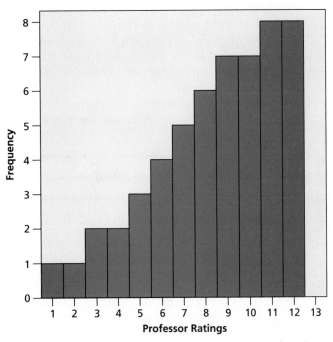

FIGURE 13 Histogram for professor ratings, showing a negatively skewed distribution.

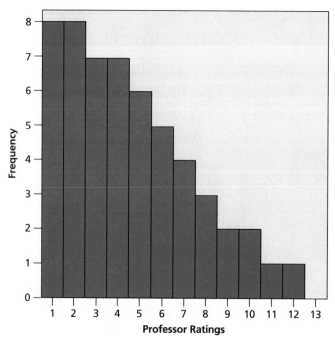

FIGURE 14   Histogram for professor ratings, showing a positively skewed distribution.

A **positively skewed** distribution is asymmetric; there are relatively few values on the high end of the scale.

Finally, data that are asymmetric in the opposite direction are **positively skewed.** In the case of positively skewed data, there are relatively few values at the high end of the scale. Figure 14 contains an example in which the data are skewed in the direction of higher values.

If a distribution has one elongated tail, that elongated tail identifies whether the distribution is positively or negatively skewed. If the tail is extended toward the lower values, the distribution is negatively skewed; if the tail is extended toward the higher values, the distribution is positively skewed.

Be Here Now

If you were a professor, which distribution shape would you prefer for your student ratings?

Answer

Assuming that you want students to rate you positively, you would prefer your ratings to assume a negatively skewed distribution. That way, most of the ratings would be high, and relatively few of the ratings would be low.

Kurtosis

Kurtosis describes the height of a distribution.

We can also describe a distribution by its height. **Kurtosis** describes the extent to which a distribution appears tall, short, or moderate in height. Although

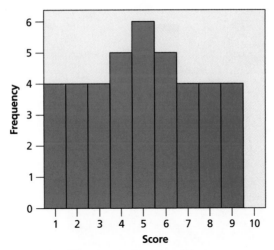

FIGURE 15 Platykurtic distribution.

Platykurtic describes a distribution that is relatively short in height.

Leptokurtic is a term used to describe a distribution that is relatively tall and slender.

Mesokurtic is a term used to describe a distribution of moderate height.

there is a statistic to compute kurtosis, here we will visually examine the relative heights of several histograms.

Shown in Figure 15 is a relatively flat distribution. When the distribution is relatively short in height, we describe the distribution as **platykurtic**.

Shown in Figure 16 is a relatively peaked distribution. When the distribution is relatively tall and slender, we describe the distribution as **leptokurtic**.

Shown in Figure 17 is a more moderate figure. When the height of the distribution is moderate (neither peaked nor flat), we consider it **mesokurtic**. We will spend considerable time in Chapter 5 describing a mesokurtic distribution called the normal curve.

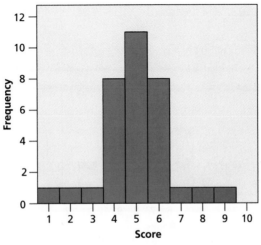

FIGURE 16 Leptokurtic distribution.

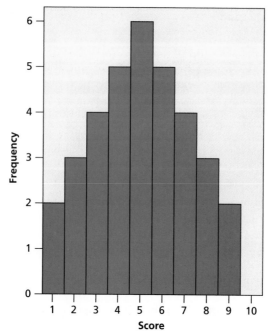

FIGURE 17 Mesokurtic distribution.

Presenting Figures in APA Style

All of the figures in this chapter are presented in APA style. A figure prepared in APA style does not have a title, but instead has a **figure caption** that explains the figure.

Figure 18, in APA style, is presented as it might appear in a journal.

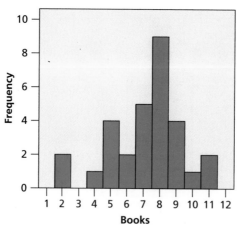

FIGURE 18 Number of books assigned in literature classes.

A figure is cited much as a table is: we use the figure number. For example, we might write

Literature classes at community colleges in the state reported varying numbers of books assigned to students (see Figure 18).

Or

As shown in Figure 18, literature classes in community colleges differ in the number of books assigned to students.

SUMMARY

We organize data in an objective way when we use frequency distribution tables or figures. Frequency distributions provide information about how frequently each score occurs in a distribution; the information presented in a frequency distribution table or figure is in a format that is easier to interpret than a listing of the individual scores. When the data span a wide range of values, we may choose a grouped frequency distribution to condense the data. Frequency distribution figures include histograms, frequency polygons, bar graphs, and pie charts. Stem-and-leaf diagrams (or stem-and-leaf plots) and box plots are additional methods used to explore data visually. SPSS can assist us in creating frequency distribution tables and figures.

PRACTICE

Frequency Distribution Tables

1. A social psychologist was interested in the number of romantic relationships experienced by college students. Students in an introductory psychology class completed a survey, one item of which asked them to report the number of romantic relationships that they had experienced in their lives. The data are as follows:

Number of Romantic Relationships Reported

1	0	3	3	7	0	2	2	4	1
3	2	6	8	6	1	6	5	10	5

individual

Create a frequency distribution table for the data reported. Include the raw frequencies, as well as proportions and percentages. Use Σp and ΣP to verify your calculations.

2. An educator was concerned that children are indirectly encouraged to play video games rather than read. She asked children how many books they owned and how many video games they owned. She found the following:

Number of Books Owned				
1	0	3	3	7
3	2	6	8	6

Number of Video Games Owned				
4	6	6	3	8
9	7	5	0	10

Create a frequency distribution for each set of data.

Frequency Distribution Tables for Grouped Data

3. In a time management exercise, students were asked to keep track of the number of minutes they spend on the cell phone during one day. Each student kept a log of his or her cell phone time on a randomly selected weekday and reported the total minutes to the professor. The data are as follows:

Total Cell Phone Minutes

54	76	33	41	71	40	72	52
98	29	61	11	69	91	66	55

Create a grouped frequency distribution table for the data, using an interval of 10 minutes.

4. Reading speed was investigated in a group of high school students. The investigator was able to calculate words per minute for a sample of 17 students. The numbers of words per minute reported for the sample were 78, 93, 110, 101, 68, 145, 117, 99, 129, 135, 80, 96, 120, 97, 140, 121, and 136. Summarize the reading speed data in a frequency distribution table for grouped data, using an interval of 10 words per minute.

Histograms and Frequency Polygons

5. Create a histogram and frequency polygon for the romantic relationship data presented in Practice Problem 1.

6. Create a histogram and frequency polygon of books owned for the sample data presented in Practice Problem 2.

Bar Graphs and Pie Charts

7. A gerontologist was interested in the leisure activities of elderly people who live in their own homes and apartments. She interviewed elderly individuals at a local grocery store, asking each person his or her favorite leisure activity. The gerontologist tallied the responses, with the following results: 48 watch television, 37 read books, 24 exercise, 16 socialize with friends, 11 volunteer, 9 knit or crochet, and 5 play with pets. Create a bar graph and a pie chart for these leisure activity data.

Stem-and-Leaf Diagrams and Box Plots

8. Use the data listed in Practice Problem 3, and create a stem-and-leaf diagram and a box plot.

Shapes of Distributions

9. Since pain management is important to dentists and oral hygienists, patients were asked about the discomfort they experienced during various routine procedures. Pain was assessed on a scale of 1 (minimum discomfort) to 5 (maximum discomfort). The raw data are as follows:

Whitening

| 2 | 1 | 2 | 4 | 2 | 3 | 2 | 1 | 5 | 3 | 1 | 3 |

Scaling

| 2 | 3 | 3 | 4 | 2 | 3 | 4 | 1 | 5 | 4 | 3 | 2 |

Injection

| 5 | 4 | 4 | 1 | 3 | 3 | 5 | 4 | 5 | 3 | 3 | 4 |

Create a histogram for each of the data sets, and identify the general shape of the distribution for each dental procedure. Which procedure is the least uncomfortable? Which is the most uncomfortable?

SOLUTIONS

Frequency Distribution Tables

1.

Number of Romantic Relationships Reported

Number of Relationships	*f*	*p*	*P*	*C%*
10	1	.05	5	100
9	0	0	0	95
8	1	.05	5	95
7	1	.05	5	90
6	3	.15	15	85
5	2	.10	10	70
4	1	.05	5	60
3	3	.15	15	55
2	3	.15	15	40
1	3	.15	15	25
0	2	.10	10	10

$$\Sigma p = 1 \qquad \Sigma P = 100$$

2.

Number of books read

Number of books	f	p	P	C%
8	1	.10	10	100
7	1	.10	10	90
6	2	.20	20	80
5	0	.00	00	60
4	0	.00	00	60
3	3	.30	30	60
2	1	.10	10	30
1	1	.10	10	20
0	1	.10	10	10

2.

Number of Video Games

Number of Video Games	f	p	P	C%
10	1	.10	10	100
9	1	.10	10	90
8	1	.10	10	80
7	1	.10	10	70
6	2	.20	20	60
5	1	.10	10	40
4	1	.10	10	30
3	1	.10	10	20
2	0	.00	00	10
1	0	.00	00	10
0	1	.10	10	10

Frequency Distribution Tables for Grouped Data

3.

Total Cell Phone Time

Minutes	f	p	P	$C\%$
90–99	2	.13	13	101
80–89	0	.00	00	88
70–79	3	.19	19	88
60–69	3	.19	19	69
50–59	3	.19	19	50
40–49	2	.13	13	31
30–39	1	.06	6	18
20–29	1	.06	6	12
10–19	1	.06	6	6
0–9	0	.00	00	00

4.

Reading Speed

Words/Minute	f	p	P	$C\%$
140–149	2	.12	12	102
130–139	2	.12	12	90
120–129	3	.18	18	78
110–119	2	.12	12	60
100–109	1	.06	6	48
90–99	4	.24	24	42
80–89	1	.06	6	18
70–79	1	.06	6	12
60–69	1	.06	6	6

Histograms and Frequency Polygons

5. Create a histogram and frequency polygon for the romantic relation-ship data presented in Practice Problem 1.

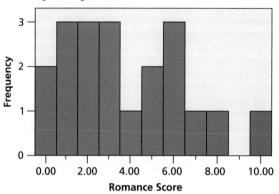

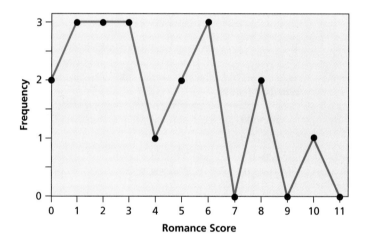

6. Create a histogram and frequency polygon of videos owned for the sample data presented in Practice Problem 2.

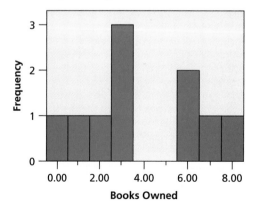

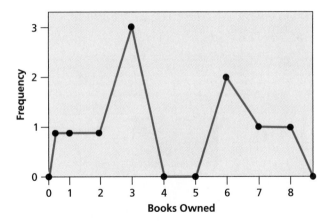

Bar Graphs and Pie Charts

7. Create a bar graph and pie chart for the leisure activity data.

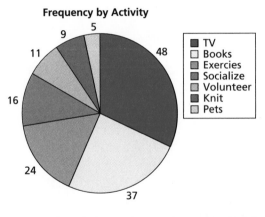

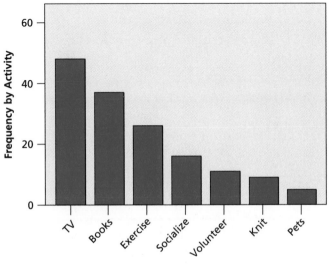

Stem-and-Leaf Diagrams and Box Plots

8. Box Plot

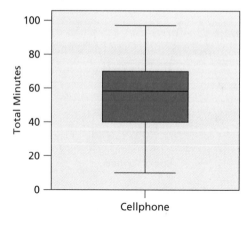

Cellphone

Stem-and-leaf

```
9 | 1 8
8 |
7 | 1 2 6
6 | 1 6 9
5 | 2 4 5
4 | 0 1
3 | 3
2 | 9
1 | 1
```

Shapes of Distributions

9. Whitening was least painful.

The most painful procedure was injection.

Whitening Histogram
Positively Skewed

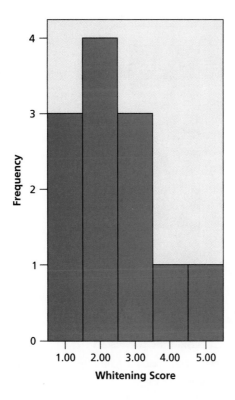

Scaling Histogram
Symmetrical

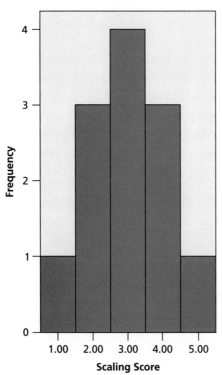

Injection Histogram
Negatively Skewed

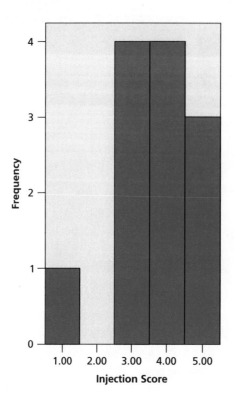

-3-

Measures of Central Tendency

In This Chapter

FOCUS

In the previous chapter, we used a frequency distribution to present data. A frequency distribution is a convenient way to organize a set of data. Sometimes, however, we want to be even more concise. For example, if someone asked you how the class did on the last quiz, you would not want to list every score: 89, 96, 79, 88, etc. Similarly, you would probably not want to create a frequency distribution table. That would be tedious.

Rather, it would be helpful to compute one number that describes the set of test scores earned by your class. For that, we need a measure of central tendency.

Measures of central tendency describe the typical score.

Measures of central tendency allow us to answer questions such as the following ones:

What is the typical reaction time to a traffic light changing to red?
What is the most popular color to paint a house?
How old is the typical university freshman?
What is the typical house price in a neighborhood?

There are several ways to describe the "typical score." In this chapter, we will present three ways to share information about the typical score: the mode, the median, and the mean.

The Mode

The **mode (*Mo*)** is the most frequently occurring score.

The **mode** (abbreviated *Mo*) is the simplest of the measures of central tendency. It denotes the most frequently occurring score. Sometimes a set of scores will have more than one mode. If a set of scores has two modes (in other words, there is a tie for the most frequently occurring score), we often say that the set of scores is *bimodal*.

TABLE 1
Frequency Distribution of Children's Ages

Age	f
9	2
8	4
7	2
6	1
5	1

Application: The Mode

Suppose that a guidance counselor is asked for the typical age of the 10 children she counsels. Their ages are as follows:

Age Scores

8	7	8	9	5	7	9	8	6	8

To obtain the mode, we need to determine the most frequently occurring age. Creating the first two columns (score and frequency) of a frequency distribution may be a helpful way to identify the mode. The age with the highest frequency is the mode. Table 1 presents the frequency distribution of children's ages.

The mode for these age scores is 8; as can be seen in Table 1, age 8 had the highest frequency.

Another example would be to determine the mode relative to the campus activities of a class of psychology students. Table 2 presents the frequency data.

As shown in Table 2, the campus activity in which students were most often engaged was Sorority/Fraternity (11 students). The mode is Sorority/Fraternity.

TABLE 2
Frequency of Campus Activity

Campus Activity	f
Sorority/Fraternity	11
Club	7
Student Government	2

Be Here Now

Ten students were asked their grades on the last test.
Their scores were as follows:
5   89   56   88   79   85   98   88   91   88
What is the mode?

Answer

Mo = 88

Computation and Interpretation: The Mode

Although computing the mode is relatively easy to do without the help of the computer, it can be done using SPSS.

Remember the age scores from our first example? Here they are again:

Age Scores									
8	7	8	9	5	7	9	8	6	8

→ **Begin by creating an SPSS file with the age scores.**

- To compute the mode, you need to select

  Analyze → *Descriptive Statistics* → *Frequencies.*

  Move *Age* **to the box labeled** *Variable(s).*

- Within the *Frequencies* menu, click on the button labeled *Statistics*, and choose *Mode.*

Figure 1 presents the SPSS results for the age data. As shown, the mode is 8. Also provided is the frequency distribution. If you "uncheck" (or "deselect") the *Display frequencies tables* option, you can avoid printing the frequency table. If you followed the instructions correctly, you should see two large output boxes.

The first output box presents the statistics you requested:
Age is the variable analyzed.

N	**Valid = 10**	Ten ages were analyzed.
	Missing = 0	No data were missing.
	Mode = 8	The mode of the 10 ages analyzed is 8.00.

TABLE 2
Frequency of Campus Activity

Campus Activity	*f*
Sorority/Fraternity	11
Club	7
Student Government	2

Let's practice computing the mode with the campus activity data presented in Table 2. As a reminder, Table 2 is presented again.

As you look at the data in the table, note that the choices of response (campus activity) were nominal (or categorical; see Chapter 1). To obtain the mode, it will be necessary to code the choices such that 1 = Sorority/Fraternity, 2 = Club, and 3 = Student Government.

You will need to enter *each* student's activity individually. For example, 11 students participated in sororities or fraternities, so the first 11 rows will

Frequencies

Statistics

age

N	Valid	10
	Missing	0
Mode		8.00

age

Valid	Frequency	Percent	Valid Percent	Cumulative Percent
5.00	1	10.0	10.0	10.0
6.00	1	10.0	10.0	20.0
7.00	2	20.0	20.0	40.0
8.00	4	40.0	40.0	80.0
9.00	2	20.0	20.0	100.0
Total	10	100.0	100.0	

FIGURE 1 SPSS output: Mode for age data.

be entered into the SPSS spreadsheet as 1; seven students participated in clubs, so the next seven rows will contain a 2; and two students participated in student government, so the last two rows will contain a 3.

➡ **Begin by creating an SPSS file with the Activity scores.**

- To compute the mode, you need to select

  Analyze → Descriptive Statistics → Frequencies.

  Move Activity to the box labeled Variable(s).

- Within the *Frequencies* menu, select *Statistics* and choose *Mode*. You may also "uncheck" *Display Frequency Table*.

Figure 2 presents the SPSS results for the campus activity data provided in Table 2. As shown, the mode is 1 (Sorority/Fraternity).

Statistics

Activity

N	Valid	20
	Missing	0
Mode		1.00

FIGURE 2 SPSS output: Mode for campus activity data from Table 2.

If you followed the instructions correctly, you should see one output box on your screen.

As shown in Figure 2, you should see that

Activity is the variable analyzed.

N	**Valid = 20**	Twenty campus activities were analyzed.
	Missing = 0	No data were missing.
	Mode = 1	The mode of the 20 campus activities analyzed is 1 (the code for Sorority/Fraternity).

The Median

*The **median** (**Mdn**) is the middlemost score.*

The **median** (sometimes abbreviated *Mdn*) is the middlemost score. Half the sample will have scores above the median and half will have scores below the median.

If you actually lined up the students in your class by the score they earned on the last quiz, the person in the middle would be at the median.

Application: The Median

Suppose that a researcher measures the perceived self-esteem of 15 students. Their scores are as follows:

Self-Esteem Scores

52	18	96	47	29	65	37	45
19	79	88	79	15	79	65	

To compute the median, you need to rearrange the scores in order of magnitude (for example, from low to high), as follows:

Self Esteem Scores Arranged in Order of Magnitude

15	18	19	29	37	45	47	52	65	65	79	79	79	88	96
							↑							

When you have an odd number of scores, the score that is in the middle is the median.

The middle score is 52, so the median is 52. When you have an odd number of scores, the score that is precisely in the middle is the median.

If you have an even number of scores, the point that best separates the two middle scores is the median. For example, following are the essay scores of 10 seventh-grade children.

Essay-Test Scores

5	8	3	7	2	9	1	6	8	10

As before, to determine the median, you must arrange the scores in order of magnitude:

Essay-Test Scores in Order of Magnitude

1	2	3	5	6	7	8	8	9	10

When you have an even number of scores, the midpoint between the two middle scores is the median.

Next, you must identify the two middle scores.

In our example, 6 and 7 are the two middle scores. You must find the midpoint between these numbers.

To find the midpoint, we add the two numbers (6 + 7) and then divide by 2. The answer is 6.5, which is the median.

Essay-Test Scores in Order of Magnitude: Median Identified

1	2	3	5	6	7	8	8	9	10

$$\frac{6 + 7}{2} = 6.5 = \text{Median}$$

6.5

Be Here Now

Eight market research participants were asked to rate a new soda on a scale of 1 (Bad) to 10 (Good).
Their ratings were as follows:
1 8 6 4 7 3 2 2 1 8
What is the median?   Remember to arrange the scores by magnitude.

Answer
The median rating is 3.5.

Computation and Interpretation: The Median

SPSS can be used to compute the median. To practice, let us use the self-esteem scores we described earlier. As a reminder, they are presented here again:

Self-Esteem Scores

52	18	96	47	29	65	37	45
19	79	88	79	15	79	65	

Statistics

Self Esteem

N	Valid	15
	Missing	0
Median		52.0000

FIGURE 3 SPSS output: Median for self-esteem.

➡ **Begin by creating an SPSS file with the Self_Esteem scores.**

- To compute the median, you need to select

  Analyze → Descriptive Statistics → Frequencies.

  Move Self_Esteem to the box labeled Variable(s).

- Within the Frequencies menu, select Statistics and choose Median. You may also "uncheck" *Display frequencies tables.*

Figure 3 presents the SPSS results for the self-esteem data. As shown, the median is 52.

If you followed the instructions correctly, you should see one output box (that is, if you remembered to "uncheck" the frequency distribution option). The output box presents the statistics you requested:

Self-Esteem is the variable analyzed.

N	**Valid = 15**	Fifteen self-esteem scores were analyzed.
	Missing = 0	No data were missing.
Median = 52.0000		The median of the 15 self-esteem scores is 52.

Next, practice computing the median with the essay-test data. As a reminder, the essay-test scores are as follows:

Essay-Test Scores

5	8	3	7	2	9	1	6	8	10

To begin computing the median, you must enter the students' scores into a data file.

➡ **Begin by creating an SPSS file with the Essay scores.**

- To compute the median, you need to select

  Analyze → Descriptive Statistics → Frequencies.

  Move Essay to the box labeled Variable(s).

- Within the Frequencies menu, select *Statistics* and choose *Median.*
- You may also "uncheck" *Display frequencies tables.*

Statistics

essay

N	Valid	10
	Missing	0
Median		6.5000

FIGURE 4 SPSS output: Median for essay data.

If you followed the instructions correctly, you should get one output box. Figure 4 presents the statistics you requested.

   Essay is the variable analyzed.

N	**Valid = 10**	Ten essay scores were analyzed.
	Missing = 0	No data were missing.
	Median = 6.5000	The median of the 10 essay scores is 6.5.

The Mean

The **mean** is the arithmetic average.

The mean is the arithmetic average. You have probably computed the mean many times. Imagine a class in which you had three tests. Rather than think about the three scores separately, you may have added the scores up and divided by three. The "average" you computed is the mean.

   To compute the mean, you add up the scores and divide by the number of scores.

   Unique to the mean, or arithmetic average, is that the magnitude of each number contributes to the average. This feature makes the mean unlike the other measures of central tendency we have described.

   The mode is based on one number: the score that has the highest frequency.

   The median is based on the order of the scores; it is the middlemost score.

   The mean is based on the magnitude of each score.

You might think of the mean as the fulcrum of a seesaw made of scores. The mean is the center of the seesaw when it is perfectly balanced. For example, if you have the five scores 1, 1, 3, 10, and 15, then

   The mode is 1 (the most frequently occurring score).

   The median is 3 (the middle score).

   The mean is 6 ((1 + 1 + 3 + 10 + 15)/5 = 6).

Imagine the seesaw:

In statistics, we distinguish between the mean of a sample and the mean of a population:

Mean	Population	Sample
Symbol	μ	M
Definitional Formula	$\mu = \dfrac{\Sigma X}{N}$	$M = \dfrac{\Sigma X}{n}$

Note that the population parameter is μ. The sample statistic is **M.** You may also see $\bar{X}$ as the symbol for the sample mean.

The formulas for the population and sample mean require the same computations. Only the symbols change.

Generally, population parameters use uppercase English letters or Greek letters. Sample statistics tend to use lowercase English letters. SPSS disregards this general rule by using **N** as the symbol for the number of scores analyzed, even though SPSS analyzes data from samples.

Application: The Mean

Suppose that a researcher studies the number of trials a rat takes to learn a maze. She has six rats. The rats' scores are as follows:

Number of Trials Required to Learn a Maze
_____

2   6   5   4   3   28
_____

To compute the mean of this sample, add the scores and divide by the number of scores:

$$M = \frac{\Sigma X}{n} = \frac{2 + 6 + 5 + 4 + 3 + 28}{6} = 8$$

Be Here Now

Imagine that you have had the following five quizzes in statistics class:
Your scores: 95   85   70   80   90
How are you doing? What is your mean score?

Answer

Your mean score is 84.

Computing and Interpreting: The Mean

SPSS can be used to compute the mean. Let's practice by using the number of trials to learn a maze already described. The data are presented again:

Number of Trials Required to Learn a Maze					
2	6	5	4	3	28

➡ **Begin by creating an SPSS file with the rats' trial scores.**

- To compute the mean, you need to select

  Analyze → Descriptive Statistics → Frequencies.

  Move Trials to the box labeled Variable(s).

- Within the Frequencies menu, select Statistics and choose Mean.
- You may also "uncheck" Display Frequency Table.

If you followed the instructions correctly, you should see one output box (if you unchecked *Display Frequency Table*).

Your output box presents the statistics you requested (see Figure 5): **Trials** is the variable analyzed.

N	**Valid = 6**	Six trial scores were analyzed.
	Missing = 0	No data were missing.
Mean = 8.0000		The mean number of trials is 8.

Weighted Mean

If you have summary data in the form of group means for two or more groups, you may need to compute the overall mean across all of the groups. If the groups are of different size, you must compute the **weighted mean**.

The **weighted mean** takes into consideration the number in each group.

For example, imagine that you are a school psychologist. The school has two third-grade classrooms. Your principal wants to know the mean achievement score of all third graders, regardless of class. But you have only the mean achievement score of each class. What to do?!

The data you are given appear in Table 3.

If you just tried to "average" the two class means ((85 + 65)/2 = 75), you would *not be correct*. Because the mean is the arithmetic average, you need to take into consideration the number of children in each class.

To compute the mean correctly—a value that will best describe the *entire* group of 30 children (number in Class A + number in Class B)—you need to compute the weighted mean.

Statistics

trials

N	Valid	6
	Missing	0
Mean		8.0000

FIGURE 5 SPSS output: Mean number of trials required to learn a maze.

TABLE 3
Achievement Data: Two Third-Grade Classes

	Class A	Class B
Number in Class	10	20
Mean	85	65

The formula for the weighted mean is

$$M_{weighted} = \frac{M_A n_A + M_B n_B}{n_A + n_B}$$

Notice that we have subscripts to distinguish information from Class A from information from Class B.

To compute the weighted mean,

1. Multiply the mean of Class A (M_A) by the number of students in Class A (n_A).
2. Multiply the mean of Class B (M_B) by the number of students in Class B (n_B).
3. Add the results of Step 1 and Step 2: $(M_A)n_A + (M_B)n_B$.
4. Add the number in Class A to the number in Class B: $n_A + n_B$.
5. Divide the result of Step 3 $((M_A)n_A + (M_B)n_B)$ by that of Step 4 $(n_A + n_B)$.

Let's apply these steps to the data in Table 3. Start with the formula:

$$M_{weighted} = \frac{(M_A n_A) + (M_B)n_B}{n_A + n_B}$$

Now replace the symbols with the numbers:

$$M_{weighted} = \frac{(85)10 + 65(20)}{10 + 20} = \frac{850 + 1300}{30} = \frac{2150}{30} = 71.67$$

The **weighted mean** will be closer to the mean of the group that contains the most scores.

Notice that the weighted mean is closer to the mean of the class with the larger number of students. Because the mean is the arithmetic average, and because it is "weighted," or affected, by *each* score, it will always be closer to the group mean having the most scores.

By the way, SPSS is not helpful in computing the weighted mean unless we have all of the original data. In that case, we would enter the entire combined set into SPSS and compute the new mean.

Reflect
The average of the averages is not necessarily the average.

Let's practice with another example:

A psychologist trained two groups and tested their short-term memories. He is given the mean memory score of each group. He would like to compute one number that describes the memory scores of all the people, regardless of the group they belong to. His data are in Table 4. Compute the weighted mean.

TABLE 4
Mean: Memory Scores

	Group A	Group B
Number in Group	30	10
Mean of Group	5	7

Begin with the formula

$$M_{\text{weighted}} = \frac{(M_A)n_A + (M_B)n_B}{n_A + n_B}$$

Replace the symbols with the numbers you know:

$$M_{\text{weighted}} = \frac{(5)30 + (7)10}{30 + 10} = \frac{150 + 70}{40} = \frac{220}{40} = 5.5$$

The mean memory score of the participants across the two groups is 5.5. Notice that the weighted mean is closer to the group that had the most people (Group A).

Be Here Now

Your town has two Girl Scout troops.

There are 8 girls in the first troop; on average, they sell 5 boxes of cookies each.

There are 20 girls in the second troop; on average, they sell 2 boxes of cookies each.

What is the average number of cookies sold by girl scouts in your town?

Answer

You need to compute the weighted mean:

$$M_{\text{weighted}} = \frac{(M_A)n_A + (M_B)n_B}{n_A + n_B}$$

$$M_{\text{weighted}} = \frac{(5)8 + (2)20}{28} = \frac{40 + 40}{28} = \frac{80}{28} = 2.86$$

Choosing the Best Measure of Central Tendency

In working with nominal data (or categories), it is common to report the most frequently occurring score or category. We would report the mode.

For example, suppose you ask your classmates what their favorite color is. Their answers are as follows:

12	Red
6	Blue
3	Yellow

Then you might report that red is the color selected most often; that is, red is the mode.

Because the mode is based on only one number, the most frequently occurring score, it does not tell us much about the distribution of colors. For this reason, unless the data are nominal, we usually want to compute and share another measure of central tendency.

For other scales of measurement, we make a choice between the mean and the median, depending upon the shape of the distribution. In particular, if our data are skewed (see Chapter 2), we are more likely to report the median. If, by contrast, our data are symmetrical (that is, we have no extreme scores at either end of the distribution), we are more likely to report the mean.

Skewness

In a symmetrical distribution, the mean = the median.

As we discussed in the previous chapter, we can describe data by the shape of the distribution. Some distributions are symmetrical.

As shown in Figure 6, when a distribution is symmetrical, the mean and the median are the same. Generally, we would report and use the mean to describe a symmetrical distribution. As you will see in later chapters, many analyses are based on the mean.

When a distribution is not symmetrical, it is said to be skewed. Recall from Chapter 2 that when there are extreme or atypical scores at the high end of the distribution, the distribution is **positively skewed.** When there are extreme or atypical scores at the low end of the distribution, the distribution is **negatively skewed.**

As shown in Figure 7, when the distribution is skewed, the mean and median are not the same. In these instances, the mean will be pulled in the direction of the extreme scores in the tail of the distribution.

In a **positively skewed** distribution, the mean > the median.

In a **negatively skewed** distribution, the mean < the median.

In a **positively skewed** distribution, the mean will be greater than the median.

In a **negatively skewed** distribution, the mean will be less than the median.

Often, when a distribution is skewed, the median will be a better description of the overall distribution than the mean. Distributions of income and housing prices are typically skewed.

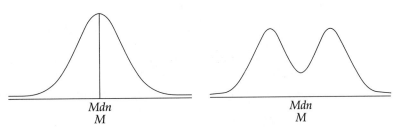

FIGURE 6 Examples of symmetrical distributions.

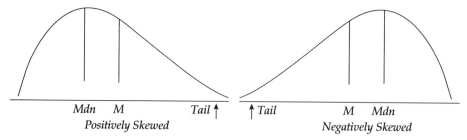

FIGURE 7 Skewed distributions.

For example, a predominantly middle-class community might have one small section that is atypically poor; this would suggest a negatively skewed distribution of income. If we reported the mean, it would be too low to provide a good estimate of the community's typical housing price. The median would be a better descriptor.

Reflect
The median provides balance in the presence of extremes.

Application: Computing and Interpreting and Measures of Central Tendency

You can now share information about the central tendency of a group of scores. As part of a study of attitudes toward divorce, 12 participants were administered a brief survey. The scores reflecting their attitudes toward divorce are as follows:

Attitude Toward Divorce Survey Scores

| 22 | 89 | 87 | 79 | 89 | 92 | 88 | 81 | 93 | 98 | 89 | 86 |

➡ Begin by creating an SPSS file with the Attitude scores.

- To compute measures of central tendency, you need to select

  Analyze → Descriptive Statistics → Frequencies.

  Move Attitude to the box labeled Variable(s).

- Within the Frequencies menu, select Statistics and choose Mean, Median, and Mode.
- You may also "uncheck" *Display frequency table.*

If you entered the data and followed the analysis instructions correctly, and you "unchecked" *Display frequency table*, you should see one output box, shown in Figure 8.

Statistics

Attitude

N	Valid	12
	Missing	0
Mean		82.7500
Median		88.5000
Mode		89.00

FIGURE 8 SPSS output for central tendency:
Attitude Toward Divorce Survey.

As you look at your output, note the following:
    Attitude is the variable analyzed.

N	**Valid = 12**	Twelve attitude scores were analyzed.
	Missing = 0	No data were missing.
Mean = 82.7500		The mean attitude score is 82.7500.
Median = 88.5000		The median attitude score is 88.5000.
Mode = 89.00		The mode attitude score is 89.00.

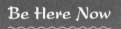

As you examine your results (Figure 8), can you determine whether the distribution is skewed?
If yes, in what direction and why?

Answers

Look at the scores. Notice that most of the scores are between 79 and 98.
However, one score (22) is an extreme, or atypical, score.
Yes, the distribution is skewed. It is negatively skewed.
The mean of 82.7500 is smaller than the median of 88.500.

Reporting Measures of Central Tendency in APA Style

For the data presented in Figure 8, you might decide to present the results in
a table. You might write something like the following:

    As shown in Table 5, the mean attitude score was 82.75 and the median
attitude score was 88.5. The mode was 89.

TABLE 5
Measures of Central Tendency: Attitude
Toward Divorce Survey Data

Score	n	M	Mdn	Mo
Attitude	12	82.75	88.50	89.00

Let's consider an additional example. Another psychologist administers the same Attitude Toward Divorce Survey to six participants.

The data collected, which we will label "Attitude2," are as follows:

Attitude2 Divorce Survey Scores

23	24	24	30	31	98

To begin calculating the three measures of central tendency, you must enter the participants' scores into a data file.

➡ **Begin by creating an SPSS file with the Attitude2 scores.**

- To compute measures of central tendency, you need to select

  Analyze ➤ *Descriptive Statistics* ➤ *Frequencies.*

  Move Attitude2 to the box labeled Variable(s).

- Within the Frequencies menu, select Statistics and choose Mean, Median, and Mode.

- You may also "uncheck" *Display frequency table.*

If you entered the data and followed the analysis instructions correctly, and if you "unchecked" *Display frequency table,* your output should look like that presented in Figure 9.

As you look at your output, note the following:

      Attitude2 is the variable analyzed.

N	**Valid = 6**	Six attitude scores were analyzed.
	Missing = 0	No data were missing.
Mean = 38.3333		The mean attitude score is 38.3333.
Median = 27.0000		The median attitude score is 27.000.
Mode = 24.00		The mode attitude score is 24.00.

Be Here Now

Review your survey results. Is the distribution skewed? If yes, in what direction and why?

Answer

Look at the scores. Notice that most scores are between 23 and 31. One score (98) was extreme, or atypical.
Yes, the distribution is skewed. It is positively skewed.
The mean of 38.3333 is larger than the median of 27.0000.

To report these data, you might create a table and write a brief description of it. You might write something like the following:

As shown in Table 6, the mean attitude score was 38.33 and the median attitude score was 27.00. The mode was 24.

Attitude2

N	Valid	6
	Missing	0
Mean		38.3333
Median		27.0000
Mode		24.00

FIGURE 9 SPSS output for central tendency: Attitude2 divorce survey data.

TABLE 6
Measures of Central Tendency: Attitude2 Survey Data

Score	n	M	Mdn	Mo
Attitude	6	38.33	27.00	24

It's Out There...

Ironsmith and Eppler (2007) examined student test scores under two conditions: traditional lecture and personalized instruction. Table 7 presents their mean results. As shown in Table 7, the mean test score for traditional lecture was 72.26, whereas that of personalized instruction was 81.34.

TABLE 7
Mean Test Scores for Traditional and Personalized Instruction

Condition	n	M	SD[1]
Traditional Lecture	278	72.26	14.70
Personalized Instruction	298	81.34	8.38

[1]SD stands for a descriptive statistic called the "standard deviation." We will discuss SD in Chapter 4.

SUMMARY

Measures of central tendency offer one number to describe a set of numbers:

The mode is the most frequently occurring score.

The median is the middlemost score.

The mean is the arithmetic average.

Most often, we report the mean. There are two exceptions to this generality: If the data are nominal, we report the mode; if the distribution is skewed, the median may be the best descriptor of central tendency. The weighted mean allows us to appropriately combine means for groups that are of different size. Selecting the best measure of central tendency depends upon whether the data are nominal. (Then we select the mode.) If ordinal, interval, or ratio data are skewed, the median is the best measure of central tendency.

PRACTICE

Mode

1. A researcher surveys 15 students to find out how many brothers and sisters they have. His data are as follows:

2	1	2	3	5	2	9	0
7	2	3	5	2	2	3	

   What is the mode?

2. The coordinator of student services wants to order sweat shirts for her student workers. The salesman is willing to give her the shirts for free. The only stipulation is that they all must be the same size. The coordinator surveys her workers and creates the following frequency distribution:

   Student Workers' Shirt Size

Size	f
Small	1
Medium	3
Large	15
Extra Large	2

   Which size should she order? What is the mode?

3. An educational psychologist tested 20 students enrolled in a certain program. The scores she obtained are as follows:

25	25	10	95	25	30	15	25	85	45
85	80	85	15	90	65	85	90	30	80

   What is the mode? Use SPSS.

Median

4. The educational psychologist described in Practice Problem 3 would also like to compute the median. What is the median?

5. An experimental psychologist tests the learning time of nine rats. Her results are as follows:

6	8	3	9	4	2	9	8	1

   What is the median learning time?

6. A personnel officer measures the typing skills of a group of 10 applicants. The results, reported as the number of words typed per minute, are below. What is the median?

35	55	45	44	25	28	35	40	55	38

Mean

7. A student is enrolled in a psychology class that has a test each week. After 8 weeks, he wants to know how he is doing. His scores are as follows:

88   96   58   75   99   85   89   79

What is his mean score?

8. A researcher measures the IQ scores of five children in a "gifted and talented" class. Their scores as follows:

115   121   135   119   129

What is their mean score?

9. A researcher wants to know the average number of hours that college students study. She asks 10 students how many hours a week they typically study. Their answers are as follows:

10   12   10   15   20   10   14   12   20   5

What is the mean number of hours studied?

Weighted Mean

10. The YMCA had two youth groups. Each was given a short community service survey. You are told the mean and the number of youngsters in each group. You need to report the overall mean, regardless of group. The data are as follows:

Group	M	n
A	23	10
B	10	5

What is the weighted mean?

11. Two classes are enrolled in a karate program. Each instructor recorded the number of sessions required to teach a new skill. The data for each class are as follows:

Class	M	n
1	5	20
2	3	4

What is the overall weighted mean?

12. A social worker runs two smoking cessation groups: one in the morning and one in the evening. After each group has met for a month, she surveys how many cigarettes each client has smoked during that month. She wants to know the overall success of her program. Her data are as follows:

Group	M	n
Morning	10	10
Evening	16	10

What is the weighted mean?

Shape of a Distribution and Choosing a Measure of Central Tendency

13. A quality control officer in a hospital recorded the number of visits to the emergency room for a group of three-year-olds. The data are as follows:

Number of Visits to an Emergency Room for 25 Three-Year-Olds

2	1	5	6	4	2	4	2	20	2
2	3	5	1	1	1	2	3	5	1
1	2	2	4	5					

Use SPSS to compute each measure of central tendency (mean, median, and mode).

Is the distribution skewed? If yes, in what direction?

Which measure of central tendency would communicate the officer's results best?

14. The counselor of a residential treatment center kept track of the number of infractions made by eight residents over a one-month period. Her data are as follows:

Resident Infractions

2	15	20	25	20	20	19	22

Use SPSS to compute each measure of central tendency (mean, median, and mode).

Is the distribution skewed? If yes, in what direction?

Which measure of central tendency would communicate the counselor's results best?

15. A psychologist examined the perceived self-esteem of 11 participants in a research study. His data are as follows:

Self-Esteem Scores of 11 Participants

50	50	75	50	55	30	70	65	35	20	50

Use SPSS to compute each measure of central tendency (mean, median, and mode).

Is the distribution skewed? If yes, in what direction?

Which measure of central tendency would communicate the psychologists' results best?

Reporting Results: Central Tendency

16. A researcher asks nine participants to study while music is playing. He then administers a test to each participant. His data are as follows:

Test Data for Nine Students Who Studied with Music

88	96	59	87	79	82	99	66	59

Use the computer to obtain each measure of central tendency (mean, median, and mode).

Develop text and a table to describe the researcher's results.

17. An evaluator collects attendance data from 15 children in an after-school program. Her data are as follows:

Attendance Data for 15 Program Participants

9	8	15	8	6	4	12	10
12	8	5	4	13	8	7	

Use SPSS to compute each measure of central tendency (mean, median, and mode).

Develop text and a table to describe the researcher's results.

18. A psychologist studied people's reaction times in responding to a change in a light. She recorded the number of seconds it took 12 participants to react. Her data are as follows:

Reaction Time, in Seconds, of 12 Participants

15	10	6	3	5	8
12	20	8	9	5	8

Use the computer to obtain each measure of central tendency (mean, median, and mode).

Develop text and a table to describe the researcher's results.

SOLUTIONS

Mode

1. The mode may be computed via a frequency distribution or via the computer:

Frequency Distribution Table

Siblings	f
9	1
8	0
7	1
6	0
5	2
4	0
3	3
2	6
1	1
0	1

The mode is 2. The number of siblings with the highest frequency is 2.

By computer, your output would be as follows:

Statistics

siblings

N	Valid	15
	Missing	0
Mode		2.00

2. The mode is Large. Size Large has the highest frequency.

3. There are two modes: 25 and 85. Each score was earned by four participants. The computer output is as follows:

Statistics

scores

N	Valid	20
	Missing	0
Mode		25.00[a]

[a] Multiple modes exist. The smallest value is shown

scores

Valid	Frequency	Percent	Valid Percent	Cumulative Percent
10.00	1	5.0	5.0	5.0
15.00	2	10.0	10.0	15.0
25.00	4	20.0	20.0	35.0
30.00	2	10.0	10.0	45.0
45.00	1	5.0	5.0	50.0
65.00	1	5.0	5.0	55.0
80.00	2	10.0	10.0	65.0
85.00	4	20.0	20.0	85.0
90.00	2	10.0	10.0	95.0
95.00	1	5.0	5.0	100.0
Total	20	100.0	100.0	

Note that SPSS lists only one mode, but alerts you that there are multiple (more than one) modes.

Median

4. The median score is 55. The computer output is as follows:

Statistics

scores

N	Valid	20
	Missing	0
Median		55.0000

5. The median learning time is 6. The computer output is as follows:

Statistics

learning

N	Valid	9
	Missing	0
Median		6.0000

6. The median number of words typed is 39. The computer output is as follows:

Statistics

typing

N	Valid	10
	Missing	0
Median		39.0000

Mean

7. The mean test score is 83.6250. The computer output is as follows:

Statistics

test_scores

N	Valid	8
	Missing	0
Mean		83.6250

8. The mean IQ score is 123.8000. The computer output is as follows:

Statistics

IQ_Score

N	Valid	5
	Missing	0
Mean		123.8000

9. The mean number of hours studied is 12.8000. The computer output is as follows:

Statistics

Hours

N	Valid	10
	Missing	0
Mean		12.8000

Weighted Mean

10. Start with the formula

$$M_{\text{weighted}} = \frac{(M_A)n_A + (M_B)n_B}{n_A + n_B}$$

Substituting values gives

$$M_{\text{weighted}} = \frac{(23)10 + (10)5}{15} = \frac{230 + 50}{15} = \frac{280}{15} = 18.67$$

The overall weighted mean is 18.67.

11. Start with the formula

$$M_{\text{weighted}} = \frac{(M_A)n_A + (M_B)n_B}{n_A + n_B}$$

Substituting values gives

$$M_{\text{weighted}} = \frac{(5)20 + (3)4}{24} = \frac{100 + 12}{24} = \frac{112}{24} = 4.67$$

The overall weighted mean is 4.67.

12. Start with the formula

$$M_{\text{weighted}} = \frac{(M_A)n_A + (M_B)n_B}{n_A + n_B}$$

Substituting values gives

$$M_{\text{weighted}} = \frac{(10)10 + (16)10}{20} = \frac{100 + 160}{20} = \frac{260}{20} = 13$$

The overall weighted mean is 13.

Shape of a Distribution and Choosing a Measure of Central Tendency

13. The mean number of ER visits is 3.44, the median is 2.0, and the mode is 2. The computer output is as follows:

Statistics

visits

N	Valid	25
	Missing	0
Mean		3.4400
Median		2.0000
Mode		2.00

The distribution is skewed. Because the mean is greater than the median, the distribution is positively skewed.

The median is the best measure of central tendency. The median is best here because the distribution is skewed.

14. The mean number of infractions is 17.8750, the median is 20.0, and the mode is 20. The computer output is as follows:

Statistics

Infractions

N	Valid	8
	Missing	0
Mean		17.8750
Median		20.0000
Mode		20.00

The distribution is skewed. Because the mean is lower than the median, the distribution is negatively skewed.

The median is the best measure of central tendency. The median is best here because the distribution is skewed.

15. The mean self-esteem score is 50.0, the median is 50.0, and the mode is 50.0. The computer output is as follows:

Statistics

Self_Esteem

N	Valid	11
	Missing	0
Mean		50.0000
Median		50.0000
Mode		50.00

The distribution is symmetrical; the mean equals the median.

The mean is the best measure of central tendency here.

Reporting Results: Central Tendency

16. The computer output is as follows:

Statistics

music

N	Valid	9
	Missing	0
Mean		79.4444
Median		82.0000
Mode		59.00

As shown in Table 16, the mean test score of participants was 79.44, whereas the median score was 82.00.

TABLE 16

Measures of Central Tendency: Music Study Achievement Data

Score	n	M	Mdn	Mo
Achievement	9	79.44	82.00	59.00

17. The computer output is as follows:

Statistics

attendence

N	Valid	15
	Missing	0
Mean		8.6000
Median		8.0000
Mode		8.00

As shown in Table 17, the mean attendance of participants was 8.60, whereas the median attendance was 8.00.

TABLE 17
Measures of Central Tendency: Participant Attendance

Score	n	M	Mdn	Mo
Attendance	15	8.60	8.00	8.00

18. The computer output is as follows:

Statistics

reaction

N	Valid	12
	Missing	0
Mean		9.0833
Median		8.0000
Mode		8.00

As shown in Table 18, the mean reaction time of participants was 9.08, whereas the median reaction time was 8.00.

TABLE 18
Measures of Central Tendency: Reaction Time

Score	n	M	Mdn	Mo
Attitude	12	9.08	8.00	8.00

-4-

Measures of Variability

FOCUS

Measures of variability describe the distance between scores within a group.

Measures of variability add to the ways in which we can describe data. Measures of central tendency (discussed in Chapter 3) provide a single number that describes the typical score. However, measures of central tendency (the mean, median, and mode) do not provide information about the *spread* of scores within a group. **Measures of variability** describe the distance between scores within a group.

For example, suppose you earned an 87 on an exam. Did most students earn a score between 80 and 87 (relatively low variability), or did the scores extend from 50 to 99 (relatively high variability)? Measures of variability quantify the spread of scores. Also, as you will discover in later chapters, measures of variability help to describe how far a score is from the average. Variability not only makes life interesting, it is the heart of many statistical analyses.

Variables vary. That is, the variables we measure take on different values. If you surveyed your friends as to the number of CD's they owned, you would get a range of responses. The number of CD's owned by different people varies. However, if you surveyed the same group as to the number of noses they had, you would not get any variability; all the answers would be "1." In this case, the number of noses is constant.

Reflect

Variables vary. Constants are constant.

Be Here Now

Consider the following examples, and decide whether each is a constant or a variable:

A. The number of noses on the faces of professional basketball players

B. The number of brothers and sisters in the families of each of your classmates

C. The SAT scores of incoming freshmen

D. A 10-point bonus received by all students

Answers

A. Constant   **B.** Variable   **C.** Variable   **D.** Constant

---

So, variability is another way to describe a distribution of scores in terms of how spread out or how clustered together the scores are. Variability also gives us information about how good the mean is as a measure of central tendency and how good one person's score is as a representation of the distribution. We all have some idea of what is average, so we also know when something has deviated from the average. For example, if you had three friends, and one had 1 sibling, one had 2 siblings, and one had 12 siblings, you would know which friend was most different from the average with regard to the number of siblings. Well, that is the idea behind variability. You know that having 1 or 2 siblings is close to average; having 12 siblings makes life a bit more exciting!

The measures of variability described in this chapter are the range, the interquartile range, the variance, and standard deviation.

The Range

The **range** is the distance between the highest and lowest scores. It is based on only two scores: the highest and the lowest.

Application and Interpretation: The Range

For example, Professor Riley tested eight students. Their scores are as follows:

Professor Riley's Test Scores

25	45	60	65	75	80	95	100

To compute the range, subtract the highest score from the lowest score. The range of the scores in Professor Riley's class is $100 - 25 = 75$.

The range may be computed with SPSS.

➡ **Begin by creating an SPSS file with Professor Riley's test scores.**

- To compute the range, you need to select

  Analyze ➡ *Descriptive Statistics* ➡ *Frequencies.*

- Within the Frequencies menu, select *Statistics* and choose *Range*.

Figure 1 presents the SPSS results for the students' scores. As shown, the range is 75.

Across the top row of the output table are the labels that define the statistics calculated.

Down the first column of the output table are the variables that were analyzed:

riley_class is the variable analyzed.

N	**Valid = 8**	There were eight records analyzed.
	Missing = 0	No data were missing.
	Range = 75.00	The range among the 8 scores analyzed was 75.

The Interquartile Range

The **interquartile range** is the distance between the 25th percentile and the 75th percentile of a distribution of scores. It is the range of the middle 50 percent

Statistics

riley_class

N	Valid	8
	Missing	0
Range		75.00

FIGURE 1 SPSS output: Range for Professor Riley's class.

of scores and can be graphically presented in a box plot (see Chapter 2). This yields a single number to represent the interquartile range.

Application and Interpretation: The Interquartile Range

We will use the computer to determine the interquartile range of the scores in Professor Riley's class. As a reminder, the scores are as follows:

Professor Riley's Test Scores

25	45	60	65	75	80	95	100

We will use SPSS to compute the interquartile range.

➡ **Begin by creating an SPSS file with Professor Riley's test scores.**

- To compute the interquartile range, you need to select

    Analyze → *Descriptive Statistics* → *Explore*

- Within the *Explore* screen, move "riley_class" to the box labeled *Dependent List*.

- Select Statistics, and choose *Descriptives* and *Percentiles*.

Presented next are the first three output boxes. Also presented (though not described here) are stem-and-leaf diagrams and box plots of Professor Riley's class (see Chapter 2).

Figure 2 presents summary information on Professor Riley's class.

First Column	The first column lists the name of the variable analyzed. We labeled our data riley_class.
Cases	Describes the cases analyzed; subdivided into three sections: Valid, Missing, and Total
Valid	Indicates the number of cases analyzed.
N	Eight cases (100 percent) were analyzed.
Missing	Indicates the number of cases with missing data.
N	No cases (0 percent) were missing.
Total	Indicates the total number of cases (valid + missing)
N	There were a total of 8 cases (100 percent).

Figure 3 presents a variety of statistics, including the interquartile range. Figure 3 has three columns. The first column labels the statistics calculated, the second column lists each statistic's value, and the last column lists the standard error (where appropriate). Many of the statistics shown we have yet to discuss. You will see many of them in the remainder of this chapter, as well as in future chapters.

Case Processing Summary

	Cases					
	Valid		Missing		Total	
	N	**Percent**	**N**	**Percent**	**N**	**Percent**
riley_class	8	100.0%	0	.0%	8	100.0%

FIGURE 2 Interquartile range output summary: Ms. Riley's class.

Descriptives

			Statistic	**Std. Error**
riley_class	Mean		68.1250	8.86090
	95% Confidence	Lower Bound	47.1723	
	Interval for Mean	Upper Bound	89.0777	
	5% Trimmed Mean		68.7500	
	Median		70.0000	
	Variance		628.125	
	Std. Deviation		25.06242	
	Minimum		25.00	
	Maximum		100.00	
	Range		75.00	
	Interquartile Range		42.50	
	Skewness		-.476	.752
	Kurtosis		-.272	1.481

FIGURE 3 Interquartile range descriptives.

Mean		The mean is 68.1250. We will describe the standard error of the mean in a future chapter.
95% Confidence	**Lower Bound**	We will discuss this statistic in a future chapter.
Interval for Mean	**Upper Bound**	
5% Trimmed Mean		We will not discuss this statistic.
Median		The median (50th percentile) is 70.0000.
Variance		We will discuss this statistic later in the chapter.

Percentiles

		Percentiles						
		5	10	25	50	75	90	95
Weighted Average (Definition 1)	riley_class	25.0000	25.0000	48.7500	70.0000	91.2500	.	.
Tukey's Hinges	riley_class			52.5000	70.0000	87.5000		

FIGURE 4   Interquartile range percentiles.

Std. Deviation	We will discuss this statistic later in the chapter.
Minimum	The lowest score was 25.
Maximum	The highest score was 100.
Range	The range was 75.
Interquartile Range	The interquartile range—the middle 50 percent of scores—spans 42.5 points.
Skewness	We discussed this concept in Chapter 2. We will not discuss the standard error of skewness.
Kurtosis	We discussed this concept in Chapter 2. We will not discuss the standard error of kurtosis.

Figure 4 presents the percentiles. Two rows of results are shown. We will describe the top row, labeled "Weighted Average (Definition 1)."

Percentile

5	The score closest to the 5th percentile is 25.0000.
10	The score closest to the 10th percentile is 25.0000.
25	The score closest to the 25th percentile is 48.7500.
50	The score closest to the 50th percentile is 70.000 (also the median).
75	The score closest to the 75th percentile is 91.2500.

As you examine your printout (see Figure 3), notice that the interquartile range is 42.5; it is the difference between the 75th and 25th percentiles (91.25 − 48.75 = 42.5).

Jt's Out There...

US News and World Report, Ultimate College Guide 2007 (2006) reports the 25th and 75th percentiles of SAT scores of students admitted to each college. For example, for Kean University, the combined SAT score at the 25th percentile was 860 and that of the 75th percentile was 1040. Thus, the interquartile range for the combined SAT scores was 180 (1040 − 860 = 180).

The Variance

The range and interquartile range are not very sensitive measures of variability. Each is based on only two numbers: the highest and lowest measurement and the 25th and 75th percentile, respectively. Moreover, neither the range nor the interquartile range tells you anything about the values of all of the other scores between their bounds. Following are the test scores for the students in Ms. Rodriguez's class and Ms. Parady's class:

Class Scores

	Ms. Rodriguez					Ms. Parady			
5	10	20	30	40	5	86	87	94	89
60	70	80	90	95	91	92	93	94	95
	Range = 95 − 5 = 90					Range = 95 − 5 = 90			

Notice that the range is the same for both classes: 90. Now look at the pattern of the scores in each of the classes; do they look the same? Not at all: In Ms. Rodriguez's class the scores are *spread out*, whereas in Ms. Parady's class they are more *uniform*. We need a measure of variability that reflects this difference. We need to compute the **variance**.

The Variance Defined

The **variance** is the average of the squared deviations from the mean.

The **variance** is a measure of variability that takes into account the distance that *each* score is from the mean. The variance is defined as the average of the squared deviations from the mean; that is, it is the average squared distance of each score to the mean of a distribution of scores. You'll see how this works a bit later in the chapter.

The definitional formula for the variance differs, depending on whether we are computing the variance of a sample or the variance of a population:

Variance	Population	Sample
Symbol	σ^2	s^2
Definitional Formula	$\sigma^2 = \dfrac{\Sigma(X - \mu)^2}{N}$	$s^2 = \dfrac{\Sigma(X - M)^2}{n - 1}$

As shown, the sample variance formula differs in several ways from that of the population.

As with measures of central tendency, population parameters are symbolized with Greek letters and English capital letters. Sample statistics are symbolized by lowercase English letters.

The formula for the sample variance requires that you divide by $n - 1$ (the number of scores, minus 1), whereas the formula for the population requires that you divide by N (the number of scores).

Although the difference between the two denominators in these formulas might seem trivial, it is important. The impact of reducing the sample number by 1 is to increase the computed sample variance. This small, but purposeful, adjustment in the size of the variance is designed to help compensate for the reduction in variability that we observe in samples (subsets of a population), compared with populations (in which we know all possible values).

Reflect

One can make an impact.

The Definitional Formula Applied

Most of the time, researchers do not have access to information about populations. Rather, we compute statistics from samples. We will apply the definitional formula for the sample variance, s^2, using the data from Ms. Rodriguez's class.

Ms. Rodriguez's Class Data

5	10	20	30	40	60	70	80	90	95

STEP 1 Compute M, the mean of the sample. Remember how to do that?

$$M = \frac{\Sigma X}{n}$$

Add the scores and divide by the number of scores.

$\Sigma X = 5 + 10 + 20 + 30 + 40 + 60 + 70 + 80 + 90 + 95 = 500$

$n = 10$ (There are ten scores)

$M = \dfrac{\Sigma X}{n} = \dfrac{500}{10} = 50$

STEP 2 Subtract M from each score $(X - M)$:

Score	X − M	Difference Scores
5	5 − 50 =	−45
10	10 − 50 =	−40
20	20 − 50 =	−30
30	30 − 50 =	−20
40	40 − 50 =	−10
60	60 − 50 =	+10
70	70 − 50 =	+20
80	80 − 50 =	+30
90	90 − 50 =	+40
95	95 − 50 =	+45
	$\Sigma(X - M) =$	0

The difference between M and each score is called the **deviation from the mean**.

If you sum the deviations from the mean, they add to zero. They should always add to zero or very close to zero, depending on how you rounded any decimals. This is a good way to check whether you did the subtracting correctly.

STEP 3 Square each deviation from the mean $(X - M)^2$.

Score	$X - M =$	Deviation from the Mean	$(X - M)^2$ Squared Deviation from the Mean
5	$5 - 50 =$	-45	2,025
10	$10 - 50 =$	-40	1,600
20	$20 - 50 =$	-30	900
30	$30 - 50 =$	-20	400
40	$40 - 50 =$	-10	100
60	$60 - 50 =$	$+10$	100
70	$70 - 50 =$	$+20$	400
80	$80 - 50 =$	$+30$	900
90	$90 - 50 =$	$+40$	1,600
95	$95 - 50 =$	$+45$	2,025
	$\Sigma(X - M) = m$	0	$\Sigma(X - M)^2 = 10,050$

STEP 4 Add up the squared deviations from the mean.

The sum of the squared deviations from the mean is 10,050. This number is called the **sum of squares (SS)**.

The **sum of squares (SS)** is the sum of the squared deviations from the mean.

$$\text{Sum of Squares} = SS = \Sigma(X - M)^2 = 10,050$$

STEP 5 Compute the sample variance by dividing the sum of squares by $n - 1$:

$$s^2 = \frac{\Sigma(X - M)^2}{n - 1} = \frac{10,050}{9} = 1,116.67$$

The variance is a useful measure of variability. It provides a means of quantifying the spread of the data in a way that is unique to each set of data.

Be Here Now

As part of a study, the time it took seven children to complete a project was measured.

Their times (in minutes) were 6, 8, 9, 10, 11, 12, and 14. Compute the variance for this sample.

(Continued)

Be Here Now

Answer

The mean is $M = 10$

Time	Deviation	Squared Deviation
6	$6 - 10 = -4$	16
8	$8 - 10 = -2$	4
9	$9 - 10 = -1$	1
10	$10 - 10 = 0$	0
11	$11 - 10 = 1$	1
12	$12 - 10 = 2$	4
14	$14 - 10 = 4$	16

The sum of squares is $SS = \Sigma(X - M)^2 = 42$

The variance is $s^2 = \dfrac{SS}{n-1} = \dfrac{\Sigma(X - M)^2}{n-1} = \dfrac{42}{7-1} = 7$

When we squared the deviations from the mean in calculating the variance, we moved away from the original numbering system. Accordingly, we need to "refine" the variance in order to be able to examine the distance that individual scores are from the mean in a way that is standardized, so that it is easily understood in any context.

To illustrate the problem, imagine you measured the height of a friend. He is 6.5 feet tall. Now measure him in the metric system; he is 2 meters tall. We need to be able to determine how far from average he is, regardless of the numbering system (English or metric).

To accomplish this refinement—to examine the distance that an individual is from the mean, regardless of the numbering system—we need to "unsquare" the variance. In other words, we need the **standard deviation**.

The Standard Deviation

The **standard deviation** is the square root of the average squared distance each score is from the mean.

The **standard deviation** is a descriptive statistic that describes the variability in terms of standard distances from the mean.

The Standard Deviation Defined

The standard deviation is the square root of the average squared distance each score is from the mean. Said another way, the standard deviation is equal to the square root of the variance.

As with the variance, the definitional formulas for the standard deviation differ, depending on whether we are computing the standard deviation of a sample or the standard deviation of a population.

Standard Deviation	Population	Sample
Symbol	σ	s or SD
Definitional Formula	$\sqrt{\sigma^2}$	$\sqrt{s^2}$
Expanded Definitional Formula	$\sqrt{\sigma^2} = \sqrt{\dfrac{\Sigma(X - \mu)^2}{N}}$	$\sqrt{s^2} = \sqrt{\dfrac{\Sigma(X - M)^2}{n - 1}}$

The Definitional Formula Applied

The standard deviation is a good approximation of the average distance of the scores from the mean. The true average difference each score is from the mean cannot be calculated. Remember that when we computed the variance, the sum of the deviation scores (each score minus the mean) was zero. We had to square the deviations to compute the variance. Now, by computing the square root of the variance, we use a mathematical "trick" to approximate the average difference a score is from the mean. This process puts the standard deviation back to the original numbering system.

Notice:

If you multiply the standard deviation by itself, you get the variance:

$$(s)(s) = s^2$$

If you take the square root of the variance, you get the standard deviation:

$$\sqrt{s^2} = s$$

In Ms. Rodriguez's class, the variance was 1,116.67.

$$\text{Remember: } S^2 = \frac{\Sigma(X - M)^2}{n - 1} = \frac{10,050}{9} = 1,116.67$$

To compute the standard deviation of Ms. Rodriguez's class, we write

$$S = \sqrt{\Sigma(X - M)^2/n - 1} = \sqrt{1,116.67} = 33.42$$

The standard deviation creates a universal language for discussing the average distance from the mean. Once you understand the concept of the standard deviation, you can apply it to just about anything. For example, if you really liked a movie, you might describe it as "clearly 2 standard deviations above the mean." By contrast, if you really disliked a certain movie, you might describe it as "2 standard deviations below the mean."

As shown in these examples, the standard deviation has universal meaning. If something is 2 standard deviations above the mean, it conveys the same message whether we are talking about test scores or ice-skating ability. The standard deviation allows us to make these comparisons in a standard format.

Computation and Interpretation: Sample Variance and Standard Deviation

We have used the definitional formulas to compute the variance and standard deviation of the students' scores in Ms. Rodriguez's class. Now let's use SPSS.

As a reminder, Ms. Rodriguez's class data are as follows:

Ms. Rodriguez's Class Data

5	10	20	30	40	60	70	80	90	95

SPSS is a great time-saver in computing the standard deviation and variance.

→   **Begin by creating an SPSS file with Ms. Rodriguez's class data.**

- To compute the standard deviation and variance, you need to select

    Analyze → *Descriptive Statistics* → *Frequencies.*

- Within the Frequencies menu, select *Statistics* and choose *Std. deviation* and *Variance.*

- You may also "uncheck" *Display frequency tables.*

If you entered the data and followed the instructions correctly, your output should look like that presented in Figure 5.

Across the top row of the output are the labels that define the statistics calculated. Down the first column of the output are the variables that were analyzed:

ms_rodriquez (Ms. Rodriguez's class scores) was the variable analyzed.

N	**Valid = 10**	Ten records (the scores of 10 students in Ms. Rodriguez's class) were analyzed.
	Missing = 0	No data were missing.
	Std. Deviation = 33.4166	The standard deviation of Ms. Rodriguez's class was 33.4166.
	Variance = 1116.667	The variance of Ms. Rodriguez's class was 11,166.67.

Note: The square root of the variance (11,166.67) is 33.4166, the standard deviation.

Statistics

ms_rodriquez

N	Valid	10
	Missing	0
Std. Deviation		33.41656
Variance		1116.667

FIGURE 5 SPSS output: Standard deviation and variance of Ms. Rodriguez's class.

Application: Computing and Interpreting Descriptive Statistics

You can now provide considerable information when you describe a group of scores. You can communicate information about the central tendency and variability of a set of scores. Let's use the scores from Ms. Parady's class to practice describing data.

Ms. Parady's Class Data

5	86	87	94	89	91	92	93	94	95

To begin, you must enter the scores into a data file.

➡ **Begin by creating an SPSS file with Ms. Parady's class data.**

- To compute descriptive statistics, you need to select

  Analyze → *Descriptive Statistics* → *Frequencies.*

- Within the *Frequencies* menu, select *Statistics* and choose *Range, Std. deviation, Variance, Mean, Median,* and *Mode.*

- You may also "uncheck" *Display frequency tables.*

If you entered the data and followed the analysis instructions correctly, your output should look like that presented in Figure 6.

As you look at your output, notice that across the top row of output are the labels that define the statistics calculated. Down the first column of the output are the variables that were analyzed:

ms_parady (Ms. Parady's class scores) is the variable analyzed.

N   Valid = 10   Ten records (the scores of 10 students in Ms. Parady's class) were analyzed.

Statistics

ms_parady

N	Valid	10
	Missing	0
Mean		82.6000
Median		91.5000
Mode		94.00
Std. Deviation		27.43558
Variance		752.711
Range		90.00

FIGURE 6 SPSS output: Central tendency and variability for Ms. Parady's class.

Missing = 0	No data were missing.
Mean = 82.6000	The mean of the class was 82.6.
Median = 91.5000	The median of the class was 91.5.
Mode = 94.00	The mode is 94.
Std. Deviation = 27.43558	The average difference of each score from the mean is 27.43558.
Variance = 752.711	The average of the squared deviations from the mean is 752.711.
Range = 90	The range in scores (high minus low) is 90.

Reporting Descriptive Statistics in APA Style

Often, descriptive statistics are presented in a table. You always describe what the reader will find in the table. Most of the time, you need not include all the statistics you computed.

You might write something like the following:

As shown in Table 1, the mean of the scores in Ms. Parady's class was 82.60, while the standard deviation was 27.44 and the variance was 752.71.

Note that, in APA format, only two decimal places are typically reported. So you may need to round the numbers when you make your tables.

Note also that, in APA format, the column heading for standard deviation is typically displayed as *SD,* but can be shown as *S* (see Table 2).

TABLE 1
Descriptive Statistics: Ms. Parady's Class

Class	*M*	*SD*	s^2
Ms. Parady	82.60	27.44	752.71

TABLE 2
Descriptive Statistics: Ms. Parady's Class

Class	*M*	*S*
Ms. Parady	82.60	27.44

Now let's consider another example. A social worker was interested in describing the income levels of 12 participants in a special program. Their incomes, in dollar amounts, are as follows:

Income					
15,000	16,500	14,600	12,000	16,000	16,500
10,000	14,900	21,000	18,000	14,500	22,000

Use SPSS to calculate the mean, range, standard deviation, and variance. Report the results in APA format.

To begin, you must enter the scores into a data file.

→ **Begin by creating an SPSS file with the income data.**

- To compute descriptive statistics, you need to select

  Analyze → Descriptive Statistics → Frequencies

- Within the Frequencies menu, select *Statistics* and choose *Range, Std. deviation, Variance, Mean, Median,* and *Mode.*

- You may also "uncheck" *Display frequency tables.*

If you entered the data correctly and followed the analysis directions, your output should be similar to that presented in Figure 7.

To share the descriptive information, you might write the following:

As shown in Table 3, the mean income for participants was $15,916.67 while the standard deviation was $3,355.28.

Take a look at the original data. The participant who earned $22,000 had above-average income. You might say that that person's income was about two standard deviations above the mean. In the next chapter, we demonstrate how the mean and standard deviation are used to describe the relative standing of an individual within a group.

Statistics

Income

N	Valid	12
	Missing	0
Mean		15916.6667
Median		15500.0000
Mode		16500.00
Std. Deviation		3355.27626
Variance		11257878.788
Range		12000.00

FIGURE 7 SPSS output: Central tendency and variability for income.

TABLE 3
Descriptive Information: Income
Data, in Dollars, for Special Program

M	Range	SD
15,916.67	12,000.00	3,355.28

It's Out There... Curhan and Pentland (2007) examined some speech characteristics of participants assigned to play the role of either middle manager or vice-president. Among the voice characteristics measured was mirroring: the number of times participants used a reciprocal utterance, such as "uh-huh." As shown in Table 4, the number of mirroring phrases uttered by middle managers was 7.52, with a standard deviation of 4.22. That of vice-presidents was 7.64, with a standard deviation of 4.55.

TABLE 4
Descriptive Statistics: Voice Characteristics of Participants
Role Playing Middle Managers and Vice-Presidents ($n = 112$)

Variable	Middle Management		Vice-Presidents	
	M	*SD*	*M*	*SD*
Mirroring	7.52	4.22	7.64	4.55

SUMMARY

Measures of variability add to the ways in which you can describe data. Measures of central tendency (discussed in Chapter 3) provide one number that describes the typical score; measures of variability describe the spread of all the scores. There are several choices as to which measure of variability you should use, depending on your data and what you want to communicate.

The **range**—the difference between the highest value and the lowest value within the distribution—gives a rough idea of the spread of the scores, but does not give any information about the scores themselves. The **interquartile range** provides a number that reflects the spread of the scores in the middle 50 percent of the distribution. The **variance**—the average of the squared deviations from the mean—is an indicator that allows you to compare how dispersed the scores are within different distributions. Finally, the **standard deviation**—the square root of the average squared distance each score is from the mean—allows you to estimate how far any one score is from the mean. An additional benefit of the standard deviation is that it is independent of the numbering system used.

Reflect
Life is Messy. Measures of variability reflect the amount of mess.

PRACTICE

Variability

1. A researcher wants to study the effect of light on mood. She uses a laboratory room at the university. For some students, red lights were

used, for others blue lights were used, and for still others, white lights were used. For each of the following items, state whether it is or is not a variable and justify your answer.

The room

The lights

The mood scores

2. For each of the following pairs of variables, if we measure 10 people, identify the item that is likely to have greater variability.
   The number of siblings each person has or the number of CD's each owns

   Their shoe sizes or SAT scores

   The number of t-shirts each owns or the number of cars each owns

Range and Interquartile range

3. A psychologist gives a self-esteem test to 10 youngsters. Following are their scores:

15	2	89	25	19	76	14	12	19	20

   What is the range of the self-esteem scores? What is the interquartile range?

4. Potential voters were asked to use a scale of 1 to 10 to indicate their approval of the mayor's performance. The observed survey scores are as follows:

7	2	5	6	4	6

   What is the range of approval ratings? What is the interquartile range?

5. A research and development (R & D) team at a cell phone manufacturing company asked a sample of individuals how many cell phone numbers were stored on their personal cell phones. The responses were as follows:

18	14	23	96	44	25	89

   What is the range of the stored cell phone numbers? What is the interquartile range?

Variance

6. A special-needs teacher records the number of trials it takes each of her students to learn a new task. The scores are as follows:

5	9	10	11	15

   Use the definitional formula to compute the sample variance for the scores.

7. Following are the achievement scores for 15 high school students enrolled in a math class:

55	61	78	96	72	55	69	89
94	88	78	83	94	85	71	

Use SPSS to compute the variance of the scores.

8. A psychologist administers a depression inventory to 10 patients diagnosed with depression. Next, she administers the inventory to eight participants not diagnosed with depression. The scores of the two groups are as follows:

Diagnosed:	86	58	59	68	59
	79	89	85	79	90
Non-Diagnosed:	15	96	42	85	
	36	49	89	12	

Use SPSS to compute the variance for each set of scores. Which group appears to have more variability? How do you know this?

Standard Deviation

9. The variance of a sample is 25. What is the standard deviation?
10. Use the definitional formula to compute the sample standard deviation of the following quiz scores in a statistics class:

10	11	14	18	21	22

11. Use SPSS to compute the standard deviation of the following prices of airline tickets:

115	125	189	152	81	56	195	103	79	88
125	119	154	189	89	96	91	139	147	151

Descriptive Statistics

12. A test developer administers a new test to a group of 15 students. He wants to describe their test scores. Using SPSS, compute the mean, median, mode, range, variance, and standard deviation of the students' test scores:

18	58	65	78	89	65	79	83
98	77	96	84	22	46	52	

13. A researcher was interested in the time (in minutes) it takes students to read a chapter in an introductory-level textbook. She collected data on 18 participants. Her data are as follows:

29	39	45	65	29	18	37	29	58
81	25	41	48	45	38	26	33	48

Using SPSS, compute the mean, median, mode, range, variance, and standard deviation of the reading times.

APA Style

14. An experimental psychologist recorded the number of trials each of 10 rats needed to learn a new maze. Her results were as follows:

Learning trials:	5	8	12	14	12
	27	10	15	19	20

Create an APA-style table that shows the mean, standard deviation, and variance for the learning trials. Write out an explanation that highlights each item in the table.

15. A test developer wanted to report the characteristics of a sample of 12 persons. One of the variables he collected data on was age. Following are his results:

Age Data

35	88	50	27	56	40
51	40	68	72	29	38

Create an APA-style table that includes the mean, median, range, and variance of the age data collected. Write out an explanation that highlights each item in the table.

SOLUTIONS

Variability

1. The room did not vary; only one room was used.
   The lights varied; three lights (red, blue, and white) were used.
   The mood scores are likely to vary; different participants will have different scores.

2. The number of CD's is likely to have greater variability than the number of siblings. For example, participants might have from 0 to 10 siblings, whereas they might have from 0 to 100 CD's.

Participants' SAT scores are likely to be more variable than their shoe sizes: SAT scores are likely to range from 200 to 800, while shoe size is likely to range from 4 to 14.

The numbers of t-shirts participants own are likely to be more variable than the number of cars they own: Participants are likely to own from 0 to 30 t-shirts, while they are likely to own from 0 to 2 cars.

Range and Interquartile Range

3. The range is 87, computed by subtracting the low score (2) from the high score (89).

   The interquartile range is 24.25, computed by subtracting the 25th percentile (13.5) from the 75th percentile (37.75).

   Interquartile range = 37.75 − 13.5 = 24.25

4. The range is 5, computed by subtracting the low score (2) from the high score (7).

   The interquartile range is 2.75, computed by subtracting the 25th percentile (3.5) from the 75th percentile (6.25).

   Interquartile range = 6.25 − 3.5 = 2.75

5. The range is 82, computed by subtracting the low score (14) from the high score (96).

   The interquartile range is 71, computed by subtracting the 25th percentile (18) from the 75th percentile (89).

   Interquartile range = 89 − 18 = 71

Variance

6. Variance = 13

   Steps:

   First, compute the mean: $M = \dfrac{\Sigma X}{n} = \dfrac{50}{5} = 10$

   Next, compute the SS (squared deviations from the mean):

Score	X − M	=	Deviation from the Mean	$(X - M)^2$ Squared Deviation from the Mean
5	5 − 10	=	−5	25
9	9 − 10	=	−1	1
10	10 − 10	=	0	0
11	11 − 10	=	+1	1
15	15 − 10	=	+5	25
				$(X - M)^2 = 52$

   $$\textbf{Variance} = s^2 = \frac{\Sigma (X - M)^2}{n - 1} = \frac{52}{5 - 1} = \frac{52}{4} = 13$$

7. Variance = s^2 = 188.838

8. Variance (diagnosed) = s^2 = 169.289
   Variance (non-diagnosed) = s^2 = 1102.857
   The score of the nondiagnosed group appears to larger than that of the diagnosed group.

Standard Deviation

9. $s = \sqrt{s^2} = \sqrt{25} = 5$

10. Standard Deviation = s = 5.09908
    Steps:

    First compute the mean: $M = \dfrac{\Sigma X}{n} = \dfrac{96}{6} = 16$

    Next, compute the SS (squared deviations from the mean):

Score	X − M	=	Deviation from the Mean	$(X - M)^2$ Squared Deviation from the Mean
10	10 − 16	=	−6	36
11	11 − 16	=	−5	25
14	14 − 16	=	−2	4
18	18 − 16	=	+2	4
21	21 − 16	=	+5	25
22	22 − 16	=	+6	36
				$(X - M)^2 = 130$

$$\textbf{Variance} = s^2 = \frac{\Sigma(X - M)^2}{n - 1} = \frac{130}{(6 - 1)} = \frac{130}{5} = 26$$

Standard deviation = $s = \sqrt{s^2} = \sqrt{26} = 5.09902$

11. Output:

Statistics

Ticket

N	Valid	20
	Missing	0
Mean		124.1500
Median		122.0000
Mode		125.00[a]
Std. Deviation		39.80118
Variance		1584.134
Range		139.00[a]

[a] Multiple modes exist. The smallest value is shown.

Descriptive Statistics

12. Output:

Statistics

new_test

N	Valid	15
	Missing	0
Mean		67.3333
Median		77.0000
Mode		65.00
Std. Deviation		24.45891
Variance		598.238
Range		80.00

13. Output:

Statistics

reading_time

N	Valid	18
	Missing	0
Mean		40.7778
Median		38.5000
Mode		29.00
Std. Deviation		15.58489
Variance		242.889
Range		63.00

APA Style

14. Output:

Statistics

learning_trials

N	Valid	10
	Missing	8
Mean		14.2000
Median		13.0000
Mode		12.00
Std. Deviation		6.42564
Variance		41.289
Range		22.00

As can be seen in Table 5 the mean number of trials was 14.20 while the standard deviation was 6.42.

TABLE 5
Descriptive Information:
Number of Trials Needed
to Learn a Maze

M	Range	SD
14.20	22	6.4.

15. Output:

Statistics

age_data

N	Valid	12
	Missing	6
Mean		49.5000
Median		45.0000
Mode		40.00
Std. Deviation		18.64745
Variance		347.727
Range		61.00

As shown in Table 6, the mean age of participants was 49.50 while the standard deviation of the age scores was 18.65.

TABLE 6
Descriptive Information:
Age of Participants

M	Mdn	Range	SD
49.50	45.00	61	18.65

-5-

Score Transformations

F O C U S | This chapter introduces you to *z* scores. *z* scores transform observed scores to a scale that has a mean of zero and a standard deviation of one. When scores have been transformed, understanding and interpreting statistical questions about life's occurrences often become easier.

Why We Need to Transform Scores

A **raw score** is an original value.

A raw score is an original data point, before anything is done to it. For example, if you take an exam that is based on 25 points and you earn a 25, your raw score would be 25. Given that the highest possible score was 25, you'd be ready to celebrate. After all, everyone knows that 25 out of 25 is a perfect score, right? However, what would happen if you told your best friend that you just got a 25 on an exam? Unless your friend knew that the exam was based on 25 points, he or she might not be ready to join in the celebration. There could be some confusion! However, if you transformed your score into a percentage and said that you got 100% on the test, then your friend (and just about anyone) would understand why you were so pleased.

A **standard score** is a score that has been transformed.

In statistics, a raw score that has been transformed into a score expressed in terms of a standard deviation is easily interpreted. Transformed scores not only allow us to understand data, but also to make comparisons among scores that may have been calculated by different numbering methods. Once a score has been transformed into standard deviation units, it doesn't matter whether the original numbering system was in inches or miles, or whether the original scores were based on a total of 25, or 100, or 62. Saying that a score is two standard deviations above the mean or one standard deviation below the mean conveys the meaning of that score, no matter what the score started out as.

Think of a hypothetical exam that was based on 50 points. Let's say that the mean on the exam was 35 and the standard deviation was 5. What would your score be if it were one standard deviation above the mean?

Your score would be 40.

If we keep the mean at 35, but change the standard deviation to 2, what will your score be if it is one standard deviation above the mean?

Your score at one standard deviation above the mean would be 37.

It is important to look at both the mean *and* the spread of the scores (the standard deviation) when you interpret data. When we use transformed scores, we do just that.

Be Here Now

Two classes have just received their scores on the same exam. The exam was based on 40 points. In Class A, the mean was 30 and the standard deviation was 4. In Class B, the mean was also 30, but the standard deviation was 3. Your score was two standard deviations above the mean. In which class would your raw score be higher? Why?

Answer

You would have earned a higher score in Class A.

Two standard deviations above the mean would be a score of 38 in Class A. Two standard deviations above the mean would be a score of 36 in Class B.

Here's why:

The standard deviation marks off the distance from the mean. The larger the standard deviation, the further from the mean will be your score.

z Scores

A **z score** is the number
of standard deviations
above or below the mean
a particular score falls.

z Scores: The Basics

A common type of score transformation results in a *z* **score**. The goal of
z-score transformations is to standardize any distribution so that the mean
is equal to zero and the standard deviation is equal to one. A z score is the
number of standard deviation units above or below the mean a particular
score falls. With z scores, we can share information about scores that are
on any scale (for example, how tall you are relative to other students, your
GPA, etc.).

It is important to remember
that, with z scores, the + and
– signs tell you whether the score
falls above or below the mean and
the actual numbers tell you the
distances from the mean. z scores
are based on a distribution with
a mean of zero and a standard deviation of one. So, -2.0 is two standard
deviations below the mean and $+1.5$ is one-and-one-half standard deviations
above the mean.

> **Reflect**
> A z score is a z score is a z score is
> a z score.
> (With apologies to Gertrude Stein.)

Let's start with the formula for the z score:

$$z = \frac{X - \mu}{\sigma}$$

In this formula,

X represents the raw score,

μ represents the mean of the population, and

σ represents the standard deviation of the population.

If we put the formula for the z score into words, it would look like this:

z = the raw score minus the population mean,
divided by
the population standard deviation

This computation is what transforms raw scores into standard scores. In
the case of z scores, the resulting mean will always be zero and the standard
deviation will always be one.

For example, let's say that you are working in a bookstore. If the average
hourly wage in the population for bookstore workers was $10.00, with a stan-
dard deviation of $2.00, and you were earning $11.00 per hour, what z score
would represent your hourly rate?

Use the formula for the z score to calculate the answer:

$$z = \frac{X - \mu}{\sigma}$$

$$X = 11.00$$

$$\mu = 10.00$$

$$\sigma = 2.00$$

$$z = \frac{X - \mu}{\sigma}$$

$$z = \frac{11.00 - 10.00}{2.00} = \frac{+1.00}{2.00} = +.5$$

Your hourly rate of $11.00 is half a standard deviation above the mean.

Let's try another example. Suppose you decide to change jobs for the summer. You decide to be a camp counselor. The pay isn't great, but at least you'll be outside. If the hourly rate in the population is $8.00, with a standard deviation of $3.00, and you are hired for $6.00 per hour, what is your pay, expressed as a z score?

Again, use the formula for the z score to calculate the answer:

$$z = \frac{X - \mu}{\sigma}$$

$$X = 6.00$$

$$\mu = 8.00$$

$$\sigma = 3.00$$

$$z = \frac{X - \mu}{\sigma}$$

$$z = \frac{6.00 - 8.00}{3.00} = \frac{-2.00}{3.00} = -.67$$

Your hourly rate of $6.00 is two-thirds of a standard deviation below the mean.

Be Here Now

Let's say that the average number of books in a student's home is 125, with a standard deviation of 35. Calculate the z scores in each of the following situations:

1. You have 150 books in your home.
2. You have 82 books in your home.

Answers

1. $z = \dfrac{150 - 125}{35} = +.71$

2. $z = \dfrac{82 - 125}{35} = -1.23$

z Scores: Comparing Scores

Sometimes we want to compare things. In high school, you may have taken both the SAT and the ACT. The SAT has a mean of 500 and a standard deviation of 100; the ACT has a mean of 20 and standard deviations ranging between 4 and 5. If you wanted to compare how well you did on these two college entrance exams, you would need to transform your scores into *z* scores. It is often said that you can't compare apples and oranges. With *z* scores, you can!

For example, imagine that your scores on the SAT and the ACT are as follows, with some information about the exams also shown.

	ACT	**SAT**
Your Score	25	450
The Tests	$\mu = 20$	$\mu = 500$
	$\sigma = 5$	$\sigma = 100$

First, compute your *z* scores for each exam.
Start with the formula for the *z* score:

ACT	**SAT**
$z = \dfrac{X - \mu}{\sigma}$	$z = \dfrac{X - \mu}{\sigma}$

Now replace the variables with numbers:

$$z = \frac{25 - 20}{5} \qquad z = \frac{450 - 500}{100}$$

$$z = +1 \qquad z = -.5$$

Were you close? Looking only at your raw scores, you might have thought you did better on the SAT than the ACT; after all, your original SAT score was higher than your original ACT score. But in fact, you did better on the ACT. Your *z* score on the ACT was +1.0, while your *z* score on the SAT was −.5. When the raw scores are transformed into the same metric (or numbering system), which, in this case, is *z* scores, it becomes evident that your performance on the ACT was higher than your performance on the SAT.

It is important to note that sometimes a negative *z* score is desirable. Time provides a good example of how this can happen. Think about scores in athletic races. Athletes want personal times that are *faster* than the population averages; their time scores must be smaller numbers than the population averages. So, their optimal *z* scores would have *negative* values.

For example, let's say that Tamsin and Esther are two female college athletes. Tamsin runs the 400 meter dash and Esther is a speed skater in the 500 meter event. They wonder who is the faster athlete. They do a little research and find out that the average speed for female college athletes is 53.35 seconds, with a standard deviation of 2.4 seconds, for the 400 meter dash and 49.13 seconds, with a 3.1 second standard deviation, for the 500 meter event in speed skating. Esther has just finished a statistics class and remembers that if she transforms their times into z scores, she can figure out who is the faster athlete. Here are the data with their best scores:

Tamsin	Esther
400 meter dash	**500 meter speed skating**
X = 54.08	X = 50.04
μ = 53.35	μ = 49.13
σ = 2.40	σ = 3.10

It looks close!

To find the answer, calculate a z score for each athlete:

Tamsin	Esther
400 meter dash	**500 meter speed skating**
$z = \dfrac{X - \mu}{\sigma}$	$z = \dfrac{X - \mu}{\sigma}$
$z = \dfrac{54.08 - 53.35}{2.40}$	$z = \dfrac{50.04 - 49.13}{3.10}$
$z = \dfrac{.73}{2.4}$	$z = \dfrac{.91}{3.1}$
$z = +.30$	$z = +.29$

Although both women's speeds are slower than the average speed for female college athletes in their respective sports (positive z scores indicate that both Tamsin and Esther took longer, more time than average), Esther can now claim that she is a slightly faster athlete than Tamsin! Her smaller z score represents a time that is faster (closer to the mean).

Let's try another example. Suppose an owner of two music stores wants to compare sales figures for the month of May. He knows that, over the past five years, his store in Westland has sold $4,600 worth of CD's, on average, with a standard deviation of $125. Over the same time, his store in Cranesford has sold $4,800 worth of CD's, on average, with a standard deviation of $135. Here are the complete data:

Westland	Cranesford
X = $4,722	X = $4,875
μ = $4,600	μ = $4,800
σ = $125	σ = $135

To begin, calculate a z score for each location:

Westland	Cranesford
$z = \dfrac{X - \mu}{\sigma}$	$z = \dfrac{X - \mu}{\sigma}$
$z = \dfrac{4722 - 4600}{125}$	$z = \dfrac{4875 - 4800}{135}$
$z = \dfrac{122}{125}$	$z = \dfrac{75}{135}$
$z = +.98$	$z = +.56$

The standardized results indicate that this May the Westland store is selling almost a full standard deviation more than in previous years. By comparison, sales in the Cranesford store are just over one-half of one standard deviation. The owner of these stores can now make some decisions about how to proceed with his businesses.

Reflect

With z scores, an apple by any other name becomes an apple.

Converting *z* Scores to a New Distribution

It is sometimes useful or desirable to transform scores from one distribution to another, new distribution. For example, you may have your score on a particular IQ test, but would like to know what that score would be if another type of IQ test were given. You can find this out by converting the z score from the original test to the scale of the new test.

Several steps are necessary:

STEP 1   First, calculate the z score for the raw-score value from the original distribution, using the formula for the z score:

$$z = \frac{X - \mu}{\sigma}$$

STEP 2   Next, algebraically rearrange the formula for the z score so that you can compute a hypothetical score on a new scale:

$$X = \sigma z + \mu$$

Be Here Now

Let's say that you eat eight apples and five oranges each week. Determine whether you eat more apples or oranges in a week, compared with the typical student.

Look at the following hypothetical data:

Apples	Oranges
$X = 8$	$X = 5$
$\mu = 7$	$\mu = 4$
$\sigma = 2$	$\sigma = 1$

Converting z scores to a new distribution.

Answer

$$z = \frac{X - \mu}{\sigma} \qquad\qquad\qquad z = \frac{X - \mu}{\sigma}$$

$$z = \frac{8 - 7}{2} \qquad\qquad\qquad z = \frac{5 - 4}{1}$$

$$z = \frac{1}{2} \qquad\qquad\qquad\quad z = \frac{1}{1}$$

$$z = +.50 \qquad\qquad\qquad z = +1.00$$

Compared with the typical student, you eat more oranges.

STEP 3   Now, compute the raw score for the new distribution by inserting the z score into the formula in step 2.

Let's try a problem:

Many IQ tests have a mean of 100 and a standard deviation of 15. Let's say that there is a new IQ test that looks promising; it has a mean of 200 and a standard deviation of 25. To communicate the meaning of the scores for the new IQ test, it may be helpful to transform them to a commonly used IQ score scale (a mean of 100 and a standard deviation of 15). What would a raw score of 235 on the new IQ test be on the original IQ test? First, set up the problem with as much information as you know:

Original IQ Test	New IQ Test
$\mu = 100$	$\mu = 200$
$\sigma = 15$	$\sigma = 25$
$X = ?$	$X = 235$
$z = ?$	

Start by calculating the z score for the raw score from the *new* IQ test:

$$z = \frac{X - \mu}{\sigma} = \frac{235 - 200}{25} = \frac{35}{25} = +1.4$$

Now, use the z score of +1.4 to calculate the raw score for the original IQ test:

$$X_{\text{original distribution}} = (\sigma_{\text{original distribution}})(z_{\text{based on the new distribution}}) + \mu_{\text{original distribution}}$$

or

$$X = \sigma z + \mu$$

$$X = 15(1.4) + 100 = 21 + 100 = 121$$

The raw score of 235 on the new IQ test would be equivalent to 121 on the original IQ test.

What is the z score corresponding to the new raw score? The z score for the new raw score is actually the same as the z score for the original distribution: z = +1.4. You can check this by calculating the z score for the original IQ score:

$$z = \frac{X - \mu}{\sigma} = \frac{121 - 100}{15} = +1.4$$

That the two z scores are the same makes sense if you remember that all you have done is convert a score from one scale to another, using a distribution that has a mean of zero and a standard deviation of one. You haven't fundamentally changed the basic score or the shape of the distribution. If, for example, you got an 87 on an exam, transforming that 87 to any number of standardized scales wouldn't change the fact that you earned an 87 on the exam. You cannot suddenly get a 98 (alas) or (thank goodness) a 42!

Often, we transform scores to a scale that is commonly used or understood. **Scale scores,** rather than z scores, are frequently used to report scores on exams to the public, parents, and students. For example, with IQ test scores we often transform z scores into a scale having a mean of 100 and standard deviation of 15.

One good reason for this kind of transformation is that it is possible to have negative z scores (scores that fall below the mean) and sharing test data in the form of negative numbers may alarm those who have yet to master z scores. For example, a z score of −1 on an IQ test would equal a scale score of 85. Scale scores eliminate the negative numbers and, many times, the anxiety!

Let's return to the SAT and the ACT for another example:

Suppose you took the SAT and your verbal score was 560. You wonder what your score would have been had you taken the ACT instead. You can use the following information to figure out what your hypothetical raw ACT score would be:

*Often, we transform z scores to more commonly used **scale scores.***

SAT Exam	ACT Exam
$\mu = 500$	$\mu = 20$
$\sigma = 100$	$\sigma = 4$
$X = 560$	$X = ?$
$z = ?$	

Start by calculating the z score for the raw score from the SAT exam:

$$z = \frac{X - \mu}{\sigma} = \frac{560 - 500}{100} = \frac{60}{100} = +.60$$

Now, use the z score of +.60 to calculate the raw score for the ACT exam:

$$X_{ACT\,distribution} = (\sigma_{ACT\,distribution})(z_{SAT\,distribution}) + \mu_{ACT\,distribution}$$

$$or$$

$$X = \sigma z + \mu$$

$$X = 4(.60) + 20 = 2.4 + 20 = 22.4$$

The original raw score of 560 on the SAT verbal exam would be 22.4 on the ACT exam.

What is the z score corresponding to the new raw score? As before, the z score is the same as the z score for the original distribution: $z = +.60$. Again, you can check this by calculating the z score for the new distribution:

$$z = \frac{X - \mu}{\sigma} = \frac{22.4 - 20}{4} = \frac{2.4}{4} = +.60$$

Be Here Now

If there were a Statistics Abilities Test and your raw score on it was 181, what would your raw score be if the mean and standard deviation were changed to those listed for the new test?

Consider the following data:

Original Statistics Abilities Test	New Statistics Abilities Test
$\mu = 150$	$\mu = 250$
$\sigma = 30$	$\sigma = 50$
$X = 181$	$X = ?$
$z = ?$	

Answer

Start by calculating the z score for the raw score from the *original test*:

$$z = \frac{X - \mu}{\sigma} = \frac{181 - 150}{30} = 1.03$$

Now use the z score of 1.03 to calculate the raw score for the *new test*:

$$X = \sigma z + \mu$$

$$X = 50(1.03) + 250 = 51.5 + 250 = 301.5$$

The original raw score of 181 would be 301.5 on the new test.

Normal Distribution

The **normal distribu-
tion** is mathematically
defined. If we measured
the population on a wide
range of variables, the distributions created would
approximate the normal
distribution.

Whether we measure the population IQ, shoe size, or attitude toward
homework, the shape of each of
these distributions will be similar.
Such distributions, as well as those
for many other variables, are said to
be **normal distributions.**

The normal distribution is symmetrical (see Chapter 2), and for each score
point X, there is a particular height. Figure 1 depicts a normal distribution.

For example, the average male in the population reports that he is able to
bench-press 135 pounds (http:www.pipeline.com/~dada3zen/average.htm).
We estimate the standard deviation to be 30. This information is used to label the normal distribution presented in Figure 2. As you look at that figure,
notice that many men can bench-press 135 pounds (the height of the curve is
greatest at $X = 135$ pounds), whereas very few can bench-press 205 pounds
(the height of the curve is small at $X = 205$ pounds).

Although we marked off the x-axis with the number of pounds bench-
pressed, we could have used z scores.

Now imagine that you are the owner of a gym. You can use the normal curve to help you determine the number of barbells to buy for each
weight. Figure 3 presents a depiction of your order for barbells. Notice that
you need to order many barbells near the mean weight bench-pressed. The
further from the mean weight bench-pressed, the fewer barbells you need
to order.

z Scores and the Normal Distribution

The Unit Normal Table in Appendix A1 may be used in combination with
z scores to answer a variety of questions about particular scores in a normal
distribution.

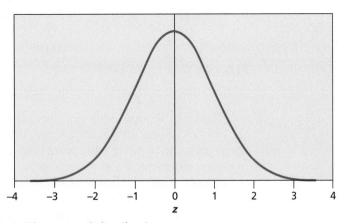

FIGURE 1 The normal distribution.

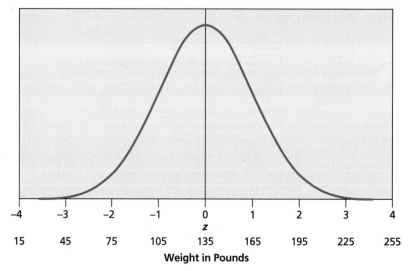

FIGURE 2 Population distribution of pounds bench-pressed by average man.

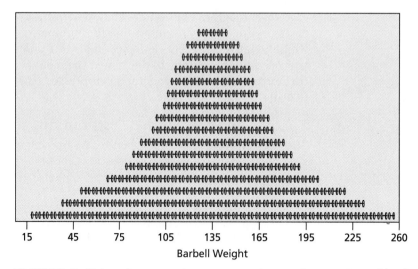

FIGURE 3 Using the normal curve to estimate the number of barbells to order.

The Unit Normal Table presents all the proportions under the normal curve—below, and above any z score. Without this table, we would need to use calculus to identify a proportion of the normal curve! When we transform raw scores into z scores, we can use the Unit Normal Table to determine the proportion of the normal distribution that is contained in an area associated with a score or a set of scores. We can also determine the probability of obtaining a z score within a particular range. Note that you cannot look up raw scores on the Unit Normal Table—the table uses z scores, not raw scores.

In fact, the universal language of z scores allows us to use the Unit Normal Table for any variable that is normally distributed—IQ scores, SAT scores, adult height, daily calorie consumption, etc.

Looking at the table, we see that the first column refers to the z scores, the second column to proportions in the *body* of the distribution, and the third column to proportions in the *tail* of the distribution. There are no negative z scores in the Unit Normal Table, so if you are looking up a z score with a negative value, just (temporarily) disregard the negative sign. Rather, as will be seen shortly, the sign of the z score will help you determine whether to use the information in the tail or body columns of the Unit Normal Table.

At this point, as you work on problems, it is a good idea to sketch the distribution and indicate the area in which you are interested on the drawing. This will help you identify the area (body or tail) of the normal curve, as well as help you check to see if your final answer looks reasonable. At first, the work may seem obvious and you may be tempted to skip the sketch, but if you get in the habit of drawing the distribution now, you'll be in a good position when the problems get a bit more difficult. And remember, once you obtain your answer, you can check your sketch to see if what you indicated initially makes sense in relation to your final answer.

Use the body column of the Unit Normal Table if the area shaded crosses the mean.

Use the tail column of the Unit Normal Table if the area shaded *does not* cross the mean.

In using the Unit Normal Table, if the area of the normal curve that you shaded includes the mean, use the body column. If the area you shaded does not include the mean, use the tail. These situations are depicted in Figure 4.

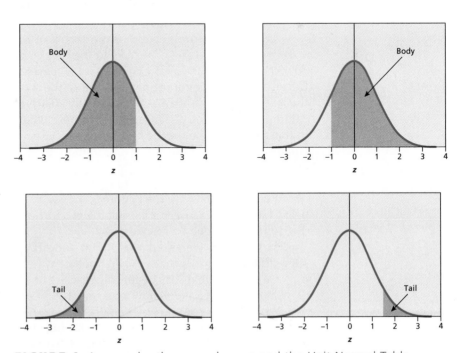

FIGURE 4 Area under the normal curve and the Unit Normal Table.

z Scores, Proportions, and Probabilities

The Unit Normal Table may be used to determine the proportion of the normal distribution above or below any *z* score, as well as between any two *z* scores.

For example, we could use the Unit Normal Table to determine the proportion of the distribution that is below the *z* score of +1.5. First, as shown next, draw the normal curve and shade in the area in which you are interested.

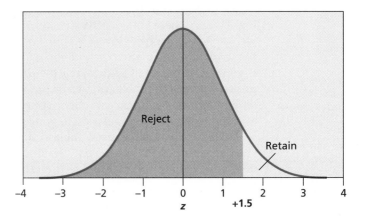

In the Unit Normal Table, find the *z* score of 1.5. Notice that the listing of *z* scores begins at 0 and continues to the *z* score of 4.00. If we scan through all of the listings in the *z* score column, we will eventually find the *z* score of 1.5. Information about proportions relative to the *z* score of 1.5 is presented in the *row* that begins with a *z* of 1.5. Directly adjacent to the *z* score column is a column labeled "body." This number represents the proportion of the distribution that is at or below the *z* scores of +1.50. So, reading this from the table, we note that .9332 of the entire distribution, or 93.32% of the distribution, is at or below the *z* score of 1.5.

If, instead, we had wanted to know the proportion or percentage of the distribution that is *above* the *z* score of 1.5, we would look in the column labeled "tail." In this case, notice that the proportion is .0668; thus, .0668 of the distribution is above a *z* score of 1.5.

The question we asked was, What was the proportion of the normal curve that is at or below $z = +1.5$? Another way to have asked the question is, What is the **probability** of getting a *z* score of +1.5 or less? The proportion of the normal curve at or below a *z* score of +1.5 is .9332. Thus, the probability of getting a *z* score of +1.5 or less is .9332.

We can also look at the proportion falling above a *z* score of +1.5. When we looked in the tail column of the Unit Normal Table for a *z* of +1.5, we found that the proportion was .0668. Hence, the probability of getting a z score above +1.5 is .0668.

We can now ask, Is it more likely to get a z score above or below $z = +1.5$? The answer is "below"; this is because a greater portion of the normal curve falls below (.9332) a z of +1.5 compared with above (.0668).

Thus, we can say that it would be more likely to earn a score at or below $z = +1.5$, compared with above. We can also say that the probability of earning a z score below +1.5 is higher (.9332) than the probability of earning a z score above +1.5 (.0668).

We can ask the question in another format. For example, if we are interested in the proportion of the normal curve that falls below a z score of 1.5, we might write:

$$p(z < 1.5)$$

The closer p is to 1, the greater is the likelihood of the event in question.

The closer p is to 0, the lower is the likelihood of the event in question.

Probability and proportion are similar concepts; both are symbolized by a lowercase *p*. Both vary from 0 to 1. The closer the probability is to 1, the greater is the likelihood of the event in question. The closer the probability is to zero, the lower is the likelihood of the event in question.

Be Here Now

1. Locate each of the following along the *x*-axis of a normal curve:
   A. $z = -1.5$
   B. $z = +1$
   C. $z = +.5$
   D. $z = -.5$

2. What proportion of the normal distribution is shown below each z.
   A. $z = +1.00$?
   B. $z = +2.00$
   C. the mean, or $z = 0$
   D. $z = -1.00$

Answers

1.

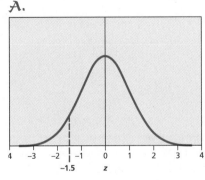

A.

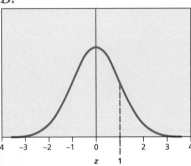

B.

(Continued)

C.

D.

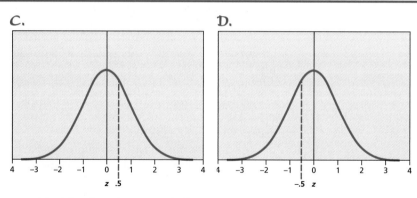

2. A. .84 of the distribution is below $z = +1.00$.
   B. .98 of the distribution is below $z = +2.00$.
   C. .50 of the distribution is below the mean.
   D. 16 of the distribution is below $z = -1.00$

Let's try another problem:

What is the probability of obtaining a score **above** 123 on an IQ test with $\mu = 100$ and $\sigma = 15$?

This question can be rewritten as

What is $p(X > 123)$?

In this example, we do not know the z score, but we need to calculate the z score in order to determine the probability (or proportion).

We begin by drawing the normal distribution and calculating the z score. Because we are interested in scores **above** $X = 123$, we shade in the portion of the curve **above** (to the right of) the z score that appears to be associated with $X = 123$.

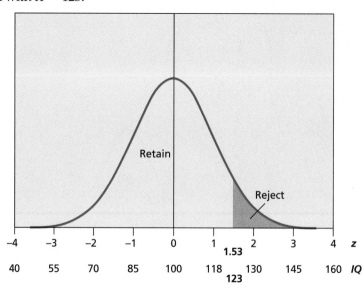

Now we compute the z score:

$$z = \frac{X - \mu}{\sigma} = \frac{123 - 100}{15} = \frac{23}{15} = +.1.53$$

Next, we go to the Unit Normal Table and look up 1.53 in the first z-score column. Look at your picture. You are interested in the proportion above +1.53. The area you shaded does *not* include the mean. Because the mean is not included in the shaded area, look in the column associated with proportions **in the tail.** The proportion associated with scores above z = 1.53 is .0630. Thus, the probability of obtaining an IQ score of 123 or above is .0630. Another way to think about the situation is that roughly 6% of the population would be expected to score above 123 on this IQ test. Conversely, about 94% would be expected to score at or below a score of 123. (See the column associated with proportions in the body of the distribution.)

With the Unit Normal Table, if you are a bit confused about when to look in the column for the tail of the distribution and when to look in the column for the body of the distribution, here are some hints:

Hints when using the Unit Normal Table

- Always draw a picture of the normal curve before trying to use the Unit Normal Table.
- Locate the z score in which you are interested. Shade in the area in which you are interested.
- If the shaded area crosses the mean (z = 0), look in the column for the body of the distribution.
- If the shaded area does not cross the mean (z = 0), look in the column for the tail of the distribution.

Be Here Now

For each of the following z scores, in which column would you look, using the Unit Normal Table to find the proportion under the curve?

1. p(z < +.71)

2. p(z > +.53)

3. p(z < −.33)

4. p(z > −.40)

Answers

1.

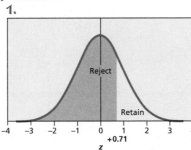

2.

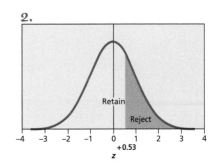

(Continued)

Be Here Now

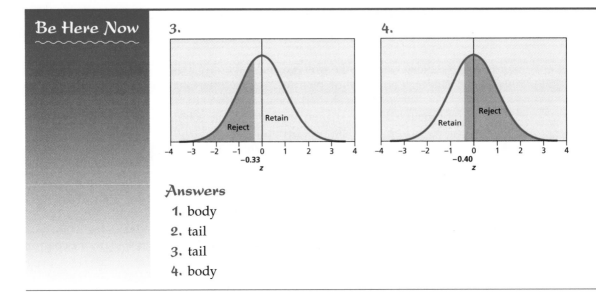

3.

Retain

Reject

−4 −3 −2 −1 0 1 2 3 4
            −0.33
            z

4.

Reject

Retain

−4 −3 −2 −1 0 1 2 3 4
            −0.40
            z

Answers

1. body
2. tail
3. tail
4. body

Let's try another example:

Imagine again a distribution of IQ scores ($\mu = 100$ and $\sigma = 15$). What is the probability of obtaining an IQ score **less than** $X = 91$?

The question can be rewritten as

What is $p(X < 91)$?

As before, begin by sketching the normal distribution and calculating the *z* score. Because you are being asked to obtain the probability associated with a score that is less than 91 ($X < 91$), shade in the distribution below (to the left of) where you locate the *z* score associated with the raw IQ score of 91 within your drawing. Then calculate *z*.

Retain

Reject

−4 −3 −2 −1 0 1 2 3 4 **z**
            −0.6
40 55 70 85 100 115 130 145 160 **IQ**
            91

$$z = \frac{X - \mu}{\sigma} = \frac{91 - 100}{15} = \frac{-9}{15} = -.60$$

Note that we obtain a negative z score (because $X = 91$ is less than our $\mu = 100$). Remember, the Unit Normal Table represents the symmetrical normal distribution. Because the top half of the distribution is a perfect mirror image of the bottom half, negative z scores do not need to be included in the Unit Normal Table, since the information is already available in the half of the distribution that corresponds to positive z scores. Note also that, in this example, you've identified (shaded in) less than 50% of the curve. Your shaded area does not cross the mean, $z = 0$. Therefore, use the column associated with the *tail* for $z < -.6$. The proportion associated with obtaining an IQ score less than 91 $(X < 91)$ is .2743, or 27.43%. Thus, the probability of selecting a score below 91 is .2743.

Be Here Now

Imagine that there is an achievement test for which $\mu = 50$ and $\sigma = 10$.

What is the probability of getting a score of 62 or less? That is, what is $p(X < 62)$?

Answer

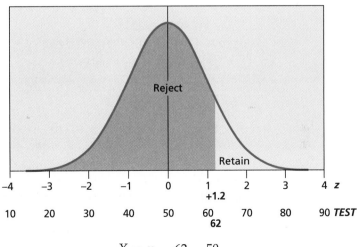

$$z = \frac{X - \mu}{\sigma} = \frac{62 - 50}{10} = +1.2$$

Go to Unit Normal Table and look up $z = +1.2$. Go to the column for the *body* of the distribution.

$$p(X < 62) = .8849$$

Area between Two z Scores

Let's try a more complex problem. Keeping our distribution of IQ scores, with $\mu = 100$ and $\sigma = 15$, what proportion of the normal distribution falls **between** the IQ score of 80 **and** the IQ score of 110. That is,

What is $p(80 < X < 100)$?

At first, this problem may look a bit difficult. But if you think about it in stages, it will be manageable.

As always, start by drawing the distribution and shading in the area of interest on the curve.

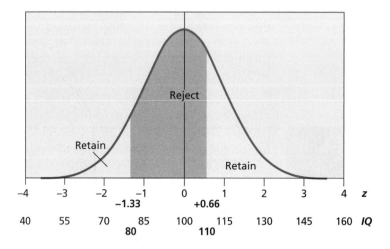

Now, calculate the z score associated with $X = 80$ and the z score associated with $X = 110$.

$$z = \frac{X - \mu}{\sigma} = \frac{80 - 100}{15} = \frac{-20}{15} = -1.33$$

$$z = \frac{X - \mu}{\sigma} = \frac{110 - 100}{15} = \frac{10}{15} = +.66$$

There are several ways to use the information from the Unit Normal Table to obtain the proportion of the normal distribution falling between the two z scores that we calculated. We present one possible method.

Start by looking at what you have shaded in on the distribution.

Look up the proportion of the curve that falls below $z = +.66$. Notice that this area does include the mean. So, using the column for the body of the distribution of the Unit Normal Curve, look up the area associated with $z = .66$. The proportion associated with z less than .66 is .7454, or almost 75% of the normal curve.

As you look at the curve, notice that you did not shade in the entire area below $z = -1.33$. You must remove the part that is not shaded. That is, you must remove the portion of the normal curve that falls below $z = -1.33$.

Look at the proportion of the curve that falls below $z = -1.33$. Notice that it does not include the mean. So, using the column for the tail of the distribution of the Unit Normal Table, look up the area associated with $z = -1.33$. The proportion associated with $z = -1.33$ is .0918, or roughly 9% of the curve.

Now look at what is shaded in the drawing. The 9% is not shaded; it is not an area that is included in the range of scores between $X = 80$ and $X = 110$.

However, the 75% associated with the z score of .66 *includes that 9%*. So, if you subtract the proportion associated with $z = -1.33$, which is .0918, from the proportion associated with $z = +.66$, which is .7454, you will have eliminated the portion that you are not interested in, leaving only what you are interested in.

$$
\begin{array}{r}
.7454 \\
-.0918 \\
\hline
.6536
\end{array}
$$

The proportion of the distribution between $X = 80$ and $X = 110$ is .6536, or about 65%. Looking at the distribution, we see that about 65% seems consistent with what was originally sketched.

This general approach will work whether the area of interest is on either side of the mean, above the mean, or below the mean. The idea is to look at what is asked for and remove any portion that is not of interest.

Let's try another example:

The Psychology Competency Examination has a mean of 500 and a standard deviation of 100. What proportion of examinees would be expected to earn scores **between $X = 585$ and $X = 670$**?

In other words,

What is $p(585 < X < 670)$?

Now calculate the z score associated with $X = 85$ and the z score associated with $X = 670$.

$$
z = \frac{X - \mu}{\sigma} = \frac{585 - 500}{100} = \frac{85}{100} = +.85
$$

$$
z = \frac{X - \mu}{\sigma} = \frac{670 - 500}{100} = \frac{170}{100} = +1.70
$$

In the following figure, we have drawn the normal curve and shaded in the area of interest on the curve:

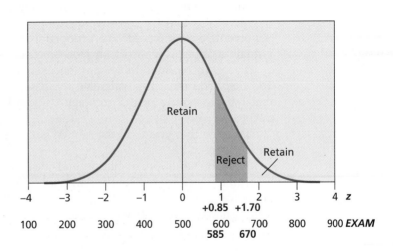

Now look up the corresponding proportions from the Unit Normal Table. As before, we will look up the proportion of the Unit Normal Table that falls below each z score. In this case, we could choose to use the proportion in the body. The proportion associated with $z = .85$ is .8023, or roughly 80% of the curve. For $z = 1.70$, the proportion is .9554, or about 95.5%. Now refer to your drawing of the curve, and subtract the proportion associated with $z = .85$ from the proportion associated with $z = 1.70$; that way, you will have eliminated the portion that you are not interested in. The result is:

$$\begin{array}{r} .9554 \\ -.8023 \\ \hline .1531 \end{array}$$

The proportion of the distribution between $X = 585$ and $X = 670$ is .1531, or about 15%. Looking at the distribution, we see that this proportion makes sense.

As you examine the preceding example, note that we could have chosen to use the proportion in the tail. The proportion in the tail associated with $z = .85$ is .1977. The proportion in the tail for $z = 1.70$ is .0446. Again, we subtract:

$$\begin{array}{r} .1977 \\ -.0446 \\ \hline .1531 \end{array}$$

Because the normal curve is symmetrical, there are several ways to approach the same problem; each will yield the same answer.

Reflect
There are many paths to a solution.

z Scores and Percentile Ranks

What we have learned about z scores can now be applied to a very useful concept: percentile ranks. We already discussed a related concept, cumulative percent, when we considered frequency distributions (see Chapter 2). A **percentile rank,** based on a proportion, tells you the percentage of individuals at or below a particular score in a distribution.

A **percentile rank** is a proportion at or below a score in a distribution.

The Unit Normal Table may be used to compute a **percentile rank.** To do so, you must first calculate the z score and then use the Unit Normal Table to determine the proportion under the curve associated with that z score.

Imagine that you took the SAT. What is the percentile rank associated with a score of 550?

First, sketch the distribution and indicate where a score of 550 would be on the distribution. Recall that, with SAT scores, $\mu = 500$ and $\sigma = 100$. Since we are interested in the percentile rank—the percent *at or below the score*—shade in the portion of the normal curve from 550 to the left (to include

Use the Psychology Competency Examination mean of 500 and a standard deviation of 100. What proportion of examinees would be expected to earn scores **between** X = 300 **and** X = 450?

Answer

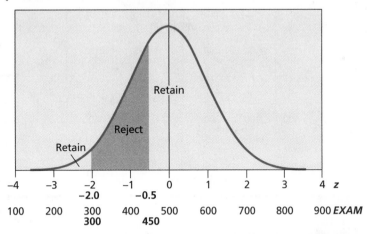

The problem is, What is $p(300 < X < 440)$?

$$z = \frac{X - \mu}{\sigma} = \frac{300 - 500}{100} = \frac{-200}{100} = -2.0$$

$$z = \frac{X - \mu}{\sigma} = \frac{450 - 500}{100} = \frac{-50}{100} = -.50$$

$$\begin{array}{r} .3085 \\ -.0228 \\ \hline .2857 \end{array}$$

The proportion between X = 300 and X = 450 is .2857, or 28.57%.

all SAT values that are lower than 550), as we have done on the following curve:

Now compute the z score:

$$z = \frac{X - \mu}{\sigma} = \frac{550 - 500}{100} = +.5$$

Your z score is +.5.

Remember, we started out by wanting to know the percentile rank of a score of 550. Now you need to use the Unit Normal Table.

As you look at your picture, notice that the area shaded includes the mean ($z = 0$). The proportion of the normal curve at or below a z score of +.5 (body) is .6915. You can convert a proportion into a percentage or percentile rank by multiplying by 100. Thus, your percentile rank is 69.15.

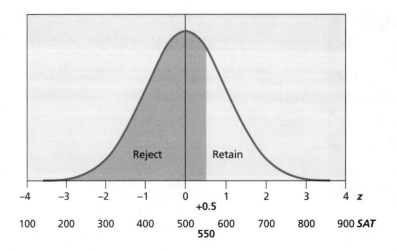

Be Here Now

For a distribution with a mean of 50 and a standard deviation of 10, what is the percentile rank that corresponds to a score of 40?

Answer

First, draw the picture:

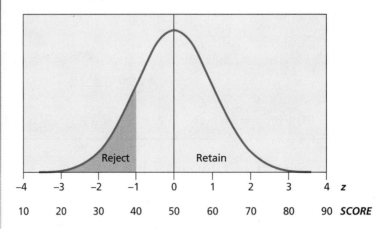

Next, compute the z score:

$$z = \frac{X - \mu}{\sigma} = \frac{40 - 50}{10} = \frac{-10}{10} = -1.0$$

The area of interest does not include the mean ($z = 0$). Therefore, the area of interest is in the tail. The proportion of the normal curve at or below a z score of -1.0 is .1587. Therefore, the percentile rank for a score of 40 is 15.87.

z Scores and Percentiles

Sometimes we know the percentile rank and we would like to find out the score associated with that percentile rank.

A **percentile** refers to the score associated with a percentile rank. This is an exact point in the distribution. For example, you may want to know which score corresponds to the 67th percentile on the Graduate Record Exam (GRE). Because you are starting out knowing the proportion—that is, the percentile rank and *not* the score (or *z* score)—the Unit Normal Table is used in a different way. To determine the score, you must first locate the *proportion* in the Unit Normal Table and then use it to find the *z* score by way of this formula:

$$X = \sigma z + \mu$$

Notice that the preceding formula is derived from our earlier-presented formula for the *z* score.

To find a score by using percentile rank information,

1. Draw the normal distribution and estimate the location of the score on the basis of the percentile rank given.

2. Locate the proportion in the Unit Normal Table, using the column for the body if the proportion is greater than 50% or the column for the tail if the proportion is less than 50%.

3. Locate the *z* score that corresponds to the proportion in question.

4. Use the formula $X = \sigma z + \mu$, **being certain to insert a negative sign for *z* scores less than 0.**

For example, the GRE has a mean of 500 and a standard deviation of 100. What score corresponds to the 67th percentile? Begin by drawing the distribution, and locate roughly where the 67th percentile would be.

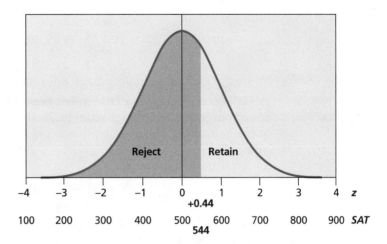

Go to the Unit Normal Table, and use the proportion in the body column, because the 67th percentile is beyond the mean (or 50th percentile). Find the number closest to .67, and then slide your finger across the row to get the z score associated with this proportion. The z score corresponding to the 67th percentile is .44. Now use the following formula:

$$X = \sigma z + \mu$$
$$X = 100(.44) + 500$$
$$X = 44 + 500$$
$$X = 544$$

The score associated with the 67th percentile is $X = 544$. Look back at the drawing of the distribution to confirm whether this score makes sense.

Let's try another example, this time using a percentile rank that is less than 50%. If a Mathematics Abilities Inventory has a mean of 75 and a standard deviation of 25, what score corresponds to the 22nd percentile?

As always, draw the distribution and locate where the 22nd percentile would be.

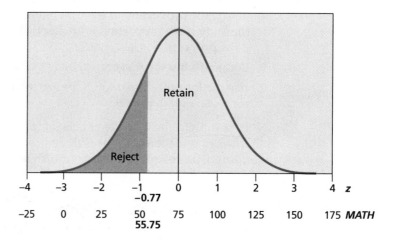

Now, using the Unit Normal Table and the column for proportion in the tail because the percentile is less than 50%, find the z score that corresponds to .22. The closest z score is .77. **However, because we are looking at the half of the distribution that is below the mean, this z score is actually −.77.**

Now use the formula:

$$X = \sigma z + \mu$$
$$X = 25(-.77) + 75$$
$$X = -19.25 + 75$$
$$X = 55.75$$

The score associated with the 22nd percentile is 55.75. Compare your results with the distribution. Had you forgotten to use a negative value for the z score, this is the point where you would most likely catch your mistake!

Be Here Now

For a distribution with a mean of 100 and a standard deviation of 20, what score corresponds to the 17th percentile?

Answer

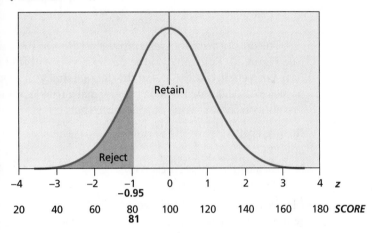

Use the column for the tail on the Unit Normal Table, and be certain to use a negative sign with the z score you obtain!

$$X = \sigma z + \mu$$
$$X = 20(-.95) + 100$$
$$X = -19 + 100$$
$$X = 81$$

Application and Interpretation: z Scores

To practice obtaining z scores with SPSS, consider the following data set, which we have obtained by asking 10 persons their age and requesting them to rate a new soda on a scale of 1 to 10 (note, these are *not* population data):

ID	Rating	Age
1	9.00	19
2	1.00	29
3	1.00	37
4	1.00	44
5	1.00	53
6	8.00	19
7	2.00	26
8	1.00	43
9	6.00	29
10	1.00	10

➜ **To compute the mean and standard deviation, after you have entered the data, you need to select**

Analyze ➙ *Descriptive Statistics* ➙ *Descriptives.*

- Check the box on the lower left that reads *"Save standardized values as variables."*

The output with the descriptive statistics for the variables will be displayed (see Figure 5). Returning to the data set, you will find that additional columns have been created at the end of the data set that will be standardized values. The name of each variable will be the original variable name, preceded by a "Z."

As you look at your output, note that

Rating and **Age** are the variables analyzed.	
N indicates that 10 records were analyzed for each variable.	
Minimum indicates the lowest score per variable.	The lowest rating was 1.00, while the lowest age was 10.
Maximum indicates the highest score per variable.	The maximum rating was 10.00, while the highest age was 58.
Mean gives the average rating.	The mean rating is 3.1. The mean for Age is 35.7000.
Std. Deviation shows the average distance each score is from the mean.	The rating standard deviation is 3.24722. For Age, the standard deviation is 13.58962.
Valid N (listwise) The number of participants with scores was 10. In this example, the N listed for "Rating" and that listed for "Valid N (listwise)" are the same. If there were any missing scores, these numbers would not be the same.	

Returning to Data View in SPSS (as shown in Figure 6), you will see that two variables have been added to the last two columns of the data set, with each of the original variable names preceded by a "Z." These are the standardized (z score) values of the original variables.

As you look at these data, note that the participant with ID number 1 has a rating of 9 for Brand A. This participant's rating expressed as a z score

	N	Minimum	Maximum	Mean	Std. Deviation
Rating	10	1.0	9.00	3.1000	3.24722
Age	10	10.0	53.00	30.9000	13.31207
Valid N (listwise)	10				

FIGURE 5 Output: Rating and age.

	ID_NUM	Rating	Age	ZRating	ZAge
1.00	1.00	9.00	19.00	1.81694	−.89393
2.00	2.00	1.00	29.00	−.64671	−.14273
3.00	3.00	1.00	37.00	−.64671	.45823
4.00	4.00	1.00	44.00	−.64671	.98407
5.00	5.00	1.00	53.00	−.64671	1.66015
6.00	6.00	8.00	19.00	1.50898	−.89393
7.00	7.00	2.00	26.00	−.33875	−.36809
8.00	8.00	1.00	43.00	−.64671	.90895
9.00	9.00	6.00	29.00	.89307	−.14273
10.00	10.00	1.00	10.00	−.64671	−1.57000

FIGURE 6 Data set with ZRating and ZAge added.

was above average ($z = 1.81694$), while his or her z score for Age was below average ($z = -.89393$).

Sometimes, you may find SPSS helpful in computing the mean and standard deviation of a set of data. These data can then be used in computing z scores for values not present in the original data set.

For example, suppose 15 students took a final exam. Their scores were as follows:

Final-Exam Scores

89	96	78	88	57	99	68	54
78	52	88	89	91	86	85	

John took the exam late. His raw score was 84. Using the class data, determine John's z score and percentile rank.

To begin, you must input the class data.

To compute the mean and standard deviation, you need to select

  Analyze → Descriptive Statistics → Frequencies.

- Within the Frequencies menu, under Statistics, choose Mean and Standard Deviation.
- Also, "uncheck" *Display frequency tables.*

Figure 7 presents the output.

If you followed the instructions correctly, you should see one large section of output on your screen (if you "unchecked" *Display frequency tables*). It presents the statistics you requested.

Score is the variable analyzed.

  N   **Valid = 15**   Fifteen scores were analyzed.

  Missing = 0   No data were missing.

Statistics

score

N	Valid	15
	Missing	0
Mean		79.8667
Std. Deviation		15.18865

FIGURE 7 SPSS output for the mean and standard deviation: Class data.

Mean = 79.8667   The mean final-exam score is 79.87.

Std . Deviation =   The standard deviation is 15.19.
15.18865

You may now compute John's z score and percentile rank:

$$z = \frac{84 - 79.87}{15.19} = +\frac{4.13}{15.19} = +.27$$

John's z score was +.27.

Now draw a normal distribution, and shade in the proportion of the curve from John's z score and below.

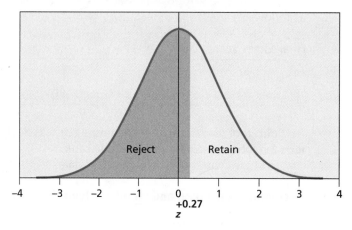

You are now ready to look up John's percentile rank in the Unit Normal Table. Since the portion you shaded includes the mean, look up +.27 in the body of the distribution. John's percentile rank is 61. The area under the curve was .6064. Thus, approximately 61% of the class scored below John on the final exam.

Reporting Results in APA Format

When reporting z scores in APA format, use an italicized lowercase z and put one space between each symbol or number. This practice simply makes the results easier to read.

Incorrect:	z=.76	(incorrect because of no italics or spacing)
	z = .76	(incorrect because of no italics)
	z=.76	(incorrect because of no spacing)
Correct:	z = .76	
	z = −.76	

Reporting percentile ranks follows similar guidelines. The *p* for the proportion should be lowercase and italicized, with spacing between the symbols and numbers.

Incorrect:	p=.8534	(incorrect because of no italics or spacing)
	p = .8534	(incorrect because of no italics)
	p = .8534	(incorrect because of no spacing)
Correct:	*p* = .8534	

Percentiles should be reported as ordinal numbers, with the appropriate superscript and the word "percentile." The symbol for a percentile is an italicized capital *P*.

Incorrect:	thirty-first percentile (incorrect because all text)
	31% (incorrect because a percentile is not a percent)
Correct:	31st percentile

Using our soda-rating study, we might create the following text:

The standardized scores for the soda ratings ranged from $z = -.65$ to $z = 1.82$. The probability of obtaining a score at or above $z = 1.82$ was $p = .0761$. The score associated with the 50th percentile was 3.10.

It's Out There...

The KDI-II is an instrument used to screen children aged 4–6 to determine their developmental readiness for kindergarten. Tables are provided in the Administration Manual to convert raw scores to scaled scores, percentiles, and labels for classification. For example, a raw score of 125 converted to a scaled score is 53, which represents the 61st Percentile. A child at this level would be classified as above average.

SUMMARY

z scores are standardized scores with a mean of zero and a standard deviation of one. Transforming a distribution of scores does not change the fundamental nature of the distribution: You cannot magically pass an exam by transforming a failing raw score into a standardized score.

Standardized scores are quite useful. They are used to determine proportions of scores under the normal curve, to compare scores from different distributions, to convert scores to new distributions, and to determine percentiles and percentile ranks.

In a later chapter, we will use standard scores to help evaluate the results of research studies.

~~~~~~
Reflect

With z scores, what is standard is standard, whether apple, orange, or orangutan.
~~~~~~

PRACTICE

z Score Basics

1. If the average height in inches for women in the population is $\mu = 64.0$, with $\sigma = 4$, what are the z scores of women with the following heights?

$$X = 71 \text{ inches}$$
$$X = 58 \text{ inches}$$

How do these women compare with the average woman in the population?

2. Let's say that the average number of points per game scored by males playing college basketball is $\mu = 12$, with $\sigma = 3$. Following are the z scores for points scored by Aaron and Lee in a recent game:

$$\text{Aaron: } X = 17$$
$$\text{Lee: } X = 11$$

How do Aaron and Lee compare with the average college-basketball-playing male in the population?

z Scores and Comparisons between Two Scores

3. The scores of one student on two IQ tests are as follows:

IQ Test A	IQ Test B
$X = 119$	$X = 89$
$\mu = 100$	$\mu = 75$
$\sigma = 15$	$\sigma = 10$

On which test did the student do better?

4. A student taking English literature and chemistry earns the following midterm scores:

English Literature	Chemistry
$X = 85$	$X = 82$
$\mu = 72$	$\mu = 65$
$\sigma = 12$	$\sigma = 10$

At the midterm, in which class is the student doing better? (Notice that the class means and standard deviations are considered population parameters for purposes of this practice problem.)

5. A student takes two math abilities tests, the data from which are as follows:

Math Abilities Test A	Math Abilities Test B
$X = 78$	$X = 82$
$\mu = 100$	$\mu = 140$
$\sigma = 20$	$\sigma = 25$

On which test did the student achieve the better score?

6. A student completes two job satisfaction measures. The data are as follows (note that the higher the score, the greater is the job satisfaction):

Job Satisfaction Measure A	Job Satisfaction Measure B
$X = 31$	$X = 47$
$\mu = 35$	$\mu = 50$
$\sigma = 3$	$\sigma = 5$

Which measure detected greater job satisfaction?

7. A student taking American history and European history receives the following scores:

American History	European History
$X = 76$	$X = 82$
$\mu = 80$	$\mu = 80$
$\sigma = 9$	$\sigma = 6$

At the end of the semester, in which class did the student have the better average? (Notice that the class means and standard deviations are considered population parameters for the purposes of this practice problem.)

z Scores and Converting to a New Scale

8. If a high school graduation exam with $\mu = 65$ and $\sigma = 15$ is revised and the new version of the exam has $\mu = 300$ and $\sigma = 100$, what would a score of $X = 91$ on the original exam now be on the new version of the exam?

9. An exam based on $\mu = 28$ and $\sigma = 6$ is being compared with a similar exam with $\mu = 70$ and $\sigma = 11$. What would a score of $X = 21$ on the original exam now be?

z Scores and Proportions

10. The SAT has $\mu = 500$ and $\sigma = 100$. What is $p(X < 395)$?

11. On an IQ test with $\mu = 100$ and $\sigma = 15$, what was $p(X < 110)$?

12. On an exam with $\mu = 80$ and $\sigma = 10$, what was $p(X > 85)$?

13. On an IQ test with $\mu = 100$ and $\sigma = 15$, what is $p(94 < X < 103)$?

z Scores and Percentiles

14. For a distribution with $\mu = 20$ and $\sigma = 2$, what score is associated with the 73rd percentile?

15. For a distribution with $\mu = 400$ and $\sigma = 75$, what score is associated with the 26th percentile?

z Scores, SPSS, and APA Reports

16. Enter the following data set into SPSS:

ID	Normal Hours of Sleep	GPA
1	8	3.44
2	4	2.08
3	6	3.11
4	11	3.53
5	7	3.61

(a) Calculate the standardized scores associated with the variables.

(b) Using the descriptive statistics from the output, state what the z scores would be for 5 hours of sleep and for a GPA of 2.24.

(c) Using the SPSS output and your calculations, write a brief summary, in proper APA format, of the results for hours of sleep.

SOLUTIONS

z Score Basics

1. $z = \dfrac{X - \mu}{\sigma} = \dfrac{71 - 64}{4} = \dfrac{7}{4} = +1.75$

$z = \dfrac{X - \mu}{\sigma} = \dfrac{58 - 64}{4} = \dfrac{-6}{4} = -1.50$

The first score is one-and-three-quarters standard deviations above the population mean of 64 inches, so the woman with that score would be considered quite tall; the second score is one-and-a-half standard deviations below the population mean.

2. Aaron:

$$z = \frac{X - \mu}{\sigma} = \frac{17 - 12}{3} = \frac{5}{3} = +1.67$$

Lee:

$$z = \frac{X - \mu}{\sigma} = \frac{11 - 12}{3} = \frac{-1}{3} = -.33$$

Aaron had a very good game compared with the population mean of 12. His score is one-and-two-thirds standard deviations above the population mean. Lee's game was not as good. His score is one-third of a standard deviation below the population mean.

z Scores and Comparisons between Two Scores

3. IQ Test A:

$$z = \frac{X - \mu}{\sigma} = \frac{119 - 100}{15} = \frac{19}{15} = +1.27$$

IQ Test B:

$$z = \frac{X - \mu}{\sigma} = \frac{89 - 75}{10} = \frac{14}{10} = +1.40$$

The higher score was obtained on IQ Test B, because the z score is further from the mean, going in the positive direction.

4. English literature:

$$z = \frac{X - \mu}{\sigma} = \frac{85 - 72}{12} = \frac{13}{12} = +1.08$$

Chemistry:

$$z = \frac{X - \mu}{\sigma} = \frac{82 - 65}{10} = \frac{17}{10} = +1.70$$

The student is doing better in chemistry, because the z score is further from the mean, going in the positive direction.

5. Math Abilities Test A:

$$z = \frac{X - \mu}{\sigma} = \frac{78 - 100}{20} = \frac{-22}{20} = -1.10$$

Math Abilities Test B:

$$z = \frac{X - \mu}{\sigma} = \frac{82 - 140}{25} = \frac{-58}{25} = -2.32$$

The higher score was obtained on Math Test A, because, with the z scores being negative, the z score closer to the mean indicates a higher score.

6. Job Satisfaction Measure A:

$$z = \frac{X - \mu}{\sigma} = \frac{31 - 35}{3} = \frac{-4}{3} = -1.33$$

Job Satisfaction Measure B:

$$z = \frac{X - \mu}{\sigma} = \frac{47 - 50}{5} = \frac{-3}{5} = -.60$$

The correct answer is Measure B, because $z_B = -.60$ is closer to the mean than $z_A = -1.33$.

7. American history:

$$z = \frac{X - \mu}{\sigma} = \frac{76 - 80}{9} = \frac{-4}{9} = -.44$$

European history:

$$z = \frac{X - \mu}{\sigma} = \frac{82 - 80}{6} = \frac{-2}{6} = +.33$$

This student had the higher average in European history, because the z score for American history is below the mean but the z score for European history is above the mean.

z Scores and Converting to a New Scale

8. Original exam to new exam:

$$z = \frac{X - \mu}{\sigma} = \frac{91 - 65}{15} = \frac{26}{15} = +1.73$$

$$X = 100(1.73) + 300 = 473$$

9. Original exam to new exam:

$$z = \frac{X - \mu}{\sigma} = \frac{21 - 28}{6} = \frac{-7}{6} = -1.17$$

$$X = 11(-1.17) + 70$$

$$X = -12.87 + 70$$

$$X = 70 - 12.87 = 57.13$$

z Scores and Proportions

10.

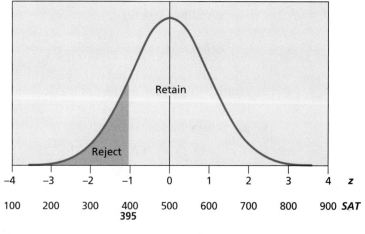

$$z = \frac{X - \mu}{\sigma} = \frac{395 - 500}{100} = \frac{-105}{100} = -1.05$$

Look in the column for the tail and look up $z = 1.05$.

$$p(X < 395) = .1469 \text{ or } 14.69\%$$

11.

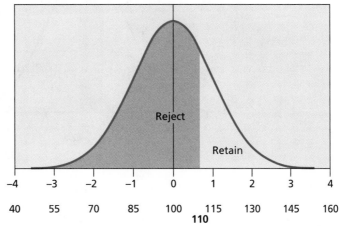

$$z = \frac{X - \mu}{\sigma} = \frac{110 - 100}{15} = \frac{10}{15} = +.67$$

Look in the column for the body and look up $z = .67$.

$$p(X < 110) = .7486 \text{ or } 74.86\%$$

12.

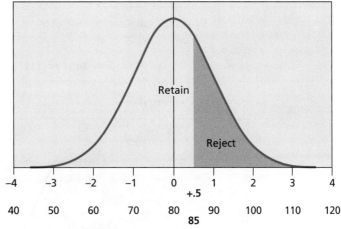

$$z = \frac{X - \mu}{\sigma} = \frac{85 - 80}{10} = \frac{5}{10} = +.50$$

Look in the column for the tail and look up $z = .50$.

$$p(X > 85) = .3085 \text{ or } 30.85\%$$

13.

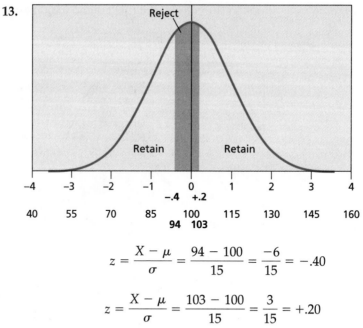

$$z = \frac{X - \mu}{\sigma} = \frac{94 - 100}{15} = \frac{-6}{15} = -.40$$

$$z = \frac{X - \mu}{\sigma} = \frac{103 - 100}{15} = \frac{3}{15} = +.20$$

For $z = -.40$, look up $z = .40$ in the column for the tail.
For $z = +.20$, look up $z = .20$ in the column for the body.
Subtract the proportion obtained for the tail from the proportion obtained for the body:

$$
\begin{aligned}
.5793 \\
-.3446 \\
\hline
.2347
\end{aligned}
$$

$p(94 < X < 103)$ is .2347 or 23.47%.

z Scores and Percentiles

14.

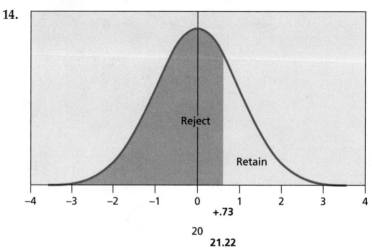

Use the column for the body on the Unit Normal Table, and obtain the z score closest to .73.

$$X = \sigma z + \mu$$
$$X = 2(.61) + 20$$
$$X = 1.22 + 20$$
$$X = 21.22$$

15. Use the column for the tail on the Unit Normal Table, and obtain the z score closest to .26. Be sure to make the z score negative for the formula!

$$X = \sigma z + \mu$$
$$X = 75(-.64) + 400$$
$$X = -48 + 400$$
$$X = 352$$

z Scores, SPSS, and APA Reports

16. (a)   **Descriptive Statistics**

	N	Minimum	Maximum	Mean	Std. Deviation
SLEEP	5	4	11	7.20	2.588
GPA	5	2.08	3.61	3.1540	.62979
Valid N (listwise)	5				

id	sleep	gpa	Zsleep	Zgpa
1	8	3.44	.30907	.45412
2	4	2.08	−1.23627	−1.70535
3	6	3.11	−.46360	−.06987
4	11	3.53	1.46807	.59703
5	7	3.61	−.07727	.72406

(b) $z = \dfrac{5 - 7.20}{2.59} = \dfrac{-2.2}{2.59} = -.85$

$z = \dfrac{2.24 - 3.15}{63} = \dfrac{-.91}{63} = -1.44$

(c) The standardized scores for hours of sleep ranged from $z = -1.24$ to $z = 1.47$. The z score associated with 5 hours of sleep was $z = -.85$. The z score associated with a GPA of 2.24 was $z = -1.44$.

-6-

Sampling

In This Chapter

Samples and Populations
- Types of Sampling

Sampling Error

Central Limit Theorem
- Mean of the Sample Means = μ
- Standard Error of the Mean

- Sampling Error is Normally Distributed
- Impact of Sample Size on Sampling Error

Probability of Getting a Mean (*M*) from a Population (*μ*): Application and Interpretation of the *z* Statistic

F O C U S | In this chapter, we discuss sampling error, which arises because samples do not match populations exactly. Most of the time, there will be a difference between a sample statistic and a population parameter. This chapter describes sampling, sampling error, and the impact of these errors.

Samples and Populations

As discussed in Chapter 1, populations comprise all of the things in which you are interested. For example, you might want to study the impact of an advertising campaign to promote dental hygiene. It is not likely that you will be able to study everyone who has been exposed to the campaign. In fact, information about the population is usually not available. For this reason, we collect data on samples. We then examine the sample data to make inferences about the population parameters of interest.

Types of Samples

A sample is a subset of the population. There are many different ways in which we might select a sample. We will discuss two of them.

In a **simple random sample,** every member of the population has an equal chance of being selected.

A **simple random sample** is a sample in which every member of the population has an equal chance of being selected. However, conducting a simple random sample is frequently impractical. More often, we use samples of convenience.

In a **sample of convenience,** the researcher uses participants who are readily available for study.

A **sample of convenience** is a sample in which a researcher uses the participants who are available for study. For example, a researcher may be interested in examining the reaction time to changes in a traffic light while an individual is actively engaged in a cell phone call. To study this phenomenon, she uses undergraduate students at her university.

The sample she obtains is a sample of convenience. The population of interest would be all adults; however, the researcher used participants easily available to her: undergraduate students. Notice that she attempted to select a sample that reflects the population of interest. She did not use 10-year-old children or 80-year-old men as participants; rather, she used participants who would be in the population of interest.

Sampling error is the difference between a sample statistic and a population parameter.

Statistics computed on samples are used to represent the population. However, regardless of the sampling method, it is likely that the sample statistic will not be exactly the same as the population parameter. This difference is **sampling error.**

It's Out There...   Meyer (2005) examined the effectiveness of air bags in reducing road fatalities in the United States. The National Highway Traffic Safety Administration used a random sample to compile data from automobile crashes. It was concluded that, through January 2004, airbags had saved over 10,000 lives.

Sampling Error

When we conduct research, we usually select a sample to study. Many times, the sample does not perfectly represent the population in which we are interested.

For example, suppose you had 10 marbles: 5 black and 5 white. Consider these marbles as your population. Your marble population is 50 percent white and 50 percent black. Now put the marbles in a brown paper bag.

Population: Bag of Marbles

Now reach into the bag and pull out a sample of 5 marbles. Do this several times. What do your samples look like? This is what ours looked like:

Samples of Marbles: Sample Size 5

Sample A
40 percent black, 60 percent white

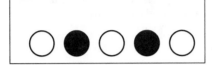

Sample B
20 percent black, 80 percent white

Sample C
60 percent black, 40 percent white

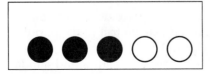

Sample D
100 percent black, 0 percent white

Within the population, 50 percent of the marbles are white and 50 percent of the marbles are black, yet none of the samples we selected were equally divided between black and white. Sample A is composed of 60 percent black marbles and 40 percent white marbles, while Sample D is 100 percent black marbles and 0 percent white marbles. None of our samples are exact replicas of the population. Each of our samples contains sampling error. Because a sample rarely serves as a perfect representation of a population, we must consider sampling error.

Certain patterns typically arise when we work with sampling error. These patterns are taken into consideration when we interpret statistics. The **Central Limit Theorem** guides our interpretation of a sample mean.

The **Central Limit Theorem** makes predictions about the characteristics of a distribution of sample means.

Central Limit Theorem

As we have already noted, when we conduct research, we usually select a sample to study, and most of the time the sample will not represent the population of interest perfectly. Sampling error is considered to happen randomly. Sometimes the sample mean *M* will be higher than the population μ; other times it will be lower than the population μ. The Central Limit Theorem has three properties that help us understand sampling error and guide the decisions we make about sample means. These properties were obtained by repeatedly sampling from a population and examining the pattern in the errors.

To examine all facets of sampling error, we would need to consider a population from which we draw *an infinite number of samples.* Carrying out this exercise is, of course, purely the-

oretical, but the Central Limit Theorem suggests that if we were to draw an infinite number of samples, we would find that

- The mean of the sample means is equal to the population μ.
- The standard deviation of the sample means is equal to the average distance a sample mean is from μ. This property is also called the standard error of the mean.
- The distribution of all sample means, of a given size, forms a normal distribution.

To explore the Central Limit Theorem, we will conduct a Monte Carlo study. In this kind of study, we generate a set of population data and then see what happens if we repeatedly select samples.

Imagine Lonely Island, where there are just 10 inhabitants. Because you are interested only in these 10 residents, they make up a population. Suppose we administered a self-esteem test to the population of 10 residents. Their scores appear in Figure 1.

Let us compute the mean, variance, and standard deviation of the Lonely Island self-esteem scores.

The mean self-esteem score of the Lonely Island population is

$$\mu = \frac{\Sigma X}{N} = \frac{500}{10} = 50$$

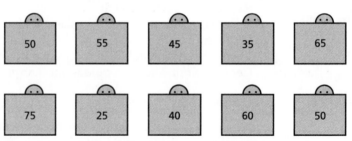

FIGURE 1 Self-esteem scores: Population of Lonely Island ($N = 10$).

The variance of the self-esteem scores of the Lonely Island population is

$$\sigma^2 = \frac{\Sigma(X - \mu)^2}{N} = \frac{1950.00}{10} = 195.00$$

The standard deviation of the self-esteem scores of the Lonely Island population is

$$\sqrt{\sigma^2} = \sqrt{\frac{\Sigma(X - \mu)^2}{N}} = 13.96$$

In Table 1, we present the descriptive parameters relative to the self-esteem scores of Lonely Island residents.

Now, suppose you took a random sample of four Lonely Island residents. By chance, you select the following residents and scores:

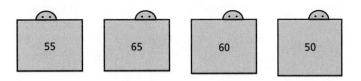

Their sample mean is $M = \dfrac{\Sigma X}{n} = 57.50$.

Suppose you did this 19 more times. Then you would have collected data for a total of 20 samples, each with $n = 4$. The means you collected are as follows:

Sample Means of $n = 4$ from Lonely Island Population

57.50	48.75	38.75	55.00	60.00	48.75	42.50	43.75	51.25	63.75
48.75	46.25	52.50	53.75	43.75	57.50	50.00	52.50	55.00	42.50

We will use these data, collected as part of our Monte Carlo study, to examine each of the three properties of the Central Limit Theorem.

The mean of the sample means is equal to the population μ.

Notice that most of the time you did not get a mean of 50, the population average (μ).

TABLE 1
Descriptive Parameters of Lonely Island Self-Esteem Scores

Descriptor		Parameter
μ	(Average)	50.00
σ^2	(Variance)	195.00
σ	(Standard Deviation)	13.96

The **grand mean** is the mean of the means.

You can compute the average of the averages. This statistic would be the mean of the means, which is also called the **grand mean.**

$$\text{Mean of the means} = M_M = \frac{\Sigma M}{n_M} = \frac{1012.50}{20} = 50.63$$

Notice how close the M_M of 50.63 is to μ of 50.00.

The Central Limit Theorem suggests that $M_M = \mu$. This assumes that an infinite number of samples were drawn. We collected only 20 samples from Lonely Island. With only 20 samples, we came very close to observing that $M_M = \mu$.

The standard deviation of the sample means is equal to the standard error of the mean.

The **standard error of the mean** (σ_M) is the standard deviation of a theoretical sampling distribution of means.

The **standard error of the mean** is the standard deviation of a theoretical sampling distribution of means. This is often interpreted as the average distance a sample mean is from μ. The symbol for the standard error of the mean is σ_M. This is the standard deviation of sample means drawn from the same population. We can use the data in our Monte Carlo study of Lonely Island (remember, this is theoretical) to compute the average distance each mean is from μ. Since we have a sample of means, we can use the formula for the standard deviation to compute the standard error of the mean. In our example,

$$\sigma_M = \sqrt{\frac{\Sigma(M - \mu)^2}{n - 1}} = \frac{123.88}{19} = 6.52$$

That is, the average mean collected in our Monte Carlo study is 6.52 units from the μ of 50.00.

Most of the time, we will not have collected repeated samples from a population. So we need another way to estimate the standard error of the mean.

We can determine the average distance a sample mean is likely to be from the population μ by using a formula based on the standard deviation of the population to compute the standard error of the mean. The standard error of the mean is equal to the population standard deviation divided by the square root of the sample size:

$$\text{Standard Error of the Mean} = \sigma_M = \frac{\sigma}{\sqrt{n}}$$

In our example, the standard deviation of the Lonely Island population is 13.96. Our samples were of size 4. We can use this information to compute the standard error of the mean:

$$\text{Standard Error of the Mean} = \sigma_M = \frac{\sigma}{\sqrt{n}} = \frac{13.96}{\sqrt{4}} = 6.98$$

As Table 2 indicates, our computed standard error of the mean based upon the data from our Monte Carlo study is very close to that computed with the formula that uses the standard deviation of the population.

The population parameters and our estimates would have been identical if we collected more samples (approaching an infinite number of samples

TABLE 2
Parameters of Lonely Island Self-Esteem Scores and Those Computed
Using the Sample of Means

Descriptor		Population Parameter	Estimate Based on Sample of Means
μ	(Average)	50.00	50.63
σ_M	(Standard Error of the Mean)	6.98	6.52

rather than only 20 samples). Remember, the basis of the Central Limit
Theorem is an infinite number of samples.

Again, the data from our Monte Carlo study support the Central Limit
Theorem.

The distribution of the sample means is a normal distribution.

Look at the 20 means sampled from the Lonely Island population. Notice that sometimes the means are higher than μ and other times they are
lower. If we had an infinite number of sample means, the distribution of the
means would be a normal distribution. This property allows us to use what
we know about the normal curve to interpret data based on samples.

Because the distribution of sample means would approximate the normal curve, we can use what we know about the normal curve to interpret
sampling error. For example, we can draw a normal curve and use the standard error of the mean to mark off standard deviation units of our sampling
distribution of means (we rounded 6.98 to 7).

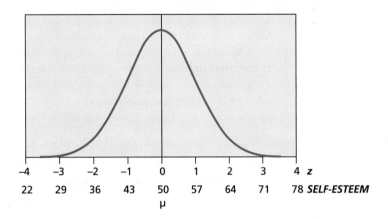

Notice that most of our sample means will be close to μ. More than half
of our sample means will be between 43 and 57. An even greater majority
of sample means will fall between 36 and 64. Conversely, very few sample
means will fall into the extreme tails of our distribution of sample means
(less than 29 or greater than 71).

Impact of Sample Size on Sampling Error

Notice that the denominator of the formula for the standard error of the mean is the square root of n, the sample size:

$$\sigma_M = \frac{\sigma}{\sqrt{n}}$$

The sample size has a direct influence on the standard error of the mean (σ_M): The larger the sample size, the closer the sample mean will be to the population μ. Also, the closer the sample mean is to the population μ, the less sampling error will be encountered. This makes sense, because the more participants we have (or the more observations or data points), the closer we are to having the total population. The closer we are to the total population, the less room there is for error.

The smaller the sample size, the larger is the standard error of the mean.

Thus, the smaller the sample size, the greater is the sampling error.

The larger the sample size, the smaller is the sampling error.

The larger the sample size, the smaller is the standard error of the mean.

Reflect

No matter how small, a sample gives insight.

Probability of Getting a Mean (*M*) from a Population (μ): Application and Interpretation of the *z* Statistic

The Unit Normal Table in Appendix A1 may be used to help determine the probability of obtaining a particular sample mean from a particular population.

We can express a sample mean as a *z* statistic, indicating the distance a particular mean is from μ.

Note that a *z* statistic is not the same as a *z* score: A *z* score refers to the location of a single score in a distribution of scores; a *z* statistic refers to the location of a mean in a distribution of sample means.

When we use the *z* statistic based on a sample, we are beginning to use **inferential statistics.** As described in Chapter 1, inferential statistics allow us to generalize from a sample to a population and make decisions about our data.

For example, the population average (μ) for IQ is 100 and the standard deviation (σ) is 15. Suppose you test a sample of 9 children and their mean (*M*) IQ is 104. How likely is it to get a mean of 104 or higher?

The *z* **statistic** is the distance an observed mean is from μ, expressed in standard units.

To begin, we can compute the *z* statistic: the distance in standard units the sample mean is from the population average.

The formula for the *z* statistic is

$$z = \frac{M - \mu}{\sigma_M}$$

In words,

$$z \text{ statistic} = \frac{\text{Sample mean } (M) - \text{Population Mean } (\mu)}{\text{Standard Error of the Mean } (\sigma_M)}$$

Remember our problem:

In our population, μ is 100 and σ is 15. Our sample data are based on $n = 9$ and $M = 104$.

We need to compute the z statistic in order to estimate the likelihood (probability) of getting a mean of 104 or higher.

Start with the formula for the z statistic:

$$z = \frac{M - \mu}{\sigma_M}$$

In order to compute the z statistic, we need to determine the standard error of the mean (σ_M):

$$\sigma_M = \frac{\sigma}{\sqrt{n}} = \frac{15}{\sqrt{9}} = \frac{15}{3} = 5$$

We can now compute the z statistic:

$$z = \frac{M - \mu}{\sigma_M} = \frac{104 - 100}{5} = +\frac{4}{5} = +.8$$

Now we draw a normal curve, as shown in Figure 2, and mark off where a z of $+.8$ would be located. Because we are interested in the probability of getting a sample mean of 104 (a z of $+.8$) *or higher*, we shade in the portion of the curve to the right of $+.8$.

Since the area shaded does not include $z = 0$, we look in the tail column of the Unit Normal Table. The proportion of the normal curve at or above $z = 0 + .8$ is .2119. Thus, the probability of getting a sample ($n = 9$) mean of 104 from a population whose mean is 100 and standard deviation is 15 is .2119, or approximately 21 percent.

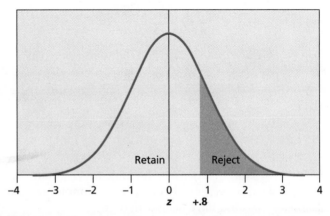

FIGURE 2 Normal curve, with the probability of getting a mean at 104 (a z of +.8) or higher shaded.

We can also use what we know about the normal curve and sampling error to estimate the proportion of means likely to fall between two sample means. For example, in a population whose average is 200 and standard deviation is 32, what is the likelihood of getting a mean between 184 and 216 if each sample contains 16 participants?

This is what we know:

$$\mu = 200$$

$$\sigma = 32$$

Two sample means are collected.
The first mean is $M = 184$, $n = 16$. The second mean is $M = 216$, $n = 16$.
To estimate the proportion of sample means likely to fall between 150 and 250, you must complete several steps:

1. Compute the z statistic for each mean.

2. Draw the normal curve.

3. Label the place on the normal curve where each z statistic (based on each sample mean) is located.

4. Shade in the area of the normal curve in which you are interested.

5. Use the Unit Normal Table to identify the proportion of means likely to fall between your sample means.

STEP 1   Compute the z statistics for each sample mean:

First Mean

$$z = \frac{M - \mu}{\sigma_M}$$

You must determine σ_M:

$$\sigma_M = \frac{\sigma}{\sqrt{n}} = \frac{32}{\sqrt{16}} = \frac{32}{4} = 8$$

Second Mean

$$z = \frac{M - \mu}{\sigma_M}$$

You must determine σ_M:

$$\sigma_M = \frac{\sigma}{\sqrt{n}} = \frac{32}{\sqrt{16}} = \frac{32}{4} = 8$$

Now compute the z statistic:

$$z = \frac{M - \mu}{\sigma_M} = \frac{184 - 200}{8} = -2$$

Now compute the z statistic:

$$z = \frac{M - \mu}{\sigma_M} = \frac{216 - 200}{8} = 2$$

Notice: Since n is the same for each sample, you needed to compute the standard error only once. If the sample sizes were different, the standard errors would have been different.

STEPS 2 TO 4   Draw the normal curve, and label where each sample mean and z statistic is located. Shade in the area between the values.

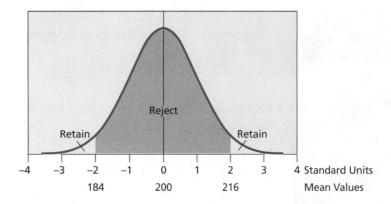

STEP 5 Use the Unit Normal Table to identify the proportion of means likely to fall between the two z statistics you computed.

We can use the Unit Normal Table to determine the proportion of the normal curve that falls to the left of a *z* statistic of +2. The proportion of the normal curve falling at or below +2 is .9772.

The area we just calculated includes not only the area of interest, but an extra portion of the curve: the proportion below −2. The area of the normal curve that falls to the left of $z = -2$ is .0228. This area has to be subtracted from the larger area of .9772.

To find the area of interest, subtract:

$$
\begin{array}{r}
.9772 \\
-.0228 \\
\hline
.9544
\end{array}
$$

Thus, from a population whose average is 200 and standard deviation is 32, approximately 95 percent of the means of sample size 16 will fall between 184 and 216.

Be Here Now

In a population, the number of words in a string of words that an adult can remember is 7. The standard deviation of the number of words remembered is 3. Thus,

$$\mu = 7 \qquad \sigma = 3$$

A researcher tests two samples, each having nine participants, to determine the number of words they can recall. Her data are as follows:

Sample A   Sample B
Mean M_A is 5; $n_A = 9$.   Mean M_B is 9; $n_B = 9$.

Given the population data, how likely is it that the researcher will get a mean less than 5 or greater than 9?

Answer

Begin with the formula: $z = \dfrac{M - \mu}{\sigma_M}$

(Continued)

Be Here Now

You know the M and μ. You do not know the standard error of the mean (σ_M).

$$\sigma_M = \frac{\sigma}{\sqrt{n}} = \frac{3}{\sqrt{9}} = \frac{3}{3} = 1.00$$

Now you can compute the z statistic for each sample mean:

Sample A

$$z = \frac{M - \mu}{\sigma_M} = \frac{5 - 7}{1} = \frac{-2}{1} = -2.00$$

Sample B

$$z = \frac{M - \mu}{\sigma_M} = \frac{9 - 7}{1} = \frac{2}{1} = 2.00$$

Now make a picture; shade the area in which you are interested.

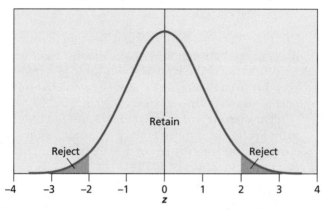

Next, use the Unit Normal Table   Sample A: Proportion below -2 = .0228
                                    Sample B: Proportion above 2 = .0228

                                    Add the two proportions: .0456

Given the population data (μ = 7 and σ = 3), approximately 5 percent of the sample means of size n = 9 will fall below 5 or above 9.

SUMMARY

When we collect information from a sample, there is likely to be sampling error. The Central Limit Theorem describes properties of sampling error. That is,

- The mean of the sample means is equal to the population μ.
- The standard deviation the sample means is equal to the standard error of the mean (the average distance a sample mean is from μ).
- The distribution of the sample means forms a normal distribution.

The standard error of the mean is the average distance a sample mean is from the population average. The standard error of the mean is influenced by the sample size; the larger the sample, the smaller is the error. We can use the Unit Normal Table to estimate the likelihood of getting a sample mean from a particular population. This will be particularly helpful as you move forward in statistics and make decisions (test hypotheses) using sample data.

PRACTICE

Samples and Population

1. A researcher is interested in the number of words three-year-olds can say. She tests 18 three-year-olds. Describe the sample and the population.

2. You are interested in researching the opinion of voters regarding whom they plan to vote for in an upcoming election. To study this phenomenon, you survey your classmates. Describe the sample and the population.

3. A marketing consultant wants to determine whether New York residents are likely to buy a new product. Her survey is drawn from the phone records of all New York residents. The computer indiscriminately selects a sample of 1000 New York residents. What sampling method does she appear to have used?

4. A researcher is interested in developing a measure of self-esteem. He tries his items out on the students in his classes. What type of sampling did he use?

Sampling Error

5. What are the three properties posited by the Central Limit Theorem?

6. A population average (μ) is 85. If you repeatedly collect samples of $n = 36$ from this sample, what would you expect to be the average of the means collected?

7. In the previous example, the standard deviation of the population is 12. What is the standard error of the mean?

8. What is the average distance a sample mean will be from μ in a population for which $\mu = 150$, $\sigma = 50$, and $n = 16$?

9. Researcher A plans on collecting a sample of size 16.
   Researcher B plans on collecting a sample of size 25.
   If both researchers sample from the same population, which researcher will have a greater standard error of the mean?

Probability of Getting a M from a Particular Population (μ): The z Statistic

10. The average population IQ test score is $\mu = 100$, and $\sigma = 15$. You decide to administer the IQ test to a sample of 25 participants. On the basis of information about the population, how likely is it for a participant to get

    a. a mean $M = 102$ or higher
    b. a mean $M = 104$ or lower

   c. a mean $M = 95$ or higher
   d. a mean $M = 97$ or lower

11. The population average score on the ABC Depression index is $\mu = 300$, and the population standard deviation is $\sigma = 70$.

   If you chose to collect two samples from this population, what is the likelihood of getting a mean between:

   a. $M = 275\ (n = 36)$ and $M = 288\ (n = 36)$
   b. $M = 280\ (n = 25)$ and $M = 315\ (n = 25)$
   c. $M = 305\ (n = 49)$ and $M = 310\ (n = 36)$

SOLUTIONS

Samples and Population

1. The population is all three-year-olds.
   The sample is the 18 three-year-olds tested.

2. The population is all likely voters.
   The sample is the classmates you surveyed.

3. Simple random

4. Sample of convenience

Sampling Error

5. First: The mean of the sample means is equal to the population μ. Second: The standard deviation of the sample means is equal to the standard error of the mean (the average distance a sample mean is from μ). Third: The distribution of the sample means is a normal distribution.

6. 85

7. What you know:

$$\mu = 85$$
$$\sigma = 12$$
$$n = 36$$

Compute the standard error of the mean:

$$\sigma_M = \frac{\sigma}{\sqrt{n}} = \frac{12}{\sqrt{36}} = \frac{12}{6} = 2$$

8. What you know:

$$\mu = 150$$
$$\sigma = 50$$
$$n = 16$$

Compute the standard error of the mean:

$$\sigma_M = \frac{\sigma}{\sqrt{n}} = \frac{50}{\sqrt{16}} = \frac{50}{4} = 12.5$$

9. Researcher A.
The larger the sample size, the less is the sampling error.

Probability of Getting a M from a Particular Population (μ): The z Statistic

10. What you know:

$$\mu = 100$$
$$\sigma = 15$$
$$n = 25 \text{ participants}$$

To begin to answer the question, you must compute the standard error of the mean. Because each sample has the same number of participants ($n = 25$), you need compute it only one time.

$$\sigma_M = \frac{\sigma}{\sqrt{n}} = \frac{15}{\sqrt{25}} = \frac{15}{5} = 3$$

a. a mean $M = 102$ or higher

$$z = \frac{M - \mu}{\sigma_M} = \frac{102 - 100}{3} = \frac{+2}{3} = +.67$$

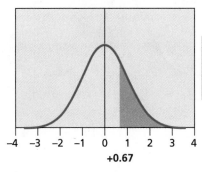

The probability of getting a mean higher than 102 (a *z* statistic of +0.67) is 0.2514, or approximately 25 percent.

b. a mean $M = 104$ or lower

$$z = \frac{M - \mu}{\sigma_M} = \frac{104 - 100}{3} = \frac{+4}{3} = +1.33$$

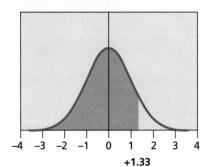

The probability of getting a mean lower than 104 (a *z* statistic of +1.33) is 0.9082, or approximately 91 percent.

c. a mean $M = 95$ or higher

$$z = \frac{M - \mu}{\sigma_M} = \frac{95 - 100}{3} = \frac{-5}{3} = -1.67$$

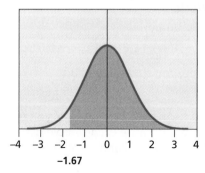

The probability of getting a mean higher than 95 (a z statistic of –1.67) is 0.9525, or approximately 95 percent.

-4 -3 -2 -1   0   1   2   3   4
          -1.67

d. a mean $M = 97$ or lower

$$z = \frac{M - \mu}{\sigma_M} = \frac{97 - 100}{3} = \frac{-3}{3} = -1.00$$

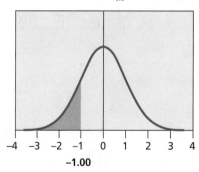

The probability of getting a mean lower than 97 (a z statistic of –1.00) is 0.1587, or approximately 16 percent.

-4 -3 -2 -1   0   1   2   3   4
          -1.00

11. a. $M = 275$ $(n = 36)$ and $M = 288$ $(n = 36)$?
What you know:

$$\mu = 300$$
$$\sigma = 70$$

To begin to answer the question, you must compute the standard error of the mean. Because each sample has the same number of participants $(n = 36)$, you need compute it only one time.

$$\sigma_M = \frac{\sigma}{\sqrt{n}} = \frac{70}{\sqrt{36}} = \frac{70}{6} = 11.67$$

Now you must compute the z statistic for each sample.

Remember, $z = \dfrac{M - \mu}{\sigma_M}$

$$M = 275$$
$$M = 288$$

$$z = \frac{275 - 300}{11.67} = \frac{-25}{11.67} = -2.14$$

$$z = \frac{288 - 300}{11.67} = \frac{-12}{11.67} = -1.03$$

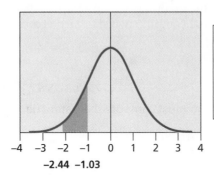

The probability of getting a mean between 275 (a z statistic of –2.14) and 288 (a z statistic of –1.03) is 0.1353 (0.1515 – 0.0162), or approximately 14 percent.

–2.44 –1.03

b. **M** = 280 (**n** = 25) and **M** = 315 (**n** = 25)? What you know:

$$\mu = 300$$
$$\sigma = 70$$

To begin to answer the question, you must compute the standard error of the mean. Because each sample has the same number of participants (**n** = 25), you need compute it only one time.

$$\sigma_M = \frac{\sigma}{\sqrt{n}} = \frac{70}{\sqrt{25}} = \frac{70}{5} = 14$$

Now you must compute the **z** statistic for each sample. Remember,

$$z = \frac{M - \mu}{\sigma_M}$$

M = 280   **M** = 315

$$z = \frac{288 - 300}{14} = \frac{-20}{14} = -1.43$$ $$z = \frac{315 - 300}{14} = \frac{15}{14} = 1.07$$

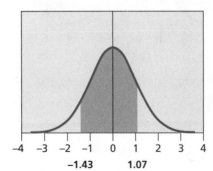

The probability of getting a mean between 280 (a z statistic of –1.43) and 315 (a z statistic of 1.07) is 0.7813 (0.8577 – 0.0764), or approximately 78 percent.

–1.43   1.07

c. $M = 305$ $(n = 49)$ and $M = 310$ $(n = 36)$?
What you know:

$\mu = 300$ $\qquad\qquad$ $\sigma = 70$

To begin to answer the question, you must compute the standard error of the mean. Because each sample is based on a different number (n), you must compute two standard errors of the mean.

$$M = 305 \ (n = 49)$$

$$\sigma_M = \frac{\sigma}{\sqrt{n}} = \frac{70}{\sqrt{49}} = \frac{70}{7} = 10$$

Now you must compute the z statistic.

Remember, $\qquad z = \dfrac{M - \mu}{\sigma_M}$

$$z = \frac{305 - 300}{10} = \frac{5}{10} = 0.50$$

$$M = 310 \ (n = 36)$$

$$\sigma_M = \frac{\sigma}{\sqrt{n}} = \frac{70}{\sqrt{36}} = \frac{70}{6} = 11.67$$

Now you must compute the z statistic.

Remember, $\qquad z = \dfrac{M - \mu}{\sigma_M}$

$$z = \frac{310 - 300}{11.67} = \frac{10}{11.67} = 0.86$$

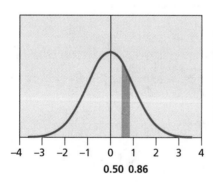

The probability of getting a mean between 305 (a z statistic of 0.50) and 310 (a z statistic of 0.86) is 0.1136 (0.8051 – 0.6915), or approximately 11 percent.

-4 -3 -2 -1 0 1 2 3 4

0.50 0.86

-7-

Hypothesis Testing

In This Chapter

Basics of Hypothesis Testing
- What Is Hypothesis Testing?
- Writing Hypotheses
- Null and Alternative Hypotheses
- Errors and Power in Hypothesis Testing
- Alpha Level; Level of Significance
- Assumptions Using the **z** Statistic

- Effect Size
- Steps in Hypothesis Testing

Application and Interpretation: Hypothesis Testing with the z Statistic

Reporting Results of the z statistic in APA Format

F O C U S | This chapter introduces hypothesis testing. Hypothesis testing is what we do when we evaluate the results of a study. There are specific rules to follow in hypothesis testing so that different researchers looking at the same results will arrive at the same conclusion. This chapter introduces the rules of hypothesis testing and applies them with the use of the z statistic.

Basics of Hypothesis Testing

What Is Hypothesis Testing?

Hypothesis testing
uses data from studies to
systematically investigate
research questions.

Hypothesis testing is a set of procedures that are used in evaluating research results. We begin with a question or an idea about what we believe is true in life. We establish a set of conditions to test our idea, gather data, and then analyze the results from the study to see if our original idea seems to be correct. We can never know for sure if we are right, but we can assign a statistical probability to the likelihood that our conclusion is correct.

Hypothesis testing involves a few concepts that need to be mastered prior to putting it into action. However, it is important to remember that this procedure is at the core of conducting research. Take some time to master these concepts; it is well worth the effort.

> **Reflect**
> Uncertainty is the very
> condition to impel man to
> unfold his powers.
> —Erich Fromm

In Chapter 1, we introduced the idea of research as investigating questions about life. We conceptualized the process as follows:

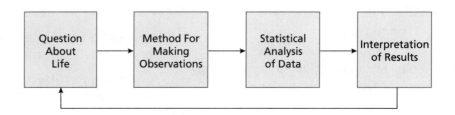

This flowchart will help you understand the underlying ideas of hypothesis testing.

> **Reflect**
> Knowledge is tentative in the
> absence of evidence to the
> contrary
> —Karl Popper

Writing Hypotheses

A **hypothesis** is a testable
statement about a possible
outcome of a study.

In research we ask questions in the form of hypotheses. A **hypothesis** is a testable statement about how a variable will behave or how the variable(s) in a study are connected. Another way of saying this is that a hypothesis is a statement of what you believe about the variable(s) you are examining. You can think of a hypothesis as a proposition about the expected outcome of whatever it is you are examining. Where do we get our hypotheses? Hypotheses can be based on prior knowledge from the literature, from previous studies, and/or from theories.

The hypothesis forms the basis of the investigation. Hypotheses must be testable.

Let's look at some examples of testable hypotheses questions:

The Question	Correctly Stated
Are boys more likely than girls to be bullies on the playground?	Boys engage in more p... bullying than girls.
Does exercise help reduce high blood pressure?	People who exercise have lower blood pressure than people who do not exercise.

Be Here Now

Write a testable hypothesis for each of the following questions:

1. Are poodles smarter than collies?
2. Is it a good idea to eat breakfast before an exam?

Answers

Some possible hypotheses are as follows:

1. Poodles learn new tricks more quickly than collies.
2. People who eat breakfast earn higher test grades than people who do not eat breakfast.

Null and Alternative Hypotheses

In hypothesis testing, you actually address two types of hypotheses: the **null hypothesis** and the **alternative hypothesis.**

In general, the **null hypothesis** proposes that there is no effect, no difference, or no relationship among the variables being studied. It may seem silly to set up a hypothesis that says, in effect, that the variables in a study will make no difference. However, the null hypothesis is actually quite important, because it is what is tested. Without an analysis of data collected to examine the null hypothesis, conclusions cannot be made about the actual relationship or effect of the variables in a study.

The **null hypothesis** states that there is no effect, no difference, or no relationship among the variables being studied.

The symbol we will use for the null hypothesis is H_0.

The **alternative hypothesis** states that there *is* an effect, a difference, or a relationship. In other words, the expected difference or relationship is confirmed.

The symbol we will use for the alternative hypothesis is H_1.

The **alternative hypothesis** states that there is an effect, a difference, or a relationship among the variables being studied.

In other contexts, you may see the alternative hypothesis symbolized as H_A.

The null and alternative hypotheses are essentially opposites.

Think about a simple experiment in which you are going to divide your class into two groups. The first group is going to sit quietly for 10 minutes. The second group is going to do 10 minutes of aerobic exercise. Then, everyone is going to measure his or her pulse. You'd expect a difference in the average pulse rates between the two groups, wouldn't you?

In our simple example, the null hypothesis would be that the average pulse rate of participants who exercise (μ_{exercise}) will be equal to the average pulse rate of participants who do not exercise ($\mu_{\text{no-exercise}}$).

In symbols, the null hypothesis would be H_0: $\mu_{\text{exercise}} = \mu_{\text{no-exercise}}$.

In this simple example, the alternative hypothesis would be that the average pulse rate (μ_{exercise}) of participants who exercise is different from the average pulse rate ($\mu_{\text{no-exercise}}$) of participants who do not exercise.

In symbols, the alternative hypothesis would be H_1: $\mu_{\text{exercise}} \neq \mu_{\text{no-exercise}}$.

A **nondirectional** alternative hypothesis does not specify the direction of the outcome.

Alternative hypotheses may be **nondirectional** or **directional**. A nondirectional alternative hypothesis states that there will be a difference, but does not specify the direction of the outcome.

A directional alternative hypothesis specifies that one mean will be greater than the other. In our example, we were looking for a difference—any difference—in pulse rate between the exercise and no-exercise groups; this is a nondirectional hypothesis.

A **directional alternative hypothesis** specifies the direction of the outcome.

Directional alternative hypotheses state a direction of the expected outcome. For example, a directional alternative hypothesis about pulse rate and physical activity might be stated as "Physical activity increases pulse rate." In other words, we would expect the group that engaged in aerobic exercise to have a higher pulse rate than the group that was asked to sit quietly.

Another example of a directional hypothesis is, "Vocabulary increases with age." Testing this hypothesis in a study, we would expect that adults would demonstrate larger vocabularies compared with children. That is, the mean vocabulary of adults is expected to be greater than the mean vocabulary of children. This directional hypothesis would be written as

$$H_1: \mu_{\text{adult}} > \mu_{\text{child}}$$

The null hypothesis would be

$$H_0: \mu_{\text{adult}} \leq \mu_{\text{child}}$$

Notice that the equality sign will always be part of the null hypothesis and never part of the alternative hypothesis. In this example, we use a "greater than" symbol ($>$) for the alternative hypothesis and a "less than or equal to" symbol ($\leq$) for the null hypothesis.

You should use a directional alternative hypothesis when you have a firm basis for predicting one direction over another. This basis may come from either the literature or previous research. Otherwise, it is advisable to use a nondirectional hypothesis.

It may seem obvious, but it is worth stating that the null hypothesis and the alternative hypothesis in a study are mutually contradictory. In other words, if one is true, the other is false, and vice versa. This feature is central to hypothesis testing. If we find evidence to reject one of the hypotheses, then we have found support for the other.

On the basis of our data, we either *reject the null hypothesis* or *fail to reject it*. If we reject the null hypothesis, we conclude that we have found support for the

> **Reflect**
>
> To see truth, contemplate all phenomena as a lie.
> —Thaganapa

Be Here Now

Match each of the following with one of the statements that follow:

a. null hypothesis

b. nondirectional alternative hypothesis

c. directional alternative hypothesis

1. There is a relationship between talking on cell phones and car accidents.

2. Listening to music decreases tension.

3. At age 10, there is no difference between boys and girls in mathematics achievement.

Answers

1.b, 2.c, 3.a

alternative hypothesis. If we fail to reject the null hypothesis, we fail to find support for the alternative hypothesis.

There are two important things to notice here. First, our conclusions are stated in relation to the null hypothesis. Second, we never *prove* either the null or alternative hypotheses. Instead, when we reject the null hypothesis, we believe we have *found evidence* that supports the alternative hypothesis.

There are many reasons that might lead to failure to reject the null hypothesis. We'll look more closely at errors in hypothesis testing later in the chapter. For now, it is important to realize that failure to reject the null hypothesis is not necessarily a bad thing: It actually can lead to more refined research.

We need some rules or guidelines so that everyone looking at the same data will come to the same conclusion. This is what the process of hypothesis testing allows us to do. It provides the guidelines we need. With that in mind, let's examine some of the guidelines we use when we engage in hypothesis testing.

Reflect
Failing to reject the null hypothesis is not necessarily failure.

Errors and Power in Hypothesis Testing

We must consider ways in which we could make an error when we test hypotheses. First, let's consider ways in which we would be correct. If we reject the null hypothesis, *and the alternative hypothesis is true,* we are correct; this is called *power.* If we fail to reject the null hypothesis, *and the null hypothesis is true,* we are again correct. However, there are two ways to make an error in our decision-making process. The first type of error occurs when we reject the null hypothesis, *when the null hypothesis is true.* This is called a *Type I error.* The second kind of error is when we fail to reject the null hypothesis, *when the alternative hypothesis is true.* This is called a *Type II error.*

		Truth	
		Null Hypothesis Is True	**Null Hypothesis Is False**
Decision Made on Basis of Sample Data	Reject H_0	Type I Error α	Correct Decision, or Power $1 - \beta$
	Fail to Reject H_0	Correct Decision	Type II Error β

FIGURE 1 Possible outcomes of hypothesis testing.

The null hypothesis may be either true or false. On the basis of the data, we decide to either reject or fail to reject the null hypothesis. Our decision may be correct or incorrect. These possibilities are depicted in Figure 1.

Let's look at these outcomes in more detail.

A **Type I error** occurs when we reject the null hypothesis, but the null hypothesis is true.

Type I errors are symbolized by alpha (α).

For example, let's say that you are comparing two methods for taking a written driving test. One method is by computer and the other is a paper-and-pencil test. The null hypothesis is that there is no difference between the methods $\left(\mu_{computer} = \mu_{paper-and-pencil}\right)$. The alternative hypothesis is that there is a difference $\left(\mu_{computer} \neq \mu_{paper-and-pencil}\right)$.

You find the group that took the test on the computer performed considerably better, on average, compared with the group that took the paper-and-pencil test. You reject the null hypothesis and conclude that the paper-and-pencil test should be eliminated in favor of the computerized test. However, perhaps what really happened is that the participants in the computer group were almost exclusively young people who were quite comfortable with computer technology. If you repeated the study with another, less technologically oriented group, you might obtain different results.

Another way of thinking about a Type I error is that it is like convicting an innocent person.

A **Type II error** occurs when we fail to reject the null hypothesis, and it is, in fact, false.

Type II errors are symbolized by beta (β).

If a Type I error is analogous to convicting an innocent person, then a Type II error is like letting a guilty person go free. For example, let's say that you are on a jury and the defendant is accused of a crime. In our legal system,

A **Type I error** occurs when the null hypothesis is rejected and it is, in fact, true.

A **Type II error** occurs when we fail to reject the null hypothesis, but the null hypothesis is false.

Reflect

Sometimes what seems to be the best decision is not correct.

there is a presumption of innocence (a legal version of the null hypothesis). You cannot convict a person unless you are assured beyond a reasonable doubt of that person's guilt. If the prosecution presents evidence that is insufficient to convict the defendant, you must deliver a verdict of "not guilty" and the person is freed. That does not mean that the person did not commit the crime; it means that there was not enough evidence—evidence beyond a reasonable doubt—to convict the person! You may have released that person to commit another crime. This would be a Type II error.

Power is the probability of correctly rejecting a false null hypothesis.

When we correctly reject a false null hypothesis, this is known as **power.** Power is symbolized by $1 - \beta$. We want our hypothesis tests to have as much power as possible, to increase the probability of finding an effect if an effect is present. And we don't want to waste time or resources conducting an investigation that has no chance of detecting an effect when one is there. So, when we design a study, we want to design a powerful study, one that is likely to reject the hypothesis if it is, in fact, false.

The more power a study has, the less likely it is that we will make a Type II error. One possible reason a study may result in a Type II error has to do with the sample size: Increasing the sample size decreases the likelihood of a Type II error and increases power. Remember the formula for the standard error of the mean:

$$\sigma_M = \frac{\sigma}{\sqrt{n}}$$

Notice that the denominator of the formula includes *n,* the sample size. The larger *n* is, the smaller will be the standard error of the mean. The smaller the standard error of the mean, the larger the test statistic will be. The larger the test statistic, the more likely we are to reject the null hypothesis. If we reject the null hypothesis, we cannot make a Type II error. Thus, the larger the sample size, the smaller is the likelihood that we will make a Type II error, and the greater is the power.

Type I and Type II errors have an inverse relationship: As we increase the probability of committing a Type I error, we decrease the probability of committing a Type II error, and vice versa. To go back to the courtroom analogy, a person can't look more innocent and more guilty at the same time. As one looks more innocent, one looks less guilty. The same applies to Type I and Type II errors.

Which type of error do we most want to avoid? The answer is not simple; it depends on what is being examined in the study. On the one hand, if you have found a cure for hepatitis, you certainly don't want to make a Type II error and say that you haven't found anything. That would allow the disease to continue to "go free." On the other hand, if you are about to invest in expensive equipment for your school district for a new program that promises to help students with learning disabilities, you want to be sure that the new program really works before spending the taxpayers' money. Otherwise, you will have made a Type I error.

Identify which of the following would present the potential for a Type I error and which would present the potential for a Type II error.

1. To fail to reject the null hypothesis when comparing the scores of participants who studied with music with the scores of those who studied without music.

2. To reject the null hypothesis upon finding that persons who use a new medication feel less pain than those who do not.

Answers

1. Type II error
2. Type I error

Alpha Level; Level of Significance

We need a set of rules to guide our decisions about hypotheses. That is, we need to establish the level of a Type I error that we are willing to accept. We need to establish the criteria by which we make the decision to either reject or fail to reject the null hypothesis.

To explore how we identify the criteria by which we decide to reject the null hypothesis, we will test a hypothesis about an expected value. Suppose you worked for a consumer protection agency. You have received complaints that a particular light bulb does not last as long as the manufacturer claims. To test the manufacturer's claim that the light bulb burns 1000 hours, with a standard deviation of 200 hours, you buy 16 light bulbs and observe how many hours each light bulb remains lit.

We can start by stating the null and alternative hypotheses:

H_0: $\mu = 1000$ **(The manufacturer's claim is true; the light bulbs burn 1000 hours.)**

H_1: $\mu \neq 1000$ **(The manufacturer's claim is not true; the light bulbs do not burn 1000 hours.)**

If you collect no data, you have only the null hypothesis: the manufacturer's claim.

If you collect data, you can make a decision. You have two choices:
There is no evidence to suggest that the manufacturer's claim is false (fail to reject H_0), or There is evidence to suggest that the manufacturer's claim is false (reject H_0).

Now let's look at the data, the number of hours each of the 16 light bulbs burned:

Number of Hours Light Bulbs Burned

950	875	1200	1100	850	850	750	1200	
950	975	1125	1200	750	855	925	1125	$M = 980$

In our light bulb study, we know that

$$M = 980$$
$$\mu = 1000$$
$$\sigma = 200$$
$$n = 16$$

We can compute the z statistic to determine the distance the observed mean (mean number of hours the light bulbs we tested burned) is from the population mean (the manufacturer's claim) in standard units.

Remember the formula:

$$z = \frac{M - \mu}{\sigma_M}$$

Now, think back to Chapter 5, and remember that samples may have error. So, we need to calculate the standard error of the mean:

$$\sigma_M = \frac{\sigma}{\sqrt{n}} = \frac{200}{4} = 50$$

We can now compute the z statistic:

$$z = \frac{M - \mu}{\sigma_M} = \frac{980 - 1000}{50} = -.4$$

Is a z statistic of −.4 units from μ far enough from μ to reject the null hypothesis? To illustrate this question, Figure 2, a normal curve, presents the sampling distribution of means when $\mu = 1000$. If in truth, $\mu = 1000$, you would expect most means to be near 1000. The further from 1000, the fewer means would be expected. It is unlikely to obtain a mean that is extremely different from 1000. How far from μ must a sample mean fall in order for you to reject the null hypothesis?

Ask your classmates if they believe that the company's claim is accurate. Some may think that the manufacturer is telling the truth; others may think

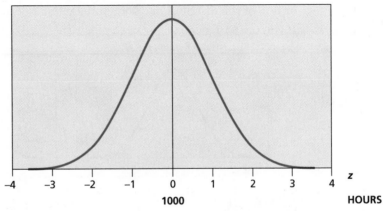

FIGURE 2 Sampling distribution of means when $\mu = 1000$.

that the number of hours is substantially below the manufacturer's claim. We need a rule!

At issue is how far away from μ is far enough to conclude that, with some particular degree of certainty, the null hypothesis is false and the data collected probably came from a different population. The probability level that we use to determine our criterion is called the **alpha level.** The alpha level is a probability level that is set prior to beginning hypothesis testing and that determines the criterion by which we will reject the null hypothesis.

The **alpha level** is the potential level of error we are willing to accept.

The **critical region** is the area in a hypothetical distribution determined by the alpha level at which we can reject the null hypothesis.

The alpha level sets what we call the *level of significance* for a hypothesis test. It determines the criterion or standard that must be met in order to reject the null hypothesis. The alpha level defines the **critical region, or region of rejection,** in a hypothetical distribution of sample statistics.

The symbol for the alpha level is α

Sometimes we use *p*—the probability of making a Type I error—to indicate the alpha level.

When we choose an alpha level, we are also setting the probability with which we are comfortable making a Type I error. By convention, the alpha level is set at .05 or .01. The smaller the alpha level, the more rigorous is our criterion for rejection of the null hypothesis. At an alpha level of .05, we are willing to tolerate a 5-in-100 chance of incurring a Type I error in deciding to reject the null hypothesis. At an alpha level of .01, we are willing to tolerate only a 1-in-100 chance of incurring a Type I error in deciding to reject the null hypothesis.

Figure 3 depicts the critical regions for each alpha level (.05 and .01). The region that is shaded is considered the critical region. As shown in Figure 3, when we have a nondirectional hypothesis, because we are not predicting a direction, we must divide the alpha level between both ends (tails) of the distribution. That means that we must divide the alpha level that we have chosen (.05 or .01) in half and designate half of the alpha level in each end, or tail, of the distribution. Therefore, with a nondirectional hypothesis, an

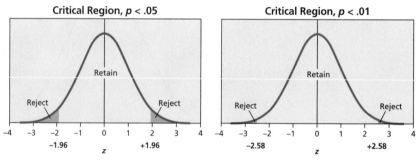

FIGURE 3 Critical region at alpha level =.05 and .01.

alpha level of .05 becomes .025 in each tail. If you look in the Unit Normal Table for .025 in the column labelled tail, you will find a z value of 1.96. Thus, ± 1.96 defines the critical region at an alpha level of .05. Similarly, an alpha level of .01 becomes .005 in each tail. Using the Unit Normal Table for .005 in the column labelled tail, you will find a z value of 2.58 (rounded). Hence, the critical region at an alpha level of .01 is ± 2.58. Notice that the critical region for an alpha level of .01 is smaller (further out in the tail) than that for an alpha level of .05. When we use .01, we are less likely to make a Type I error.

Now, consider our light bulb example. The z statistic of $-.4$ that we computed does *not* fall into the critical region. Therefore, we *fail to reject the null hypothesis*; we have no evidence to conclude that the manufacturer's claim was incorrect.

As shown in Figure 4, when we have a directional hypothesis, we place the entire value of the alpha in one tail. The critical region is placed in the tail in which we would expect to find the z statistic if the null hypothesis is rejected. To illustrate, we have stated possible alternative hypotheses concerning our light bulb study.

When a result falls within the critical region, it is considered **significant.**

As you work on problems, it is helpful to draw a picture of the distribution and indicate the alpha level that you have chosen by shading in the region of rejection. When you obtain your results and make your decision about whether to reject or fail to reject the null hypothesis, it is a good idea to refer to your illustration and check to see if your decision makes sense.

A **significant** statistic indicates that the null hypothesis is rejected.

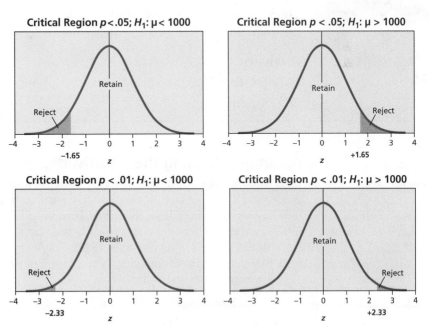

FIGURE 4 Alpha levels of .05 and .01 for directional hypothesis tests.

		Type of Test	
		Directional	**Nondirectional**
Alpha Level	**$p < .05$**	$z = 1.65$ or -1.65	$z = 1.96$ and -1.96
	$p < .01$	$z = 2.33$ or -2.33	$z = 2.58$ and -2.58

FIGURE 5 Critical values of the z statistic.

So, when we set an alpha level, we consider

- Whether to use an alpha level of .05 or an alpha level of .01
- Whether to use a directional or a nondirectional hypothesis

Although you can look up the critical values in the Unit Normal Table, Figure 5 presents the critical regions of the z statistic for alpha levels of .05 and .01 for directional and nondirectional hypotheses.

Reflect
The beginning (alpha) can determine the end.

Be Here Now

Identify which situation would increase the likelihood of rejecting the null hypothesis:

1. Alpha level of .05 or alpha level of .01.
2. Your study is based on 20 participants or 200 participants

Answers
1. alpha level of .05
2. 200 participants

Assumptions Using the z Statistic

Four assumptions using the z statistic underlie hypothesis testing. If we seriously violate these assumptions, we may weaken the validity of our research (see for example, Hays, 1994; Kirk, 1999; Nunnally, 1975; Pedhazur & Schmelkin, 1997). However, statistical tests typically hold up well despite moderate violations of the assumptions (Stevens, 1996). There are also alternative statistical tests that may be considered, but they are for a more advanced statistics class.

The four assumptions are

- Random sampling
- Independence of observations
- Normal distribution
- Homogeneity of variance

In **random sampling,** participants are selected at random from a population.

Independence of observations means that an observation is not related to any other observations within or between conditions.

The **normal distribution** permits us to use the Unit Normal Table to determine the critical region for a hypothesis test.

Homogeneity of variance means that the variance of the sample is the same as the variance in the population.

Effect size refers to the magnitude of a treatment effect.

In *random sampling,* we make the assumption that the participants in a sample were drawn at random from the population. This assumption reduces the chance for bias and so facilitates generalization, or being able to make statements about the population of interest on the basis of results from a sample.

Independence of observations means that an observation is not related to any other observations within or between conditions.

The assumption that we have a *normal distribution* of sample means permits us to use the Unit Normal Table and the probabilities associated with the normal distribution to determine the critical region for a hypothesis test.

Homogeneity of variance means that the variance of the sample is the same as the variance in the population.

Effect Size

Sometimes, even though we reject the null hypothesis, our finding may be small or not meaningful. The American Psychological Association (APA, 2001) suggests that we calculate **effect size.** Measures of effect size are not affected by the size of the sample. Rather, they use a standard scale to show the magnitude or the strength of a treatment effect. Effect size is a good way to further examine significant results. Later in this chapter, we will describe a commonly used method of estimating effect size called Cohen's *d*.

Steps in Hypothesis Testing

Here, then, are the steps to follow for hypothesis testing:

- State the null and alternative hypotheses for the study.
- Choose the alpha level, draw the distribution, and label the critical values.
- Collect the data for the study and calculate the appropriate test statistic.
- From the results you obtain, make a decision whether to reject the null hypothesis or fail to reject the null hypothesis.
- If you reject the null hypothesis, calculate the effect size.

Application and Interpretation: Hypothesis Testing Using the *z* Statistic

We have already seen an example of hypothesis testing with the *z* statistic: our light bulb study. What follows are two more examples.

Let's say you are teaching introductory statistics. You have $n = 25$ students in your class. You hypothesize that their level of motivation is greater than that of the typical student. If the average score on a motivation scale is $\mu = 100$, with a standard deviation of $\sigma = 10$, and your class average is $M = 105$, is there enough evidence for you to conclude that your students are more motivated than the typical student?

To answer the preceding question, begin by laying out what you already know, and then follow the steps in hypothesis testing. You know that

$$\mu = 100$$

$$\sigma = 10$$

$$M = 105$$

$$n = 25$$

STEP 1   State the null and alternative hypotheses for the study.

In words, the null hypothesis is: "The motivation of your class is similar to or less than that of the typical college class (population)."

The alternative hypothesis is: "The motivation of your class is greater than that of the typical college class (population)."

Now you state the null and alternative hypotheses in statistical notation:

$$H_0: \mu \leq 100$$

$$H_1: \mu > 100$$

Because you are testing whether the class has greater-than-average motivation, you use a greater-than sign in the alternative hypothesis. You then use a less-than-or-equal-to sign in the null hypothesis, to indicate that anything equal to the known mean of 100 or less than the known mean of 100 will not be significant.

Notice that you are employing a directional test. Directional tests are often called one-tailed tests because the critical region falls into only one tail of the hypothetical distribution of sample means. Here, you are using a one-tailed test because you have stated an expected direction.

STEP 2   Choose the alpha level and draw the distribution

In this example, you will use an alpha level of .05. Refer to the chart in Figure 5 of this chapter or look up the appropriate value for the critical region of .05 in the Unit Normal Table (see Appendix A1); be sure to use the column related to the tail of the distribution. The table identifies the critical value of the test statistic to be $z = 1.65$. When you draw a picture of the distribution, only one tail—the one at the positive end of the distribution—is shaded in. In this study, you will reject the null hypothesis if your computed z statistic falls within the critical region of rejection.

STEP 3   Collect the data for the study and calculate the appropriate test statistic. The calculations are as follows:

$$z = \frac{M - \mu}{\sigma_M}$$

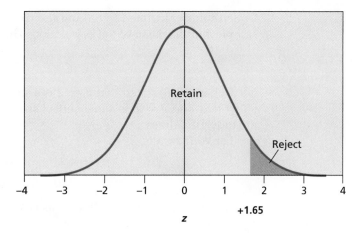

where

$$\sigma_M = \frac{\sigma}{\sqrt{n}}$$

$$\sigma_M = \frac{10}{\sqrt{25}} = \frac{10}{5} = 2$$

$$z = \frac{105 - 100}{2} = \frac{5}{2} = 2.50$$

STEP 4   On the basis of the results you obtain, make a decision whether to reject the null hypothesis or fail to reject the null hypothesis.

The question here is whether the result $z = 2.50$ falls into the critical region. If so, you can reject the null hypothesis and conclude that your class is significantly greater than average in motivation. The critical region begins at $z = 1.65$. $z = 2.50$ falls within the critical region. Your decision, then, is to reject the null hypothesis and conclude that your class is significantly more motivated than average.

Might you have made an error? Yes, you may have rejected the null hypothesis when the null hypothesis was really true. This would be a Type I error. What is the probability that you made a Type I error? Remember, you set the probability of a Type I error when you choose the alpha level for your study. So, the probability that you made a Type I error is 5% or less. Could you have made a Type II error? The answer is no. In a Type II error, you incorrectly fail to reject the null hypothesis. Since you rejected the null hypothesis, you could not have made a Type II error.

Because you obtained a significant result, it is important to calculate the effect size associated with that result. The effect size tells you the magnitude of the effect and is not dependent on the size of the sample. You can use a basic formula that is widely used: Cohen's *d.*

STEP 5   Calculate the effect size that results upon rejection of the null hypothesis.

Cohen's *d* is calculated by taking the difference between the mean of the sample and the mean of the population. The formula looks like this:

$$d = \frac{M - \mu}{\sigma}$$

Cohen (1988) stated that *d* = .2 is a small effect size, *d* = .5 is a medium effect size, and *d* = .8 is a large effect size; however, he cautioned that these are only guidelines.

For your results,

$$d = \frac{105 - 100}{10} = \frac{5}{10} = .50$$

According to Cohen's guidelines, this is a medium effect size.

Let's try another example, this time using a nondirectional test. Imagine that you are interested in the writing skills of people who use e-mail. You have found no theoretical or empirical basis for forming a directional hypothesis. You know that, on an established writing test, the mean score is $\mu = 350$, with a standard deviation of $\sigma = 75$. The mean score of a sample of $n = 144$ people who use e-mail regularly is 335. Is this sample significantly different from the population?

Again, organize what you already know, and then follow the steps in hypothesis testing. You know that

$$\mu = 350$$
$$\sigma = 75$$
$$M = 335$$
$$n = 144$$

STEP 1 State the null and alternative hypotheses for the study.

In words, the null hypothesis would be "Regular e-mail users' writing ability is similar to that of the general population."

The alternative hypothesis would be "Regular e-mail users' writing ability is different from that of the general population."

In statistical notation, these hypotheses would be

$$H_0: \mu = 350$$
$$H_1: \mu \neq 350$$

Notice that because this is a nondirectional, or two-tailed test, you use an equals sign and an inequality sign in the hypotheses.

STEP 2 Choose the alpha level and draw the distribution.

For purposes of demonstration, let us use an alpha level of .01 in this example. Refer to the chart in Figure 5 of this chapter, or look up the appropriate value for the critical region on the Unit Normal Table (Appendix A1). Notice that the critical region has been divided between the two tails and is shaded. The alpha level of .01, therefore, becomes .005 if you are looking in the Unit Normal Table. Critical values associated with alpha .01, two tailed, are ±2.58. If the calculated z statistic falls into either of the shaded areas, you will be able to reject the null hypothesis and conclude that, for this sample, those who regularly use e-mail have different writing skills than the general population.

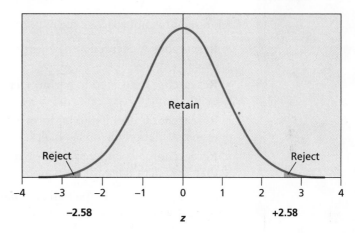

STEP 3   Collect the data for the study and calculate the appropriate test statistic. The calculations are as follows:

$$z = \frac{M - \mu}{\sigma_M}$$

where

$$\sigma_M = \frac{\sigma}{\sqrt{n}}$$

$$\sigma_M = \frac{75}{\sqrt{144}} = \frac{75}{12} = 6.25$$

$$z = \frac{335 - 350}{6.25} = \frac{-15}{6.25} = -2.40$$

STEP 4   On the basis of the results obtained, make a decision whether to reject the null hypothesis or fail to reject the null hypothesis. Notice that the value of z that you obtained has a negative sign ($z = -2.40$) The critical regions are the areas beyond $z = +2.58$ and $z = -2.58$. Your calculated value of z does not fall into either critical region. So, you must fail to reject the null hypothesis; there was no evidence to conclude that the writing ability of persons who regularly use e-mail is different from that of the general population.

Be Here Now

Four researchers wanted to test whether a certain accelerated class' IQ was greater than average. There were 25 youngsters in the class; their mean IQ score was 108. In the population, the mean IQ is 100 and the standard deviation is 16. As the research assistant, you compute the z statistic. To begin, you must compute the standard error of the mean:

$$\sigma_M = \frac{\sigma}{\sqrt{n}} = \frac{16}{\sqrt{25}} = \frac{16}{5} = 3.2 \quad z = \frac{M - \mu}{\sigma_M} = \frac{107 - 100}{3.2} = \frac{7}{3.2} = 2.5$$

Each researcher approaches the hypothesis from a different vantage point:

- **Researcher A** decided to perform a nondirectional test at an alpha level of .05.
- **Researcher B** decided to perform a nondirectional test at an alpha level of .01.
- **Researcher C** decided to perform a directional test (class mean greater than population mean) at an alpha level of .05.
- **Researcher D** decided to perform a directional test (class mean greater than population mean) at an alpha level of .01.

For each researcher, state the null and alternative hypotheses, draw the critical region, and make a decision.

Answers

Researcher A	Researcher B

Researcher A

$H_0: \mu = 100$ $H_0: \mu \neq 100$

Alpha level of .05; Critical Value: $z \pm 1.96$

Decision: Reject H_0; Compute Cohen's $d = .44$

Researcher B

$H_0: \mu = 100$ $H_0: \mu \neq 100$

Alpha level of .01; Critical Value: $z \pm 2.58$

Decision: Fail to Reject H_0

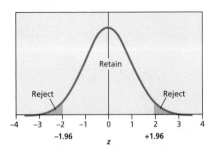

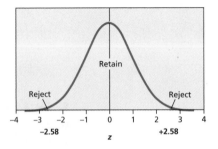

Researcher C

$H_0: \mu \leq 100$ $H_0: \mu > 100$

Alpha level of .05; Critical Value: $z + 1.65$

Decision: Reject H0; Compute Cohen's $d = .44$

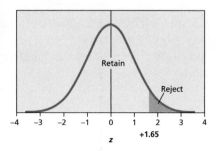

Researcher D

$H_0: \mu \leq 100$ $H_0: \mu > 100$

Alpha level of .01; Critical Value: $z + 2.33$

Decision: Reject H_0; Compute Cohen's $d = .44$

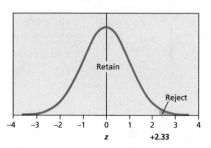

Because you failed to reject the null hypothesis, you did not make a Type I error. However, because you failed to reject the null hypothesis, you may have made a Type II error. You did not obtain a significant result, so you would not calculate an effect size.

Reporting Results of the *z* statistic in APA Format

When we present our results in APA format, we make a statement about our decision to reject or fail to reject the null hypothesis and then provide the statistical documentation to support our decision. Here are some examples of how to write results in APA format, based on the examples presented in this chapter.

When you reject the null hypothesis:

The statistics class showed greater-than-average motivation, $z = 2.50$, $p < .05$. The effect size was moderate, Cohen's $d = .50$.

When you fail to reject the null hypothesis:

There was no evidence to suggest that the light bulb manufacturer's claim was false, $z = -0.40$, $p = .98$. or Writing skills of people who use e-mail regularly were not significantly different in this sample from those of people in the general population, $z = -2.40$, *ns*.

Reflect

In hypothesis testing, as in criminal justice, we report the results, the whole results, and nothing but the results.

Note that *ns* is used to indicate a finding that is not significant.

If you have more than one test statistic, consider how to best report your findings so that you communicate them to your reader in a clear and easy-to-understand manner. If you have more than four test results to report, consider placing them in a table. More than 20 test results may be best represented in a graph or a figure. Remember, part of the scientific method is sharing results in such a way that they can be easily comprehended.

It's Out There...

Layton, Heeley, and Shakir (2004) examined the adverse side effects of drugs used for colitis. Relative to a known database, they found that there was a greater potential signal of colitis when the drug rofecoxib was present ($z = 5.6$, $p < .05$).

SUMMARY

In research, we formulate two hypotheses: the null hypothesis (H_0) and the alternative hypothesis (H_1). On the basis of our data, we either reject the null hypothesis or fail to reject it. If we reject the null hypothesis and the alternative hypothesis is true, we are correct; this state of affairs is called power. If we fail to reject the null hypothesis and the null hypothesis is true, we also are correct. However, there are two ways to make an error in our decision-making process. The first type of error occurs when we reject the null hypothesis, but the null hypothesis is true. This is called a Type I error, or alpha. The researcher sets the level of alpha he or she is willing to tolerate; usually, alpha is set to .05 or .01. The second kind of error occurs when we fail to reject the null hypothesis and the null hypothesis is false. This is called a Type II error. Measures of effect size show the magnitude or the strength of a treatment effect, using a standard scale that is not influenced by the number of participants.

PRACTICE

Writing Hypotheses

1. Write a testable hypothesis for each of the following questions:
   a. Are people who study statistics brighter than other people?
   b. Is blue a more soothing color than yellow to look at when studying?
   c. Does a six-cylinder car use more gas than a four-cylinder car?
   d. Do tall parents have tall children?

2. Write the null and alternative hypotheses for each of the following situations, using the appropriate symbols. For each state whether it is directional or nondirectional.

a. Males and females differ in how quickly they can learn computer games.

b. Poor people commit more crimes than wealthy people do.

c. Smokers have shorter life spans than the average population's 78 years.

Errors and Power in Hypothesis Testing

3. Place the appropriate term in each box. Use the following terms: Type I error; Type II error; Correct decision: power; and correct decision.

<table>
<tr><td rowspan="2"></td><td colspan="2">Situation in the Population</td></tr>
<tr><td>Null Hypothesis Is True</td><td>Null Hypothesis Is False</td></tr>
<tr><td rowspan="2">Decision Made on Basis of Sample Data</td><td>Reject the Null Hypothesis</td><td></td><td></td></tr>
<tr><td>Fail to Reject The Null Hypothesis</td><td></td><td></td></tr>
</table>

Alpha Level—Level of Significance

4. Which alpha level should you use if you want to set your critical region farther out in the tails of the distribution, alpha = .05 or alpha = .01?

5. You have a directional hypothesis. Is your critical region in one tail of the distribution or both tails of the distribution?

Hypothesis Testing Using the z Statistic

6. Historically, the number of crimes committed in the town of XYZ is 141 per month (μ), with a standard deviation (σ) of 20. You believe that more violent crimes are committed when there is a full moon than during other phases of the moon. The average number of violent crimes committed over 12 months of full moons is $M = 156$. Using the steps for hypothesis testing, test the hypothesis that more violent crimes are committed when the moon is full.

7. A researcher is interested in whether short-term memory in the elderly is different from short-term memory in the general population. The literature suggests that short-term memory declines with age. The mean on a short-term memory task for the general population is $\mu = 125$, with a standard deviation of $\sigma = 10$; the mean for a sample of $n = 36$ people over the age of 80 is $M = 121$. Using the steps for hypothesis testing, test the hypothesis that the elderly sample has significantly less short-term memory than the general population.

8. A teacher thinks that it does not really matter if children have recess. She knows that the average score in the population on a standardized achievement test is $\mu = 300$, with a standard deviation of $\sigma = 50$. She decides to eliminate recess for a term and then give her class the achievement test. The mean for her sample of $n = 25$ students is $M = 280$. Using the steps for hypothesis testing, test the null hypothesis that there is no difference between her class and the population on the achievement test.

9. Suppose you want to know if only children have larger vocabularies than children with siblings. You know that the average five-year-old child's vocabulary has a mean of $\mu = 2,500$ words, with a standard deviation of $\sigma = 200$. The mean number of vocabulary words for a sample of $n = 9$ five-year-old only children is $M = 2,600$. Using the steps for hypothesis testing, test the hypothesis that five-year-old only children have a significantly higher vocabulary than five-year-olds in the general population.

Writing Results for the z Statistic in APA Format

10. Write your results for practice problems 8 and 9 in APA format.

SOLUTIONS

Writing Hypotheses

1. **a.** There is a relationship between intelligence and people who study statistics.
   b. Studying in a blue room produces higher exam scores than studying in a yellow room does.
   c. A six-cylinder car uses more gasoline than a four-cylinder car in running 100 miles.
   d. There is a positive relationship between the height of parents and the height of their children.

2. **a.** $H_0: \mu_{males} = \mu_{females}$
   $H_1: \mu_{males} \neq \mu_{females}$
   This is nondirectional.
   b. $H_0: \mu_{poor} \leq \mu_{wealthy}$
   $H_1: \mu_{poor} > \mu_{wealthy}$
   This is directional.
   c. $H_0: \mu_{smokers} \geq 78$
   $H_1: \mu_{smokers} < 78$
   This is directional.

Errors and Power in Hypothesis Testing

3.

		Situation in the Population	
		Null Hypothesis Is True	Null Hypothesis Is False
Decision Made on Basis of Sample Data	**Reject the Null Hypothesis**	Type I Error	Correct Decision: Power
	Fail to Reject The Null Hypothesis	Correct Decision	Type II Error

Alpha Level; Level of Significance

4. $alpha = .01$

5. One tail of the distribution

Hypothesis Testing Using the z Statistic

6. Step 1: H_0: $\mu_{\text{violent crimes}-\text{full moon}} \leq \mu_{\text{violent crimes}-\text{other phases}}$

H_1: $\mu_{\text{violent crimes}-\text{full moon}} > \mu_{\text{violent crimes}-\text{other phases}}$

$alpha = .05$, one tailed directional test

Step 2:

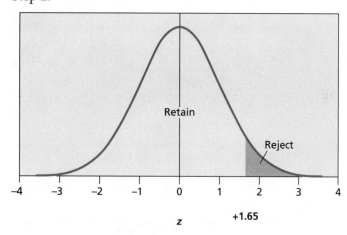

Step 3: $z = \dfrac{M - \mu}{\sigma_M}$

$$\sigma_M = \frac{\sigma}{\sqrt{n}} = \frac{20}{\sqrt{12}} = \frac{20}{3.47} = 5.76$$

$$z = \frac{M - \mu}{\sigma_M} = \frac{156 - 141}{5.76} = \frac{15}{5.76} = 2.60$$

Step 4: We reject the null hypothesis: The z statistic we calculated, 2.60, falls into the critical region, or region of rejection. For this sample, we conclude that more violent crimes were committed when the moon was full than in any of the other phases.

Step 5: Because we rejected the null, we calculate the effect size:

$$d = \frac{M_1 - M_2}{\sigma}$$

$$\frac{156 - 141}{20} = \frac{15}{20} = .75$$

This is a medium effect size.

7. Step 1: H_0: $\mu_{\text{elderly}} \geq \mu_{\text{general population}}$

H_1: $\mu_{\text{elderly}} < \mu_{\text{general population}}$

alpha = .05, one tailed/directed test

Step 2:

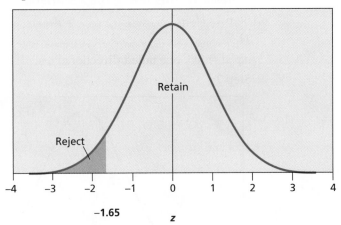

Step 3: $z = \dfrac{M - \mu}{\sigma_M}$

$$\sigma_M = \frac{\sigma}{\sqrt{n}} = \frac{10}{\sqrt{36}} = \frac{10}{6} = 1.67$$

$$z = \frac{M - \mu}{\sigma M} = \frac{121 - 125}{1.67} = \frac{-4}{1.67} = -2.40$$

Step 4: We reject the null hypothesis: The z statistic we calculated, −2.40, falls into the critical region, or region of rejection. We conclude that the elderly sample has significantly less short-term memory than the general population has.

Step 5: Because we rejected the null hypothesis, we calculate the effect size:

$$d = M1 - M2 = \frac{121 - 125}{10} = \frac{-4}{10} = -.40$$

This is a small effect size.

8. Step 1: H_0: $\mu_{\text{no recess}} = \mu_{\text{recess}}$

   H_1: $\mu_{\text{no recess}} \neq \mu_{\text{recess}}$

   alpha = .05, two-tailed nondirectional test

   Step 2:

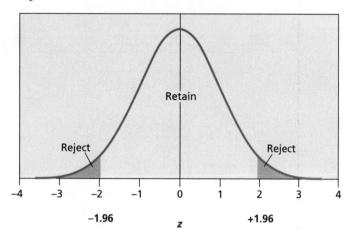

   Step 3: $z = \dfrac{0 - \mu}{\sigma_M}$

   $$\sigma_M = \frac{\sigma}{\sqrt{n}} = \frac{50}{\sqrt{25}} = \frac{50}{5} = 10$$

   $$z = \frac{M - \mu}{\sigma_M} = \frac{280 - 300}{10} = \frac{-20}{10} = -2.00$$

   Step 4: We reject the null hypothesis: The z statistic we calculated, 2.00, falls into the critical region, or region of rejection. We conclude that the students in this sample who did not have recess scored significantly below the population on this achievement test.

   Step 5: Because we rejected the null hypothesis, we calculate the effect size:

   $$d = \frac{M_1 - M_2}{\sigma} = \frac{280 - 300}{50} = \frac{-20}{50} = -.40$$

   This is a small effect size.

9. Step 1: H_0: $\mu_{\text{onlychildren}} \leq \mu_{\text{children}}$

   H_1: $\mu_{\text{onlychildren}} > \mu_{\text{children}}$

   alpha = .05, one-tailed/directional test

Step 2:

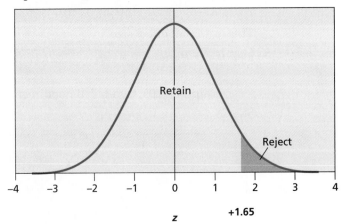

+1.65

Step 3: $z = \dfrac{M - \mu}{\sigma_M}$

$$\sigma_M = \dfrac{\sigma}{\sqrt{n}} = \dfrac{200}{\sqrt{9}} = \dfrac{200}{3} = 66.67$$

$$z = \dfrac{M - \mu}{\sigma_M} = \dfrac{2,600 - 2,500}{66.67} = \dfrac{100}{66.67} = 1.50$$

Step 4: We fail to reject the null hypothesis: The z statistic we calculated, 1.50, does not fall into the critical region, or region of rejection. We conclude that the average vocabulary of the five-year-old only children in this sample was not significantly higher than that of the five-year-old children in the population.

Note that, because we failed to reject the null hypothesis, we do not calculate an effect size.

Writing Results for the z Statistic in APA Format

10. Students in this sample who did not have recess scored significantly below the population on this achievement test, $z = 2.00$, $p < .05$. The effect size was moderate (Cohen's $d = -40$).

The average vocabulary of the five-year-old only children in this sample was not significantly higher than that of the five-year-old children in the population, $z = 1.50$, *ns*.

-8-

The Single-Sample *t*-test: Hypothesis Tests for a Single Sample

In This Chapter

Why *t*?

The Single-Sample (or One-Sample) *t*-test
- Degrees of Freedom
- Assumptions of Hypothesis Testing Using the *t* Statistic
- Confidence Intervals

Application and Interpretation: Hypothesis Testing with a Single-Sample *t*-Test

Reporting Nondirectional Results in APA Format

Application and Interpretation: Directional Hypotheses for a Single-Sample *t*-Test

Reporting Directional Results in APA Format

FOCUS

When we know the population standard deviation, we can use it to compute the standard error of the mean and the *z* statistic. When we do not know the population standard deviation, we must estimate the standard error of the mean. When we estimate the standard error of the mean, we use the *t*-test to test hypotheses. With the *t*-test, we abandon the Unit Normal Table and use a table of critical values of *t*.

The single-sample **t-test** compares a single sample mean with a population mean.

The **single-sample** (or **one-sample**) *t*-test is nearly identical to the *z* statistic. The single-sample *t*-test allows us to compare a sample mean with a population or hypothetical mean *when the standard deviation of the population is not known.*

Why *t*?

In order to use the Unit Normal Table and the normal curve to examine sampling error, we need population parameters, including the standard deviation (σ). However, in most scientific investigations, whether in the field of psychology, education, medicine, agriculture, or business, we do not know the standard deviation of the population. We can use the sample standard deviation to estimate the standard error of the mean when we do not know the population standard deviation.

Despite the fact that we are typically limited to using sample data to estimate the standard deviation of the population, we are not limited in our ability to test hypotheses. Indeed, *t*-tests allow us to use the information from our sample in such a way that we can generalize the results to the population of interest.

The Single-Sample (or One-Sample) *t*-Test

Recall the basic formula for the *z* statistic that we use to test hypotheses about a single sample when the population standard deviation is known:

$$z = \frac{M - \mu}{\sigma_M}$$

The **standard error of the mean** is an index of variability based on the *population* standard deviation.

Recall also how we calculate the denominator in the formula for the **standard error of the mean**:

$$\sigma_M = \sigma / \sqrt{n}$$

With a *t*-test, we do not know the population standard deviation; thus, we use a formula that reflects the fact that we are working with a sample standard deviation. The formula for the estimate the standard error of the mean is

$$s_M = s / \sqrt{n}$$

The **estimate of the standard error of the mean** is an index of variability based on the *sample* standard deviation.

This new formula is called the **estimate of the standard error of the mean.** With it, we can create a statistical formula that is comparable to our *z* statistic. Because we are using sample data to estimate of the standard error of the mean, we can no longer call this inferential statistic *z*; instead, we call it *t*:

$$t = \frac{M - \mu}{s_M}$$

z and *t* behave similarly. As with *z*, if there is no difference between the sample mean *M* and population mean *μ*, *t* will equal 0. Again as with *z*, we can conduct directional or nondirectional tests of hypotheses using *t*.

Using the same logic that we learned when we employed the *z* statistic in evaluating hypotheses, we will need to compare the *t* we calculate (or let SPSS calculate it for us—but more on that later) with a critical value of *t* to ascertain whether the *t* value based upon our sample is sufficiently large to reject the null hypothesis. A table of critical values for *t* appears in Appendix A2.

As you look in Appendix A2, across the top of the table, notice two rows of column headings. The top row of column headings presents values of alpha for directional, or one-tailed, tests. In the row beneath are values of alpha for nondirectional, or two-tailed, tests. Conveniently, the table divides the alpha level for us! Thus, if we were using a nondirectional test with $\alpha = .05$, we would use the column labelled .05 for "proportion in two tails combined. "Notice that this is the same column of critical values for .025 for "proportion in one tail."

Let's continue to examine the table. Notice that the extreme left of the table contains a column labelled *df*, the abbreviation for **degrees of freedom.** This column contains information about the *size* of our sample. For the one-sample *t*-test,

$$df = n - 1$$

Thus, the critical value of *t* is partially based on the size of the sample.

When we have a very small sample, we will need a relatively large observed *t* (based on our sample) to reject the null hypothesis. Conversely, the larger the sample size, the smaller is the observed *t* necessary to reject the null hypothesis.

Let us suppose that we have a sample size of $n = 2$ and we are using a nondirectional test with $\alpha = .05$. Then our degrees of freedom equal 1 ($n - 1 = 2 - 1 = 1$), and the critical value of *t* that we will need to exceed in order to reject the null hypothesis is 12.706.

Notice what happens if we increase our sample size to $n = 3$ ($df = 2$). The critical value then drops to 4.303. As you scan down the column labeled *df*, notice that the larger the size of the sample (the larger the value of *df*), the smaller will be the critical value of *t*. Notice also that, as the sample size increases to infinity ("∞" is the symbol for infinity) the critical *t* value becomes 1.960. Hence, when $df = \infty$, the critical values for *t* and *z* are identical. Thus, when we have a very large sample, we are approaching population parameters and the normal distribution associated with a population.

We use Table A2 to look up critical values of *t*. The critical value will vary as a function of the alpha level, the sample size, and whether we conduct a directional (one-tailed) or nondirectional (two-tailed) test. In a directional test, if we hypothesize *M* to be less than μ, we use a negative critical value. Conversely, if we hypothesize *M* to be more than μ, we use a positive critical value. When our hypothesis is nondirectional, we use both a positive and a negative ($\pm$) critical *t* value to establish the critical region.

Reflect

Where does infinity begin?

Recall that the more rigorous the test of our hypotheses, the smaller is the alpha level we use. Thus, if we were extremely concerned about making a Type I error (rejecting the null hypothesis when it is true), we would use a small alpha level (.01). This will result in selecting a larger critical value of *t*. So, for example, if $df = 10$, the critical (nondirectional) *t* value at $\alpha = .05$ is 2.28, whereas at $\alpha = .01$, the critical *t* value is 3.169.

Be Here Now

For each of the following sets of information, identify the critical value of *t* from the table of *t* values:

1. Nondirectional test for $\alpha = .05$ and *df* = 9.
2. Directional test for $\alpha = .01$ and *df* = 23.
3. Directional test for $\alpha = .05$ and *df* = 23.
4. Nondirectional test for $\alpha = .01$ and *df* = 25.

Answers

1. $t = \pm 2.262$
2. $t = 2.500$ (or $t = -2.500$ if the critical region is in the lower tail)
3. $t = 1.714$ (or $t = -1.714$ if the critical region is in the lower tail)
4. $t = \pm 2.787$

Degrees of freedom

From a practical perspective, it is important that you know that degrees of freedom reflect the size of the sample for the *t*-test. However, there is a bit more to the concept of degrees of freedom.

Degrees of freedom refer to the degree to which measurements are free to vary.

For every set of descriptive statistics that we calculate from a sample, we lose a degree of freedom. **Degrees of freedom** refer to the degree to which measurements are "free" to vary. So, if we know that a sample mean is based on 10 test scores, and we know 9 of the 10 scores, then the tenth score is not "free" to vary. We have "lost" a degree of freedom.

An intuitive way to conceptualize the loss of a degree of freedom is to imagine serving tea and pastries to four friends and yourself. You have a plate of five pastries. If each of your guests selects a pastry, do you have any "freedom" in your own selection? No. A degree of freedom was lost.

Reflect
Freedom's just another word for nothing left to lose.
Janice Joplin

Let's consider another example of how we lose a degree of freedom. Suppose you want to form a student group to enter a math contest. The rules of the contest require that the mean score of each group be 50. You randomly select the first four students. You select:

Mary, who earned a 10;

George, who earned a 30;

Kiwana, who earned a 60; and

Diego, who earned an 80.

The scores for these four children were free to vary. The fifth child selected must "lock in" the group mean of 50. To determine what the fifth child's score must be, we use the formula for the mean and X to represent the fifth child's score:

$$M = \frac{\Sigma X}{n} = \frac{10 + 30 + 60 + 80 + X}{5} = 50$$

Now we can solve for X:

Begin by multiplying each side of the equation by 5

$$M = \frac{\Sigma X}{n} = \left(\frac{10 + 30 + 60 + 80 + X}{5}\right)5 = (50)5$$

$$10 + 30 + 60 + 80 + X = 250$$
$$180 + X = 250$$
$$X = 250 - 180$$
$$X = 70$$

In order to create a group with a mean score of 50, the last child selected *must* have a score of 70. We were free to select any four children, whose scores could be any values, as long as the last child selected created a group whose mean was 50. There were $5 - 1 = 4$ *df* in this case.

Assumptions of Hypothesis Testing Using the *t* Statistic

As with *z*, there are assumptions that underlie hypothesis testing using the *t* statistic (see Chapter 7):

- Random sampling
- Independence of observations
- Observations sampled from a normally distributed population
- Homogeneity of variance

Random sampling states that we make the assumption that the participants in a sample were drawn at random from the population.

Independence of observations indicates that an observation is not related to any other observations within or between conditions.

The *normality* assumption states that, to use the *t* distribution to test our hypotheses, the population from which the observations are sampled must be normal.

Homogeneity of variance indicates that the variance of the sample is the same as the variance in the population.

Confidence Intervals

Let's say that your oceanography professor asks, "How deep is the Marianas Trench in the Pacific Ocean?" Would you be more confident in giving a response if you could give a potential range—say, between 35,000 and 40,000 feet—than if you had to state an exact depth? You probably would be. (By the way, the Marianas Trench is 11,033 meters, or 36,201 feet, deep.) You'd be more confident because you have room for error. As long as the interval you provided had the correct response, you would not be wrong. However, if you stated "exactly 36,000 feet," you would not be correct.

Confidence intervals involve placing a band (an interval) around a test statistic such that the band tells us how confident we are in our findings at a specific level (or probability). For example, we might want to ensure that we are 95% or 99% confident that our sample mean falls between the values that we calculate as the confidence interval.

In the next section, we will use SPSS in an example of the single-sample *t*-test. The SPSS output will show the confidence interval, and we will return to our discussion of confidence intervals at that point.

Be Here Now

The mean for a set of 10 quiz scores is 7. The instructor has recorded nine of the student's grades as follows: 10, 4, 7, 9, 8, 8, 5, 4, and 5. The instructor has misplaced the tenth student's score, but that is not a problem. What is the tenth student's score?

Answer

The tenth student's score has to be 10. It is not "free" to be any other score. How do we know this? Remember our formula for the mean? $M = \Sigma X / n$. Now substitute what we know into that formula:

$$7 = \frac{10 + 4 + 7 + 9 + 8 + 8 + 5 + 4 + 5 + X}{10}$$

Next, we solve for X:
Multiply each side by 10

$$(10)7 = \left(\frac{10 + 4 + 7 + 9 + 8 + 8 + 5 + 4 + 5 + X}{10}\right)10$$

$$70 = 60 + X$$

$$10 = X$$

Once the nine scores are determined, the tenth is fixed. It is not free to take any other value!

Application and Interpretation: Hypothesis Testing with a Single-Sample (or One-Sample) *t*-Test

Let's begin with a question: Do people who practice yoga show a level of depression that is different from that of the general population? A researcher uses a depression survey that has an average of 25. The population standard deviation is not known. A yoga class with an enrollment of 14 individuals agrees to be tested with the depression inventory. The depression scores for the yoga class participants are as follows:

Depression Scores

15	19	24	27	18	22	23	25	17	16	20	21	17	18

Do the yoga students show a level of depression that is different from that of the general population? To begin answering this question, start with the null and alternative hypotheses:

STEP 1   State the null and alternative hypotheses:
    In words, the null hypothesis is "Those who practice yoga experience symptoms of depression similar to those of the general population."
    The alternative hypothesis is "Those who practice yoga do not experience symptoms of depression similar to those of the general population."
    Stated symbolically, the null and alternative hypotheses would read,

$$H_0: \mu = 25$$
$$H_1: \mu \neq 25$$

Notice that this is a two-tailed, or nondirectional, test.

STEP 2   Choose the alpha level, determine the critical value of *t*, and draw the distribution.
    For this example, we will use an alpha level of .05. To determine the critical value, we need to compute *df*. Remember, since $n = 14$, $df = n - 1 = 13$. Now that we have selected an alpha level and know *df*, we can look up the critical value of *t* in the table of *t* values (Table A2). We look under the two-tailed proportion column of .05 and then look across the degrees-of-freedom row labeled 13, and we find that the critical value of *t* is 2.160.
    We draw the hypothetical distribution of *t* values when the null hypothesis is true.

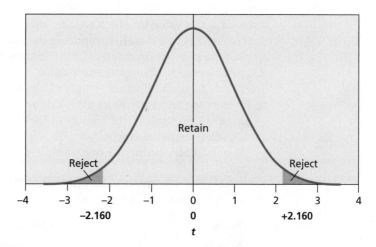

In a perfect world, if the null hypothesis were true, we would expect our sample mean to coincide precisely with the population mean of 25. The upper-right half of our distribution corresponds to the possibility that our sample mean will be greater than the population mean (suggesting increased depression). Similarly, the lower-left half of our distribution corresponds to the possibility that our sample mean will be less than the population mean (suggesting decreased depression).

As we noted with the z statistic, it is more likely to obtain a sample mean close to the population mean when the null hypothesis is true. Also, it becomes increasingly less likely to obtain a sample mean that is very different from the population mean when the null hypothesis is true.

In our figure, we have labeled hypothetical values of t. When our sample mean coincides precisely with the population mean, $t = 0$. When our sample mean is greater than the population mean, $t > 0$. When our sample mean is less than the population mean, $t < 0$.

STEP 3   Calculate t.

SPSS can calculate the value of t for us. It also will calculate df and p. If you would like to compute the one-sample t-test without the aid of a computer, the formula appears in the Synthesis Section (Appendix D).
Let's go through our example using SPSS.

➡ **Begin by creating an SPSS file with the data.**

To perform a single-sample t-test, you need to select

Analyze → Compare Means → One-Sample t-test.

- SPSS will calculate the value of t for the depression scores that we labeled "score." Move the variable "score" on the left to the box labeled "Test Variable(s)" on the right.

We are nearly ready to let SPSS do the work for us, but there is one more important task: You must change the value in the "Test Value" box to the value of μ, the value to which we are comparing our sample mean. Remember that in this case we are comparing the sample mean of depression scores against the population mean of 25. SPSS uses 0 as the default test value, and if we neglected to change the test value, our test result would be incorrect. To change the default value, simply click on the box and change the value 0 to 25. Now we are ready to let SPSS do the work.

Figure 1 presents the SPSS output for the results of the one-sample t-test for the set of depression scores.

When SPSS conducts a one-sample t-test, it provides two sections of output.

SPSS first presents an output box of "One-Sample Statistics" that includes the following information:

Score	Our sample's depression scores
N = 14	The size of the sample is 14.
Mean = 20.1429	The mean of the sample is 20.1429.
Std. Deviation = 3.63439	The standard deviation is 3.63439.
Std. Error Mean = .97133	The estimate of the standard error of the mean is .97133.

T-Test

One-Sample Statistics

	N	Mean	Std. Deviation	Std. Error Mean
Score	14	20.1429	3.63439	.97133

One-Sample Test

	Test Value = 25					
					95% Confidence Interval of the Difference	
	t	df	Sig. (2-tailed)	Mean Difference	Lower	Upper
Score	−5.000	13	.000	−4.85714	−6.9556	−2.7587

FIGURE 1 SPSS output for depression scores: one-sample *t*-test.

The next output box is labelled "One-Sample Test." It contains the following information:

Score	The variable analyzed is score.
t = −5.000	The *t* statistic computed is −5.00.
df = 13	There are 13 degrees of freedom.
Sig (2 − tailed) = .000	The alpha at which the result is significant is so small that, when rounded, it equals 0.
Mean Difference = −4.85714	The difference between the sample mean and the population is −4.85714.
95% Confidence Interval **Lower = −6.9556** **Upper = −2.7587**	Given an alpha of .05, we are 95% sure that the mean difference between our sample and the population ranges from −2.7587 to −6.9556.

SPSS computed the confidence interval by adding to and then subtracting from the mean difference, the product of the critical *t* value and the estimate of the standard error of the mean.

In words, the confidence interval is

Mean Difference ± (Critical *t* Value) (Estimate of the Standard Error of the Mean)

In our example, we have

$$-4.85714 \pm (2.160)(.97133) = -4.85714 \pm 2.0981$$

Thus, the confidence interval is

$-4.85714 - 2.0981 = -6.95524$ to $-4.85714 + 2.0981 = -2.75904$ (differences in rounding result in slightly different outcomes relative to the SPSS printout).

Note that when this range (95% confidence interval) includes zero, we fail to reject the null hypothesis.

When this range (95% confidence interval) does not include zero, as in our example, we reject the null hypothesis.

STEP 4   Evaluate the hypotheses on the basis of the results obtained.

Since our t of -5.01 falls into the critical region of rejection, we reject the null hypothesis and conclude, on the basis of our sample, that those who practice yoga experience a level of depression different from that observed in the general population.

Notice that we arrive at the same decision if we use the confidence interval. That is, 0 does not fall within the 95 percent confidence interval, which ranged from -2.7587 to -6.9556. Accordingly, we reject the null hypothesis.

Depending on the application or discipline, researchers choose to report their results relative to either the critical t value or the confidence interval.

STEP 5   Calculate the effect size.

Because we obtained a significant result, it is important to calculate the effect size. We again employ Cohen's d, a formula that is widely used.

Recall that Cohen's d is calculated by taking the difference between the: means. Since we are basing our conclusions on a sample and have no information about the population variability, the formula looks like this:

$$d = \frac{M - \mu}{s}$$

Cohen (1988) stated that $d = .2$ is a small effect size, $d = .5$ is a medium effect size, and $d = .8$ or greater is a large effect size; however, he cautioned that these are only guidelines.

For our results,

$$d = \frac{20.14 - 25}{3.63}$$

$$d = \frac{-4.86}{3.63}$$

$$d = -1.34$$

According to Cohen's (1988) guidelines, this is a large effect size. For this sample we conclude that those who practice yoga exhibit significantly less depression than the population.

Remember, it is possible that we have made a Type I error: that we rejected the null hypothesis when, in fact, the null hypothesis is true. When we set our alpha level, we indicated that we were comfortable with .05 as the probability of making a Type I error. Another way to think about this is that we are stating how confident we are that our results are correct and would be replicated. With an alpha level of .05, we indicated that we are willing to tolerate a 5% probability of rejecting the null hypothesis when it is true; simultaneously, we are 95% confident that our decision is the correct one.

Reporting Nondirectional Results in APA Format

As you learned with the *z* statistic, we make a statement about our decision to reject or fail to reject the null hypothesis and then provide the statistical documentation in support of that decision. When reporting the results of a *t*-test, we report *t* and the value we calculate on the basis of our sample. We also must include the degrees of freedom, by specifying the value in parentheses.

In our example, the correct APA format for documenting our results would be as follows:

$$t(13) = -5.00, \ p < .05; d = -1.34$$

It is also acceptable to report the exact probability for a *t*-test. Recall in our current example that the SPSS output box indicated that the significance level for a two-tailed test was .000. We would **never** indicate that our probability of making a Type I error is zero (even if SPSS rounds it to 0). In this case, it is more prudent to report our probability as less than our alpha level or as less than .001, as we have done:

$$t(13) = -5.00, p < .001$$

Now that we know how to format our results, we can consider the various statements we might make when we reject the null hypothesis. For our example, we might write one of the following statements:

Those who practiced yoga did not experience the same level of depression as was observed in the general population, $t(13) = -5.00, p < .05$. A large effect size was evident ($d = -1.34$).

We can also state: Fewer symptoms of depression were found in a sample of yoga students, compared with those exhibited by the general population, $t(13) = -5.00, p < .05, (d = -1.34)$.

Reflect

Life, language, and APA present many choices.

 It's Out There... Cheung, Brogan, Pilla, Dillon, and Redington (2003) compared arterial disease in children with arthritis with what they would expect in a normal population. Single-sample *t*-test results indicated that children with arthritis are more likely ($p < .05$) than the population to have arterial distension.

Be Here Now

A high school guidance counselor wants to examine whether students accurately predict how well they will do on the SAT test. The counselor is not interested in the actual SAT score, only whether the students underestimated their actual scores or overestimated them. A class of high school juniors participates in the study. The differences between the predicted score and the actual SAT quantitative scores are recorded, where positive numbers indicate that students overestimated how well they would do and negative numbers indicate underestimates.

Prediction Accuracy

+10	+30	+90	+50	+10	−20	+20	−30	0	−10	−50	−10
−20	+40	−30	+10	−20	−70	+10	+10	+30	0	+30	+40

Do students accurately predict their SAT scores? Test with $\alpha = .05$, using a two-tailed (nondirectional) test.

Answer

In words, the null hypothesis might be "Students accurately predict their SAT scores; there is no difference between the score that is predicted and the score that is obtained." The alternative hypothesis is that there is a difference between the predicted and obtained scores.

Stated symbolically, the null and alternative hypothesis would read,

$$H_0: \mu = 0$$
$$H_1: \mu \neq 0$$

We have chosen an alpha level of .05, and with $df = 24 - 1 = 23$; the critical value for t is ± 2.069.

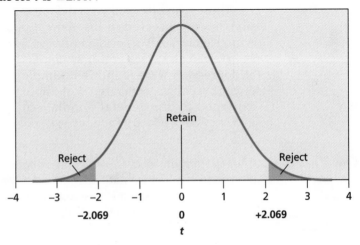

Be Here Now	We enter the data into SPSS and make certain that Test Value is set to 0.

We enter the data into SPSS and make certain that Test Value is set to 0.

On the basis of our sample, SPSS calculated *t* to be .710. This value does not exceed the critical value of *t*, so we fail to reject the null hypothesis. Our decision is confirmed by SPSS, since the results of the *t*-test show *p* (sig 2 tailed) to be .485. Because *p* is not less than .05, we fail to reject the null hypothesis.

Given our sample of high school juniors, we failed to find a difference between predicted and actual SAT scores, $t(23) = 0.710$, $p = .485$.

We did not need to calculate Cohen's *d*, because we failed to find a statistically significant result.

Application and Interpretation: Directional Tests Using *t*

As we have noted with the *z* statistic, on occasion we can make a prediction about the *direction* of the difference between a sample mean and a population mean. When we can make a prediction about the specific direction of difference, we perform a directional (or one-tailed) test, and we can do this by using *t*. Only some of the details of the hypothesis-testing procedure differ—and even then, just slightly—when we perform a one-tailed test.

Let's consider an example. Suppose a psychologist wants to examine whether young adults underestimate the amount of time that passes when they play a computer game. A sample of university students is asked to play a computer video game. At the end of the 10-minute session, the students are asked to estimate the amount of time they spent, in seconds, playing the video game.

Their time estimates are as follows:

Time Estimates (in seconds)

610	575	545	600	630	675	500	580	455	560
520	490	685	530	575	625	555	590	635	665

Is there any evidence that young adults *underestimate* the time they spend playing a video game?

STEP 1   State the null and alternative hypotheses.

In words, the null hypothesis is "Young adults do not underestimate the time they spend playing a video game."

The alternative hypothesis is, "Young adults underestimate the time they spend playing a video game."

Stated symbolically, the null and alternative hypotheses would read,

$$H_0: \mu \geq 600 \text{ seconds (10 minutes)}$$
$$H_1: \mu < 600 \text{ seconds}$$

Notice that this is a one-tailed, or directional, test.

STEP 2 Choose the alpha level and draw the distribution.

In this example, we will use an alpha level of .05. Since $n = 20$ and $df = 19$, we can look up the critical value of t in the table of t values (Table A2). We look under the one-tailed proportion column of .05 and then look across the degrees-of-freedom row labeled 19, and we find that the critical value of t is 1.729.

We draw the hypothetical distribution of t values, when the null hypothesis is true. In a perfect world, if the null hypothesis were true, we would expect our sample mean to coincide precisely with the hypothetical mean of 600 seconds. This would result in a t value of 0. The upper-right half of our distribution corresponds to the possibility that our sample mean will be greater than the population mean (suggesting overestimates of time and positive values of t). The lower-left half of our distribution corresponds to the possibility that our sample mean will be less than the population mean (suggesting underestimates of time). In this example, we are predicting underestimates of time; therefore, our critical region is in the lower tail (below $t = -1.79$).

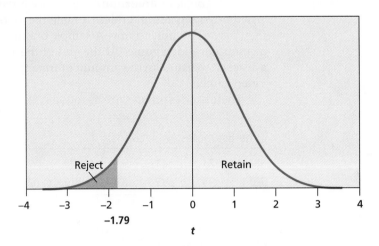

STEP 3 Calculate the appropriate test statistic, using a one-sample t-test.
First, we enter the data into SPSS. We will name the variable "estimate" and enter our 20 time estimates.

To perform a single-sample t-test, you need to select

Analyze → Compare Means → One-Sample t-test.

- SPSS will calculate the value of *t* for the time estimates that we labeled "estimate." Move the variable "estimate" on the left to the box labeled "Test Variable(s)" on the right.

Remember to change the value in the "Test Value" box to the expected value of μ, the value to which we are comparing our sample mean. In this case, we are comparing the sample mean of time estimates against the hypothetical mean of 600 seconds. Change the default value by clicking on the box, and change the value of 0 to 600.

Figure 2 presents the SPSS output for the results of the one-sample *t*-test for the observed set of time estimates.

- SPSS provides two sections of output for a one-sample *t*-test.
- SPSS first presents an output box of "One Sample Statistics" that includes the following information:

estimate	Our sample's time estimates were analyzed
N = 20	The size of the sample is 20.
Mean = 580.0000	The mean of the sample is 580.0000.
Std. Deviation = 62.72329	The standard deviation is 62.72329.
Std. Error Mean = 14.02535	The estimate of the standard error of the mean is 14.02535.

The next output box is labeled "One-Sample Test." It contains the following information:

estimates	Our sample's time estimates were analyzed.

T-Test

One-Sample Statistics

	N	Mean	Std. Deviation	Std. Error Mean
Estimate	20	580.0000	62.72329	14.02535

One-Sample Test

	Test Value = 600					
					95% Confidence Interval of the Difference	
	t	df	Sig. (2-tailed)	Mean Difference	Lower	Upper
Estimate	−1.426	19	.170	−20.00000	−49.3554	9.3554

FIGURE 2 SPSS output for time estimates: one sample *t*-test.

t = −1.426	The *t* statistic computed is −1.426.
df = 19	There are 19 degrees of freedom.
Sig (2-tailed) = .170	The alpha at which our result is significant is .170.
	Note: The default test for SPSS is two-tailed.
Mean Difference = −20.00000	The difference between the sample mean and the population mean is −20.00000.
95% Confidence Interval **Lower = −49.3554** **Upper = 9.3554**	Given an alpha of .05, we are 95% sure that the mean difference between our sample and the population ranges from −49.3554 to 9.3554.

STEP 4   Evaluate the hypotheses on the basis of the results obtained. Since our *t* of −1.426 fails to fall into the critical region, we fail to reject the null hypothesis and conclude, on the basis of this sample, that we failed to find evidence that young adults underestimate the amount of time they spend playing a video game, $t(19) = -1.426$, ns.

SPSS provides us with an actual value of *p* in the column labeled "Sig. (2-tailed)," but since we conducted a one-tailed test and not a two-tailed test, we cannot report that $p = 17$. If we wish, we can divide *p* by 2 to obtain the one-tailed value of *p* (.170/2 = .085).

Also, because we conducted a one-tailed test, we are not able to use the confidence interval computed.

STEP 5   Effect Size

Because our test did not produce a significant result, we do not calculate Cohen's *d*.

Presenting Directional Results in APA Format

When we fail to reject the null hypothesis, we can report our findings in the following format:

$$t(19) = -1.43, ns$$

Using APA format, we might make the following statement:

Young adults did not significantly underestimate the amount of time they spent playing a video game, $t(19) = -1.43$, *ns*.

Be Here Now

A social psychologist wants to test the assumption that people attribute greater confidence to others. She asks participants to indicate their best guess about how much more or less confident the person is who is seated next to them in relation to their own self-confidence. She asks students to use a positive number (1 to 5) to indicate that the person has greater self-confidence, and a negative number (–1 to –5) to indicate that the person has less self-confidence. The scores obtained are as follows:

Compared Self-Confidence

| 1 | 2 | −5 | 5 | 5 | −1 | 4 | −3 |
| 4 | 1 | 3 | 4 | −2 | 2 | 5 | 3 |

Is greater confidence attributed to others? Test with $\alpha = .05$, using a one-tailed (directional) test.

Answer

In words, the null hypothesis might be "Greater confidence is not attributed to others." The alternative hypothesis might state "Greater confidence is attributed to others."

Stated symbolically, the null and alternative hypotheses would read,

$$H_0: \mu \leq 0$$
$$H_1: \mu > 0$$

We have chosen an alpha level of .05, and with $df = 16 - 1 = 15$, the critical value for t is +1.753.

We enter the data into SPSS and make certain that Test Value is set to 0.

On the basis of the sample, SPSS calculated t to be 2.283. This value exceeds the critical value of t, so we reject the null hypothesis. Thus, the social psychologist has found that greater confidence is attributed to others, $t(9) = 2.283, p < .05$.

$$\text{Cohen's } d = \frac{1.75 - 0}{3.06} = .57; \text{ there is a moderate effect size.}$$

SUMMARY

In the absence of information about the population variance, we must estimate it in order to engage in hypothesis testing. When we do not know the population standard deviation, we use the sample standard deviation to estimate the standard error of the mean. We then use our estimate of the standard error of the mean together with the *t*-test to test hypotheses. We no longer use the Unit Normal Table; rather, we use the table of critical values of t, which takes into account the fact that we are working with sample data rather than population data. Using the single-sample t-test, we can test hypotheses about whether a sample mean is different from a population, or hypothetical, mean. We can use a one-tailed, or directional, test when the direction of difference is specified. Cohen's d can be used to measure the size of the effect when the null hypothesis is rejected.

PRACTICE

Degrees of freedom and critical values of t

1. For each of the following, identify the critical value of t:
   a. a nondirectional test for $n = 22$, $\alpha = .05$
   b. a directional test for $n = 22$, $\alpha = .05$
   c. a nondirectional test for $n = 37$, $\alpha = .05$
   d. a nondirectional test for $n = 37$, $\alpha = .01$

Nondirectional tests using t

2. Hospital administrators want to investigate the effect of background classical music on the duration of patients' hospital stay. On one floor of the hospital, the average stay for patients is 4.3 nights. Classical music is piped into all rooms on the floor, and the length of hospital stay is noted for each of the patients.

Hospital Stay (nights)

4	3	2	1	8	4	5	2
6	7	9	2	3	6	1	3

Is the duration of hospital stay different for patients who are exposed to classical music? Test with $\alpha = .05$. Report the results in APA format.

3. A new university art gallery director wishes to measure the success of her premier exhibit. The former gallery director reported 38 visitors per day. The new director records the number of visitors each day for three weeks.

Daily Visitors to the Gallery

49	41	63	52	31	38	44	57	45	32	47
62	65	27	39	28	53	60	41	37	49	

Is the visitor rate under the new gallery director different from that of her predecessor? Test at $\alpha = .05$.

Directional tests using t

4. A social psychologist is interested in the phenomenon of white lies—relatively minor misrepresentations of truth—and whether they are universally told. She asks a group of 12 students from her class to record the number of white lies they tell for one week. The data are as follows:

White Lies

2	3	1	9	0	7
6	0	5	2	4	0

Are the number of white lies that students tell over the course of a week significantly greater than 0? Test with $\alpha = .05$. Present the results in APA format.

5. Anxiety-like behavior can be examined in mice by measuring the relative amount of time they spend in an open section of a maze instead of an enclosed section. A researcher makes the assumption that if the mice spend less than 50% of the time in the open section of the maze, they are exhibiting anxiety-like behavior. Ten mice are individually placed in the maze, and the amount of time they spend in the open is reported as follows:

Time Spent in Open (expressed as percentages)

59	36	41	38	43	27	52	49	46	39

Can the researcher conclude that the mice engage in anxiety-like behavior? Test with $\alpha = .05$. Present the results in APA format.

SOLUTIONS

Degrees of freedom and critical values of t

1. **a.** The critical value of *t* is ±2.074.
   b. The critical value of *t* is *either* +1.717 or −1.717.
   c. The critical value of *t* is ±2.030, with *df* = 35 (the closest lower value for *df* in the table).
   d. The critical value of *t* is ±2.750, with *df* = 30 (the closest lower value for *df* in the table).

Nondirectional tests using t

2. In symbols, the null and alternative hypotheses read as follows:

$$H_0: \mu = 4.3 \text{ nights}$$
$$H_1: \mu \neq 4.3 \text{ nights}$$

We have chosen an alpha level of .05, and with $df = 16 - 1 = 15$, the critical value for t is ± 2.132.

We enter the data into SPSS and make certain that Test Value is set to 4.3.

T-Test

One-Sample Statistics

	N	Mean	Std. Deviation	Std. Error Mean
Nights	16	4.1250	2.47319	.61830

One-Sample Test

			Test Value = 4.3			
					95% Confidence Interval of the Difference	
	t	df	Sig. (2-tailed)	Mean Difference	Lower	Upper
Nights	−.283	15	.781	−.17500	−1.4929	1.1429

On the basis of our sample, SPSS calculated t to be $-.283$. This value fails to exceed the critical value of t, so we fail to reject the null hypothesis.

On the basis of this sample, there is no evidence that hospital duration is different from 4.3 nights, $t(15) = -.28$, *ns*. We do not calculate Cohen's *d*, because we failed to reject the null hypothesis.

3. In symbols, the null and alternative hypotheses are:

$$H_0: \mu = 38 \text{ daily visitors}$$
$$H_1: \mu \neq 38 \text{ daily visitors}$$

We have chosen an alpha level of .05, and with $df = 21 - 1 = 20$, the critical value for t is ± 2.086.

We enter the data into SPSS and make certain that Test Value is set to 38.

T-Test

One-Sample Statistics

	N	Mean	Std. Deviation	Std. Error Mean
Visitors	21	45.7143	11.57645	2.52619

One-Sample Test

			Test Value = 38			
					95% Confidence Interval of the Difference	
	t	df	Sig. (2-tailed)	Mean Difference	Lower	Upper
Visitors	3.054	20	.006	7.71429	2.4447	12.9838

On the basis of the three-week sample, SPSS calculated *t* to be 3.054. This value exceeds the critical value of *t*, so we reject the null hypothesis. There is evidence that the new gallery director is experiencing a degree of success different from that of her predecessor, $t(11) = 3.77, p < .01$.

Cohen's $d = \dfrac{45.714 - 38}{11.574} = .67$; there is a moderate effect size.

Directional tests using t

4. In symbols, the null and alternative hypotheses are:

$$H_0: \mu \le 0$$
$$H_1: \mu > 0$$

We have chosen an alpha level of .05, and with $df = 12 - 1 = 11$, the critical value for *t* is +1.796

We enter the data into SPSS and make certain that Test Value is set to 0.

T-Test

One-Sample Statistics

	N	Mean	Std. Deviation	Std. Error Mean
Lies	12	3.2500	2.98861	.86274

One-Sample Test

	Test Value = 0					
	t	df	Sig. (2-tailed)	Mean Difference	95% Confidence Interval of the Difference	
					Lower	Upper
Lies	3.767	11	.003	3.25000	1.3511	5.1489

On the basis of our sample, SPSS calculated *t* to be 3.767. This value exceeds the critical value of *t*, so we reject the null hypothesis.

The number of white lies that students told was significantly greater than 0, $t(11) = 3.77, p < .01$.

Cohen's $d = \dfrac{3.25 - 0}{2.989} = 1.09$; there is a large effect size.

5. In symbols, the null and alternative hypotheses read as follows:

$$H_0: \mu \ge 50$$
$$H_1: \mu < 50$$

We have chosen an alpha level of .05, and with $df = 10 - 1 = 9$, the critical value for t is -1.833.

We enter the data into SPSS and make certain that Test Value is set to 50.

T-Test

One-Sample Statistics

	N	Mean	Std. Deviation	Std. Error Mean
OpenPrcnt	10	43.0000	9.01850	2.85190

One-Sample Test

	Test Value = 50					
					95% Confidence Interval of the Difference	
	t	df	Sig. (2-tailed)	Mean Difference	Lower	Upper
OpenPrcnt	−2.455	9	.036	−7.00000	−13.4514	−.5486

On the basis of our sample, SPSS calculated t to be -2.455. This value exceeds the critical value of t, so we reject the null hypothesis.

Cohen's $d = \dfrac{43.00 - 50}{9.02} = \dfrac{-7.00}{9.02} = .78$; there is a large effect size.

The mice show evidence of anxiety-like behavior, spending significantly less than 50 percent of the time in the open, $t(9) = -2.46, p < .05$.

The Independent-Samples *t*-Test: Hypothesis Tests for Two Samples

F O C U S

Independent-samples *t*-tests examine differences between two independent groups.

The *t*-statistic is a versatile statistical test. The single-sample *t*-test can be expanded to test for differences between two groups in the **independent-samples *t*-test**. With the independent-samples *t*-test, the same logic behind hypothesis testing is applied to two groups; however, there are important differences between the two *t*-tests. With the independent-samples *t*-test, we incorporate *two population means* into our null and alternative hypotheses. We also look at the hypothetical distribution of *sample mean differences* when we test our hypotheses. For these reasons, we have to modify our estimate of the standard error of the mean to include two measures of variability and two sample sizes. And, of course, our degrees of freedom must reflect the *n* of each group.

Furthermore, for the independent-samples *t*-test, data entry into SPSS is different from when we use a single-sample *t*-test, because we now must identify the *group* associated with each measurement.

In life, there are many potential questions about differences between two groups. We can compare outcomes of two substance abuse programs, two different forms of reading instruction, two university curricula, two chemotherapy treatments, or two vitamin supplements. The hypotheses that can be tested with the independent-samples *t*-test are endless. We will begin by examining the differences in anxiety between males and females.

Hypotheses about Two Groups

Again, imagine that your professor asks you and everyone in your class about your anxiety associated with taking this statistics course. Suppose that each of you is asked to use a number from 1 to 10 to represent your personal level of anxiety, where 1 would mean that you experience very little anxiety and 10 would indicate that you are extremely fearful. We can pose a variety of questions in relation to the data we will collect. Is anxiety different in males and females? Are juniors more anxious than seniors? Do summer students experience as much anxiety as students who take a statistics course during a regular academic term? Do students who have previous math experience suffer less anxiety? Each of these questions can be examined with an **independent-samples *t*-test**.

For example, consider the following sets of anxiety-level data from a class:

Anxiety-Level Scores for Males

7	10	2	3	7	3	6	6	9	1	7	5

Anxiety-Level Scores for Females

9	5	7	8	6	1	7	8	2	6	5	3

Our question is:

Do males differ from females in their anxiety level?

Let's begin by formulating a null and an alternative hypothesis:

H_0: Males and females do not differ in anxiety level in a statistics class.

H_1: Males and females differ in anxiety level in a statistics class.

Our question about same or different levels of anxiety is nondirectional. In other words, in this case, our question is *not* whether males' anxiety levels are greater than females' or whether females' anxiety levels are greater than males'. Our question is whether a difference in *any* direction exists between males' and females' anxiety levels.

In expressing our hypotheses, we might initially write

$$H_0: \ \mu_\delta = \mu_\varphi$$

$$H_1: \ \mu_\delta \neq \mu_\varphi$$

Notice that we use subscripts to identify the particular population mean (in this case, male or female). It happens that our example is about gender differences, so we can make use of the universal symbols for male and female as our subscripts. Alternatively, we could have used the numbers 1 and 2 as subscripts, or letters of the alphabet as subscripts, or even the descriptors "male" and "female" to help us in referring to the two groups that we are interested in comparing.

Another way to express the hypotheses about the potential difference between males' and females' anxiety levels is the following:

$$H_0: \ \mu_\delta - \mu_\varphi = 0$$

$$H_1: \ \mu_\delta - \mu_\varphi \neq 0$$

The two versions of our null and alternative hypotheses mean the same thing even though they look different. The null hypothesis stipulates that males' and females' anxiety levels are the same; in other words, there is no difference (the difference is zero) between the two populations ($H_0: \ \mu_\delta - \mu_\varphi = 0$). The alternative hypothesis stipulates that males' and females' anxiety levels differ; in other words, there *is* a difference (the difference is nonzero) between the two populations ($H_1: \ \mu_\delta - \mu_\varphi \neq 0$).

We will use our two sample means to make inferences about males' and females' anxiety levels in the population. We will reject our null hypothesis ($H_0: \ \mu_\delta - \mu_\varphi = 0$) if we observe a difference in our sample means that is sufficiently large.

Let's explore this idea a bit more fully. Let's consider some possible patterns of sample mean differences:

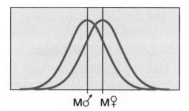

M♂ M♀

The distribution of male and female anxiety may overlap and the mean male and female anxiety levels may be similiar. This depicts a non-significant difference between groups.

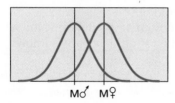

M♂ M♀

The distribution of male and female anxiety may overlap and the mean male and female anxiety levels may be different. This depicts a non-significant difference between groups.

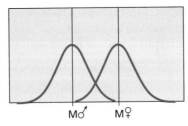

$M\male$ $M\female$

The distribution of male and female anxiety may have little overlap and the mean male and female anxiety levels may be very different. This depicts a significant difference between means.

The question of interest is how much of a difference between the two distributions is necessary for us to conclude that there is a significant difference? To answer that question, we use the independent-samples t-test. In particular, the independent-samples t-test examines the difference between the two group means relative to the overlap of the two distributions.

Reflect
To appear different does not mean that a difference exists.

Assumptions of the Independent-Samples t-test

There are some things we assume to be true when we use the independent-samples t-test:

- The variable being measured is normally distributed;
- The variances of the groups being assessed are equivalent (i.e., the variances are homogeneous); and
- Sample 1 is randomly sampled from population 1, and sample 2 is randomly sampled from population 2.

Although these assumptions underlie the use of the independent-samples t-test, research suggests that violations of the assumptions are likely to have minimal effects on the statistical outcomes. In particular, if the numbers of participants in each group are equal, the assumption of equal variances is not an issue (Box, 1954). Violations of the assumption of a normal distribution have also been found to have a minimal effect on the outcome of the independent-samples t-test (Boneau, 1960).

Calculating the Independent-Samples t

By now, you may have guessed that calculating t for two groups is going to be a bit more complex than when we calculated t for a single sample. Our test statistic will have to include two sample means, two measures of variability, and two sample sizes. Let's consider the inferential statistical tests that we have been mastering.

To compare a mean with a population value when the population variance (and the standard error of the mean) is known, use

$$z = \frac{M - \mu}{\sigma_M}$$

To compare a mean with a population value when the population variance (and the standard error of the mean) is not known, use

$$t = \frac{M - \mu}{s_M}$$

where

$$s_M = \frac{s}{\sqrt{n}}$$

In order to accommodate two groups, we "stretch" these basic formulas.

$$t = \frac{M_1 - M_2}{s_{M1-M2}} = \frac{\text{Difference between Means}}{\text{Standard Error of the Mean Difference}}$$

Notice the resemblance in the formulas: In each one, the numerator consists of the difference between means and the denominator is an estimate of error. With two groups, our numerator is the difference between the two sample means and the denominator is the estimate of the standard error of the difference between the means. As with *z* and the one-sample *t*-test, if the difference between the two sample means is 0, our test statistic, *t*, will be 0 and the null hypothesis will be retained.

The definitional formula for the independent-samples *t*-test can be more broadly stated as

$$t = \frac{\text{Difference between Means}}{\text{Standard Error of the Mean Difference}} = \frac{\text{Variance between Means}}{\text{Variance within Groups (Error)}}$$

The entire computational formula is given in Appendix D.

Degrees of Freedom for the Independent-Samples *t*-test

The number of degrees of freedom that we calculate for the independent-samples *t*-test must reflect the number in each sample, minus 1.

Thus, for the independent-samples *t*-test,

$$df = n_1 + n_2 - 2$$

or

$$df = (n_1 - 1) + (n_2 - 1)$$

In our example,

$$df = (n_\delta - 1) + (n_\female - 1)$$
$$df = (12 - 1) + (12 - 1)$$
$$df = 11 + 11$$
$$df = 22$$

Now we are ready to tackle our example of the independent-samples *t*-test.

Application: Hypothesis Testing Using the Independent-Samples *t*-Test

Let's look again at the anxiety-level data that we have already collected from a sample of males and females in the class.

Anxiety-Level Scores for Males

7	10	2	3	7	3	6	6	9	1	7	5

Anxiety-Level Scores for Females

9	5	7	8	6	1	7	8	2	6	5	3

STEP 1   State the null and the alternative hypotheses.

$$H_0: \ \mu_\delta - \mu_\female = 0$$
$$H_1: \ \mu_\delta - \mu_\female \neq 0$$

In words, the null hypothesis states that anxiety levels between males and females are the same. The alternative hypothesis stipulates that anxiety levels between males and females differ. Notice that this is a nondirectional test, because we are not specifying whether males have more or less anxiety than females.

STEP 2   Choose the alpha level and draw the distribution.

For this example, we will use an alpha level of .05. We have already calculated our *df* (*df* = 22). Next, we can look up the critical value of *t* in the table of *t* values. We look under the column labeled .05 for two tails, look across the degrees of freedom row labeled 22, and find that the critical value of *t* is 2.074.

We draw the hypothetical distribution of values of *t* when the null hypothesis is true.

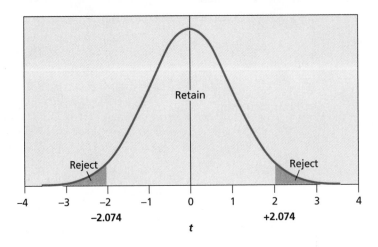

When the null hypothesis is true, and if it were a perfect world, we would expect no difference between our sample means and a t value of 0.

Since this is a two-tailed test, we have two critical regions defined by our critical value of $t(t = 2.074)$. To reject the null hypothesis, our calculated value of t must fall beyond $+2.074$ or -2.074.

STEP 3   Calculate the appropriate test statistic.

Data entry for the independent-samples *t*-test is different from that fowr the single-sample *t*-test, because here we must identify which values belong to each group. In using SPSS, it is important to remember that each row in the "Data View" page represents a participant and each column represents a variable. In our example, we must indicate which scores were from males and which were from females.

Click on the tab for "Variable View," and define one variable to identify the group associated with each score and one variable to identify our anxiety-levelmeasurements.Inthisexample,wewilluse"gender"and"anxiety" as our variable names.

Now click on the tab labeled "Data View." Before entering our gender information, we will code it. Because many statistical procedures in SPSS require numerical data, we will use a numerical code for nominal data such as gender. We will use "1" for males and "2" for females. We can enter each anxiety-level number under the column heading "anxiety."

Once data entry is complete, there should be 24 gender designations in the first column and 24 anxiety scores in the second. Each row presents the information for one participant. Each column presents the information for one variable. A sample of what the data look like in SPSS "Data View" is presented in Figure 1.

We are now ready for SPSS to calculate t.

➡ **To conduct an independent-samples *t*-test, you need to choose**

Analyze ➜ *Compare Means* ➜ *Independent-Samples t-Test.*

- "Anxiety" is the "Test Variable" because we are "testing" differences in anxiety between males and females, so move "Anxiety" over to the box labeled "Test Variable."

- "Gender" is the "Grouping Variable," because the information that we put in the column labeled "Gender" identifies group membership. Move "Gender" over to the box labeled "Grouping Variable." However, we still have to define "Gender" a bit more fully by clicking on the button labeled "Define Groups." In this new dialog box, we must provide SPSS with information about what we called each of our two groups. For Group 1, we will enter the number "1" and for Group 2 we will enter the number "2." Once this information is provided, we are ready to let SPSS perform the independent-samples *t*-test.

	Gender	Anxiety
1	1.00	7.00
2	1.00	10.00
3	1.00	2.00
4	1.00	3.00
5	1.00	7.00
6	1.00	3.00
7	1.00	6.00
8	1.00	6.00
9	1.00	9.00
10	1.00	1.00
11	1.00	7.00
12	1.00	5.00
13	2.00	9.00
14	2.00	5.00
15	2.00	7.00
16	2.00	8.00
17	2.00	6.00
18	2.00	1.00
19	2.00	7.00
20	2.00	8.00
21	2.00	2.00
22	2.00	6.00
23	2.00	5.00
24	2.00	3.00

FIGURE 1 SPSS "Data View" sheet.

Figure 2 presents the SPSS output for the results of the independent-samples t-test for anxiety levels among males and among females.

In the first box of output, SPSS presents the descriptive statistics. The top row labels the statistics computed; below each label are the values for each of our samples (1 and 2).

Gender   Two levels of anxiety were examined, by gender: Sample 1.00 (Male) and Sample 2.00 (Female).

T-Test

Group Statistics

Gender		N	Mean	Std. Deviation	Std. Error Mean
Anxiety	1.00	12	5.5000	2.77980	.80246
	2.00	12	5.5833	2.50303	.72256

Independent Samples Test

		Levene's Test for Equality of Variances		t-test for Equality of Means					95% Confidence Interval of the Difference	
		F	Sig.	t	df	Sig. (2-tailed)	Mean Difference	Std. Error Difference	Lower	Upper
Anxiety	Equal variances assumed	.201	.659	−.077	22	.939	−.08333	1.07983	−2.32277	2.15610
	Equal variances not assumed			−.077	21.762	.939	−.08333	1.07983	−2.32418	2.15752

FIGURE 2 SPSS output for the independent-samples *t*-test comparing anxiety levels in males and females.

N	There were 12 participants in Sample 1 and 12 participants in Sample 2.
Mean	The mean anxiety level for Sample 1 was 5.5000, and the mean anxiety level for Sample 2 was 5.5833.
Std. Deviation	The standard deviation for Sample 1 was 2.77980, while the standard deviation for Sample 2 was 2.50303.
Std. Error Mean	The estimate of the standard error of the mean for Sample 1 was .80246, while that for Sample 2 was .72256.

The preceding information is useful, because we can use these results for our report in APA format.

The second output box contains the results of the *t*-test. Notice that there is much more information in this output box than when SPSS produced results for the one-sample *t*-test.

The top row labels the statistics computed; below each label are the values calculated by SPSS.

Before we describe the results, notice that the first column is divided into two rows. The first row is labeled "Equal variances assumed," the second "Equal variances not assumed." As you may recall, one of the assumptions of *t* is equal variances. When this assumption is violated, we have the option of using a more conservative estimate of the number of degrees of freedom (Behrens–Fisher solution). Levene's test is used to determine whether the variances of the two groups (male and female) are equivalent.

Levene's Test for Equality of Variances

F   Levene's test of homogeneity of variance computes a statistic called *F*. For our data, $F = .201$.

Sig.   In this column, SPSS reports the significance of Levene's *F*. If the significance level is less than or equal to .05, then we conclude that the variances of the two groups differ significantly. The alpha associated with Levene's *F* is .659. Since .659 is greater than .05, the difference between the variances is not significant. Thus, we do not have evidence that we have violated the assumption of equal variances.

> *Reflect*
> *Strive for equality.*
> *If necessary, accept*
> *imperfection.*

t-Test for Equality of Means

Because Levene's test was not significant, we use the top row of output labeled "Equal variances assumed." In that case, we have

t   The *t* value is $-.077$.

df   There are 22 degrees of freedom (*df*).

Sig (2-tailed)   The *p* at which this *t* value is significant is .939. Since we use $p < .05$, and .939 is greater than .05, the difference between anxiety levels in males and females is not significant.

Mean difference   The difference in anxiety between males and females is $-.08333$.

Std. Error Difference   The standard error of the mean difference $(S_{M1-M2}) = 1.07983$.

95% Confidence Interval of the Difference

Another way of determining whether there is a significant difference between the two means is to compute confidence intervals, or bands, around the observed *t*. If 0 falls within the interval, we do not have a significant difference; if the interval does not include 0, the difference is significant.

Lower   The lower point of the interval is -2.32277.

Upper   The upper point of the interval is 2.15610. Since 0 falls within the 95 percent confidence interval, the difference in anxiety levels between males and females is not significant.

STEP 4   Evaluate the hypothesis on the basis of the results obtained.

There are several ways to evaluate our hypotheses with SPSS. We can use the value of *p* in the column labeled "Sig (2-tailed)" and reject the null hypothesis if $p < .05$. We can use the confidence interval and reject the null hypothesis when it does not include 0. Finally, we can compare our calculated value of *t* with the critical value of *t*.

On the basis of our sample of males and females, $t = -.077$. Our calculated *t* of $-.077$ does not exceed the critical value of $t = \pm 2.074$. Thus, we fail to reject the null hypothesis. In other words, we did not find a significant difference between the anxiety levels of males and females in our statistics class. Thus, we have no evidence to suggest a difference by gender in the population. This conclusion is illustrated graphically as follows:

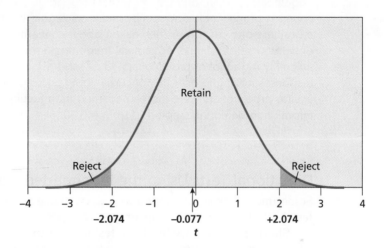

As described in the previous chapter for the single-sample *t*-test, had the results indicated a significant difference, we would also have computed the effect size, using Cohen's *d*. When the two samples have equal values of *n*, we can compute the pooled standard deviation (s_p) by finding the square root of the average of the two sample variances (Cohen, 1988).

$$\text{Cohen's } d = \frac{M_1 - M_2}{S_p}$$

where

$$S_p = \sqrt{\frac{S_1^2 + S_2^2}{2}}$$

In Appendix D, a formula is presented that can be used when the numbers in each group are different.

> **Reflect**
> Just as the song is sweeter when voices combine, the estimate is better when variances are pooled.

Reporting the Results in APA Style

We have already learned how to report the results of a *t*-test in the previous chapter. Here, we can review the format and consider the various statements we might make when we fail to reject the null hypothesis, as in our example. We might write any one of the following:

Mean male anxiety level was 5.50 ($SD = 2.78$), while that of females was 5.58 ($SD = 2.50$). Anxiety levels between males and females did not differ significantly within this study sample, $t(22) = -.08$, *ns*.

No significant difference was found in anxiety levels between males and females, $t(22) = -.08$, $p = .939$.

It's Out There...

Ragothaman, Lavin, and Davies (2007) compared the perceptions of accounting practitioners and faculty members regarding their concern about e-business and security. When asked about concerns relative to the delivery of items ordered on line, educators' mean response (3.61) did not differ significantly from that of practitioners' (3.67), $t(167) = .029$, *ns*. This is unlike responses having to do with concerns about personal information. In particular, practitioners were more concerned than faculty about their personal information being intercepted, $t(167) = 3.10$, $p < .05$.

Directional Tests Using the Independent-Samples *t*-Test

Sometimes there is sufficient information to use a directional (or one-tailed) test. Let's illustrate with an example.

Short-term memory has been tested with the use of a technique called "digit span"—the number of randomly presented digits that a person can remember immediately after their verbal presentation. A cognitive psychologist wants to test whether short-term memory is impaired when an individual is under stress. One group of students is tested while waiting to purchase basketball tickets, and another group is tested while waiting to receive an influenza vaccine. The total number of digits recalled accurately for the two groups are reported as follows:

Stress Condition (vaccine group)

5	6	6	5	4	7	8	7	6	5	6	7	5	6	5

Nonstress Condition (ticket group)

6	7	6	7	8	8	5	4	6	7	8	9	6	7	8

Is short-term memory impaired by stress?

STEP 1   State the null and the alternative hypotheses.

$$H_0: \mu_S - \mu_N \geq 0 \text{ or } H_0: \mu_S \geq \mu_N$$
$$H_1: \mu_S - \mu_N < 0 \text{ or } H_1: \mu_S < \mu_N$$

Be Here Now

A scientist wishes to compare the effects of two different diets on the weight of mice. Diet A is fed to a group of 8 mice and Diet B is fed to a different group of 8 mice. At the end of a four-week period, the mice are weighed. The data (in grams) are as follows:

Diet A:	28	32	31	29	33	34	32	30
Diet B:	29	27	28	28	30	29	25	22

Is there a significant difference between the two diets? Test at $a = .05$.

Answer

In words, the null hypothesis is "There is no significant difference between the two diets." The alternative hypothesis is "There is a significant difference between the two diets."

Stated symbolically, the null and alternative hypothesis would read as follows:

$$H_0: \mu_A = \mu_B$$
$$H_1: \mu_A \neq \mu_B$$

Notice that this is a two-tailed, or nondirectional, test.

$df = 14[df_A = (8 - 1) = 7; df_B = (8 - 1) = 7; df_{total} = df_A + df_B = (7 + 7) = 14]$

Given an alpha level of .05, the critical value of *t* is ±2.145.

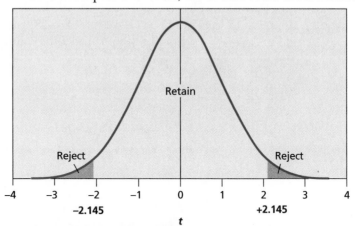

Enter the data into SPSS, and name two variables: weight and diet.

As we look at the output, we see that the mean for Diet A was 31.1250 while that of Diet B was 27.2500.

Levene's test verifies homogeneity of variance $(p > .05)$.

The *t*-test result of 3.318 exceeds the critical value of *t*. Thus, we conclude that, on the basis of this sample, there is a significant difference in the weight of mice as a function of diet, $t(12) = 3.318, p = .005$.

Because the test result is significant, we will calculate Cohen's *d*.

First we will compute the pooled standard deviation. To do so we need the variance of each sample. We squared the standard deviation provided by the output to determine the variance, as needed in the formula. As shown below, the effect size was large ($d = 1.66$).

Begin by computing the pooled standard deviation:

$$S_p = \sqrt{\frac{S_1^2 + S_2^2}{2}} = \sqrt{\frac{2.03^2 + 2.60^2}{2}} = \sqrt{\frac{4.12 + 6.76}{2}} = \sqrt{\frac{10.88}{2}} = \sqrt{5.44} = 2.33$$

$$\text{Cohen's } d = \frac{M_1 - M_2}{S_p} = \frac{31.125 - 27.25}{2.33} = \frac{3.875}{2.33} = 1.66$$

In APA style we might report the following:

A significant difference in the weight of mice was found as a function of diet, $t(12) = 3.32$, $p = .005$; the effect size was large ($d = 1.66$).

In words, the null hypothesis states that stress does not impair the recall of digits (there is no difference between the two groups, or the difference is positive). The alternative hypothesis stipulates that those under stress will recall fewer digits than those who are not under stress. Thus, the difference between the two means will be less than 0.

STEP 2 Choose the alpha level and draw the distribution.

For this example, we will use an alpha level of .05. We calculate $df [df = (15 - 1) + (15 - 1) = 14 + 14 = 28]$ and look up the critical value of t in the table of t values. We look under the column labeled .05 for one tail, and look across the degrees of freedom row labeled 28, to find the critical value of t, which is 1.701 in the table.

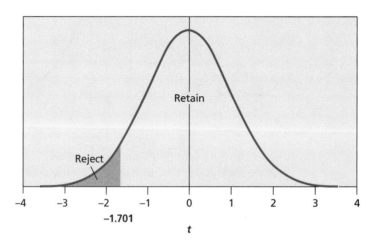

Because this is a one-tailed test, we have one critical region, in the lower tail of the distribution, bounded by our critical value of $t(t = -1.701)$. If we are to reject the null hypothesis, our calculated value of t must be negative.

STEP 3 Calculate the appropriate test statistic.

Remember that each row in the data entry page in SPSS represents a participant and each column represents a variable. In our example, we must indicate which scores were obtained under conditions of stress and which were obtained under conditions in which stress was absent.

- Click on the tab for "Variable View," and define one variable for our memory measurements and another variable to identify the group associated with each. In this example, we will use "Recall" and "Condition" as our variable names.

When we now click on the tab labeled "Data View," we can enter each number under the column heading "Recall" in the SPSS data entry page. Each recall number is followed by an entry in the column headed "Condition." We must "code" our condition information; we can use "1" for the stress condition and "2" for the nonstress condition.

Once we have entered all of the data, our data entry page consists of 30 recall scores in the first column and 30 condition designations in the second column.

We are now ready for SPSS to calculate *t*.

- To conduct an independent-samples *t*-test, we choose

Analyze → Compare Means → Independent-Samples t Test.

"Recall" is the "Test Variable," because we are testing differences between the digit span in both groups, so move "Recall" over to the "Test Variable" box.

"Condition" is the "Grouping Variable," so move it over to the "Grouping Variable" box. We still have to define "Condition" a bit more fully by clicking

T-Test

Group Statistics

Condition		N	Mean	Std. Deviation	Std. Error Mean
Recall	1.00	15	5.8667	1.06010	.27372
	2.00	15	6.8000	1.32017	.34087

Independent Samples Test

		Levene's Test for Equality of Variances		t-test for Equality of Means							
		F	Sig.	t	df	Sig. (2-tailed)	Mean Difference	Std. Error Difference	95% Confidence Interval of the Difference	
									Lower	Upper
Recall	Equal variances assumed	.700	.410	−2.135	28	.042	−.93333	.43716	−1.82882	−.03785
	Equal variances not assumed			−2.135	26.752	.042	−.93333	.43716	−1.83071	−.03596

FIGURE 3 SPSS results for recall under stress and nonstress conditions.

on the button labeled "Define Groups." In this new dialog box, we will provide SPSS with the specific information about what we called each of our two groups. For Group 1, we will enter the number "1" to designate the stress condition. For Group 2, we will enter the number "2," denoting the nonstress condition. We are now ready to let SPSS perform the independent-samples t-test.

Figure 3 presents the SPSS output for the results of the independent-samples *t*-test of memory under stress and nonstress conditions.

Recall	Memory for each of two conditions—Sample 1 (stress) and Sample 2 (nonstress) was examined, with the following results:
N	There were 15 participants in Sample 1 and 15 participants in Sample 2.
Mean	The mean recall for Sample 1 (stress) was 5.8667, and the mean recall for Sample 2 (nonstress) was 6.8000.
Std. Deviation	The standard deviation for Sample 1 (stress) was 1.06010, while the standard deviation for Sample 2 (nonstress) was 1.32017.
Std. Error Mean	The estimate of the standard error of the mean for Sample 1 (stress) was .27372, while that for Sample 2 (nonstress) was .34087.

Levene's Test for Equality of Variances

F	Levene's test of homogeneity of variance computes $F = .700$.
Sig	In this case, the alpha at which F is significant is .410. Since .410 is greater than .05, the difference between the variances is not significant. Thus, we do not have evidence that we have violated the assumption of equal variances.

T-test for Equality of Means

Because Levene's test was not significant, we use the top row of output, labeled "Equal variances assumed":

t	The *t* value is -2.135.
df	There are 28 degrees of freedom (*df*).
Sig (2-tailed)	The *p* at which the *t* value is significant is .042 (two tailed). We do not report this value, since we conducted a one-tailed directional test. If we wish, we may divide *p*, as reported by SPSS, by 2 to obtain the one-tailed value of $p(.042/2 = .021)$.
Mean Difference	The difference in recall under stress and nonstress conditions is $-.9333$. The negative value indicates that fewer digits were recalled under stress.

Be Here Now

A researcher believes that a new formula will enhance the flavor of a diet drink. To test her hypothesis, she has 10 participants drink a cup made with the new formula and 10 drink a cup made with the old formula. Participants are asked to rate the flavor of the drink. Drink ratings ranged from 1 to 7 and were as follows:

New:	7	5	6	3	4	5	2	4	2	5
Old:	2	5	3	5	2	4	2	3	5	1

Does the new formula increase the flavor? Test at $a = .05$.

Answer

In words, the null hypothesis is "The new formula did not enhance the flavor." The alternative hypothesis is "The formula enhanced the flavor."

Stated symbolically, the null and alternative hypothesis would read as follows:

$$H_0: \mu_{new} \leq \mu_{old}$$
$$H_1: \mu_{new} > \mu_{old}$$

Notice that this is a one-tailed, or directional, test.

$$df = 18 \left[df_{new} = (10 - 1) = 9; df_{old} = (10 - 1) = 9; df_{total} = (9 + 9) = 18 \right]$$

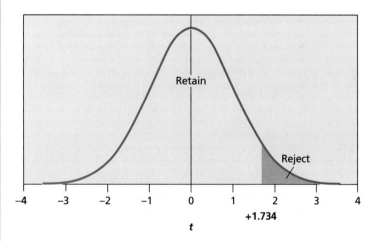

Given an alpha level of .05, the critical value of *t* is +1.734.

Enter the data into SPSS, and name two variables: rating and formula.

As you look at the output, the mean new formula rating was 4.3 while that of the old formula was 3.2.

Levene's test verifies homogeneity of variance ($p > .05$).

The *t*-test result of 1.579 fails to exceed the critical value of *t*.

Thus, we conclude that based on this sample, there is no evidence that the new formula increased flavor, $t(18) = 1.579$, *ns*.

Because the test result was not significant, we do not calculate Cohen's *d.*

Std. Error Difference.   The standard error of the mean difference (S_{M1-M2}) = 43716.

95% Confidence Interval of the Difference   SPSS assumes a two-tailed test. Because this example is a directional test, we cannot use the confidence interval printed.

STEP 4   Evaluate the hypothesis on the basis of the results that we obtained.

The t-test result of -2.135 exceeds the critical value of t, of -1.701. Thus, we conclude that short-term memory is impaired under conditions of stress, as opposed to nonstress, $t(28) = -2.135$, $p < .05$.

In a directional test, it would not be appropriate to use the confidence interval reported by SPSS, because the SPSS default assumes a nondirectional (two-tailed) test.

STEP 5   Compute the effect size.

Because we rejected the null hypothesis, we will compute Cohen's d, the effect size. The formula requires a pooled estimate of the standard deviation. To compute the pooled standard deviation, we need the variance of each group (we squared the standard deviation provided by the output to determine the variance). As the following calculations show, the effect size was moderate $(d = .78)$:

$$S_p = \sqrt{\frac{S_1^2 + S_2^2}{2}} = \sqrt{\frac{1.12 + 1.74}{2}} = \sqrt{\frac{2.86}{2}} = 1.19$$

$$\text{Cohen's } d = \frac{M_1 - M_2}{S_p} = \frac{5.8667 - 6.8000}{1.20} = .78$$

SUMMARY

For questions regarding *differences* between two independent groups, we use **the independent-samples t-test.** Just as with the single-sample t-test, we can use Cohen's d to measure effect size. We can use the independent sample t-test to examine directional and nondirectional hypotheses.

PRACTICE

The t Statistic

1. In words, what is the formula for the independent-samples t-statistic?
2. What is the definitional formula of the independent-samples t-statistic?

3. A large difference between group means and little variability within groups is likely to result in a *t*-statistic of what size?

4. A large difference between group means and large variability within each group is likely to result in a *t*-statistic of what size?

5. A small difference between group means and large variability within each group is likely to result in a *t*-statistic of what size?

Stating Hypotheses

6. A researcher is interested in determining whether there is a difference in achievement between participants who are given an outline before studying, compared with those who are not given an outline. What are the null and alternative hypotheses?

7. A researcher wants to test whether rats that are deprived of stimulation in their first month of life are less social than those which are afforded stimulation during that period. What are the null and alternative hypotheses?

df and Critical Values

8. A researcher tests 20 participants who are told that their task is difficult and 20 participants who are told that their task is easy. The researcher tests for differences between the groups. How many *df* are there? What is the critical value at alpha = .05?

9. A researcher is examining the impact of a new drug on the speed at which rats learn a new maze. Fifteen rats are given the drug and 15 are not. The speed of learning is compared for the two groups. How many *df* are there? What is the critical value at alpha = .01?

10. A researcher wants to test whether watching TV **reduces** accuracy in a homework exercise. Twelve children are asked to study while a TV is playing, and 12 are asked to study without a TV playing. The performance of the two groups of children is compared. How many *df* are there? What is the critical value at alpha = .05?

Nondirectional Tests Using the Independent-samples t-Test

For each problem that follows, be sure to

State the null and alternative hypotheses

Chose the alpha and critical value

Compute the statistic

Evaluate the hypothesis on the basis of the test results

If appropriate, compute Cohen's *d*

11. A psychologist wishes to test whether judgments of physical attractiveness differ as a function of whether one is raised in a rural or urban

setting. Pictures of faces of fashion models are presented to participants, who judge the physical attractiveness of each face on a scale of
1 to 20. Two samples are selected for testing: one from the New York
metropolitan area and one from a rural setting in central Indiana. Judgments of attractiveness for each sample are as follows:

New York City

11	18	14	17	15	20	15	13	12	16	15	19	10	14	17

Central Indiana

13	15	11	16	15	14	13	17	18	9	12	15	14	15	18

Do the groups differ in their judgments of attractiveness?

12. Vocational choice may be influenced by various personality dimen-
sions. A social psychologist wishes to examine whether shyness
influences the choice of college major. University theater students
were compared with psychology students on a shyness inventory.
The inventory scores reported are as follows:

Theater Majors

455	525	654	345	591	615	569	497	575	553

Psychology Majors

533	625	446	544	559	515	531	617	609	493

Do the college majors differ in shyness?

Directional Tests Using the Independent-samples t-test

For each problem that follows,

State the null and alternative hypotheses

Chose the alpha and critical value

Compute the statistic

Evaluate the hypothesis on the basis of the test results

If appropriate, compute Cohen's *d*

13. A gerontologist wishes to assess aspects of the mental health of elderly women who regularly participate in organized social activities,
compared with those who do not. A psychological "hardiness" scale
is administered to a sample of elderly women; higher scores indicate
greater hardiness. Some respondents indicate that they participate
regularly in organized community activities (socially active), while

some respondents indicate that they do not (socially inactive). The hardiness scores obtained are as follows:

Socially Active

| 45 | 52 | 65 | 44 | 59 | 61 | 56 | 49 | 57 | 55 | 63 |

Socially Inactive

| 33 | 52 | 46 | 44 | 49 | 55 | 51 | 37 | 40 | 43 | 39 |

Do socially active elderly women show greater psychological hardiness compared with socially inactive elderly women?

14. A study examined the relationship, if any, between immune function and humor. Moviegoers were asked to answer a brief survey about how well they felt after watching either a humorous movie or a dramatic movie. The number of physical symptoms reported for each of the two groups is summarized as follows:

Humorous Movie

| 1 | 3 | 1 | 0 | 1 | 2 | 1 | 3 | 2 | 1 | 1 | 1 | 0 | 5 | 1 | 0 | 0 | 3 | 1 |

Dramatic Movie

| 3 | 5 | 1 | 1 | 5 | 4 | 3 | 1 | 1 | 0 | 2 | 5 | 4 | 5 | 4 | 2 | 4 | 2 | 3 |

Did the group that watched the humorous movie report fewer symptoms than those who watched the dramatic movie?

Report your results in APA style.

SOLUTIONS

1. $t = \dfrac{\text{Difference between Means}}{\text{Standard Error of the Mean Difference}}$

   $= \dfrac{\text{Variance between Means}}{\text{Variance within Groups (Error)}}$

2. $t = \dfrac{M_1 - M_2}{S_{M1-M2}}$

3. A large *t* value, likely to be a significant difference

4. A small *t* value, likely not to be a significant difference

5. A small *t* value, likely not to be a significant difference

6. $H_0: \mu_{\text{outline}} = \mu_{\text{no outline}}$

   $H_1: \mu_{\text{outline}} \neq \mu_{\text{no outline}}$

7. H_0: $\mu_{\text{deprived}} \geq \mu_{\text{not deprived}}$

   H_1: $\mu_{\text{deprived}} < \mu_{\text{not deprived}}$

8. **dft** = 38 at $p < .05$, = 2.042 (due to table limitation, 30 **df** used)
9. **dft** = 28 at $p < .01$, = 2.763
10. **dft** = 22 at $p < .05$, = 1.717 (one-tailed test)

Nondirectional Tests Using the Independent-samples t-test

11. In words, the null hypothesis is "There is no significant difference in judgments of attractiveness between participants in the two geographic locations. The alternative hypothesis is "There is a significant difference between the participants in the two locations."

    Stated symbolically, the null and alternative hypotheses would read as follows:

$$H_0: \mu_{\text{NY}} = \mu_{\text{IN}}$$
$$H_1: \mu_{\text{NY}} \neq \mu_{\text{IN}}$$
$$df = 28[df_{\text{NY}} = (15 - 1) = 14; df_{\text{IN}} = (15 - 1) = 14;$$
$$df_{\text{total}} = df_{\text{NY}} + df_{\text{IN}} = (14 + 14) = 28]$$

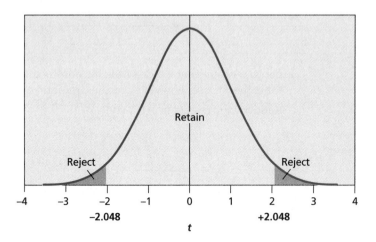

Given an alpha level of .05, the critical value of *t* is ±2.048.

Enter the data into SPSS, and make certain to name two variables (Judgment and Location).

Levene's test verifies homogeneity of variance $(p > .05)$, and the *t*-test result of 0.747 fails to exceed the critical value of *t*. Thus, we conclude that, on the basis of this sample, there is no significant difference in the judgments of attractiveness as a function of geographic location, $t(28) = 0.747$, *ns*.

Because the test result is not significant, we do not calculate Cohen's *d*.

12. In words, the null hypothesis is "There is no significant difference in shyness of students in two college majors. The alternative hypothesis is "There is a significant difference in shyness."

Stated symbolically, the null and alternative hypotheses would read as follows:

$$H_0: \mu_T = \mu_P$$
$$H_1: \mu_T \neq \mu_P$$
$$df = 18[df_T = (10 - 1) = 9; df_P = (10 - 1) = 9;$$
$$df_{total} = df_T + df_P = (9 + 9) = 18]$$

Given an alpha level of .05, the critical value of *t* is ±2.101.

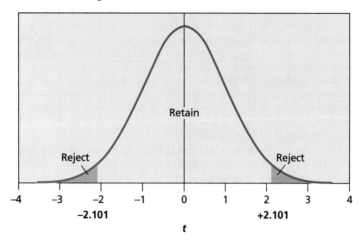

Enter the data into SPSS, and make certain to name two variables (Shyness and Major).

Levene's test verifies homogeneity of variance ($p > .05$), and the *t*-test result of $-.279$ fails to exceed the critical value of *t*. Thus, we conclude that, on the basis of this sample, there is no significant difference in measurements of shyness for theater and psychology majors, $t(18) = -0.279$, *ns*.

Because the test result is not significant, we do not calculate Cohen's *d*.

Directional Tests Using the Independent-samples t-test

13. In words, the null hypothesis might be "Socially active elderly women do not show greater psychological hardiness than socially inactive elderly women." The alternative hypothesis might say "Socially active elderly women show greater psychological hardiness than socially inactive elderly women."

Stated symbolically, the null and alternative hypotheses would read as follows:

$$H_0: \mu_A \leq \mu_I$$

$$H_1: \mu_A > \mu_I$$
$$df = 20[df_A = (11 - 1) = 10; df_I = (11 - 1) = 10;$$
$$df_{total} = df_A + df_I = (10 + 10) = 20]$$

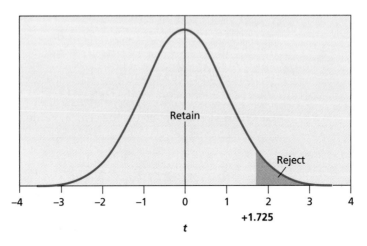

Given an alpha level of .05, the critical value of t is +1.725.

Enter the data into SPSS, and make certain to name two variables (Hardiness and Activity Level).

Levene's test verifies homogeneity of variance ($p > .05$), and the t-test result of 3.601 exceeds the critical value of t. Thus, we conclude that, on the basis of this sample, socially active elderly women show greater psychological hardiness than inactive elderly women, $t(20) = 3.601$, $p < .05$.

Because the test result is significant, we calculate Cohen's d, which equals 1.53; the effect was large.

14. In words, the null hypothesis is "Those who view a humorous movie do not have fewer symptoms of illness than those who view a dramatic movie." The alternative hypothesis is "Those who view a humorous movie have fewer symptoms of illness than those who view a dramatic movie."

Stated symbolically, the null and alternative hypotheses would read as follows:

$$H_0: \mu_A \geq \mu_1$$
$$H_1: \mu_A < \mu_1$$
$$df = 20[df_H = (19 - 1) = 18; df_D = (19 - 1) = 18;$$
$$df_{total} = df_H + df_D = (18 + 18) = 36]$$

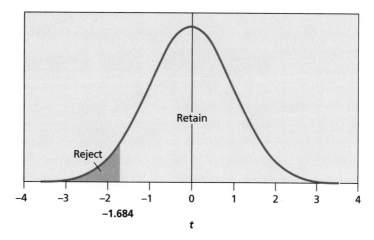

Given an alpha level of .05, the critical value of t is -1.684.

Enter the data into SPSS, and make certain to name two variables (Movie Type and Symptoms).

Levene's test verifies homogeneity of variance $(p > .05)$, and the t-test result of -3.08 exceeds the critical value of t. Thus, we conclude that, on the basis of this sample, those who view a humorous movie have fewer symptoms of illness than those who view a dramatic movie, $t(36) = -3.08$, $p < .05$.

Because the test result is significant, we calculate Cohen's d. The effect size of 1.00 is large.

In APA style:

It was found that those who viewed a humorous movie had significantly fewer symptoms of illness than those who viewed a dramatic movie, $t(36) = -3.08$, $p < .05$. The effect size was large (Cohen's $d = 1.00$).

-10-

Two or More Independent Samples

One-Way ANOVA

FOCUS

Analysis of variance
is used when mean
differences between
two or more groups are
compared.

Analysis of variance is often used to compare mean differences among two or more groups. As you will see in future chapters, there are many different forms of analysis of variance. To analyze differences in one variable among two or more groups, we use **one-way analysis of variance**, often called one-way ANOVA (ANOVA is an acronym for "<u>AN</u>alysis <u>O</u>f <u>Va</u>riance") or, simply, ANOVA.

One-Way ANOVA

Often, we want to study the differences between more than two groups. For example, we might want to examine the extent to which music stimulates babies. To examine this, babies might be randomly assigned to one of three conditions: No Music, Rock Music, and Classical Music. The number of kicks each baby makes would be recorded. In particular, we want to compare the number of kicks made when no music, rock music, or classical music is played. Thus, we are interested in studying the difference(s) in kicks among music groups.

Notice that we used random assignment in the preceding example. Random assignment allows us to determine whether one or more music conditions "caused," or is responsible for, a change in the number of kicks made. Random assignment is necessary to make statements about causation.

To examine differences among music conditions, we might consider doing a series of *t*-tests. That is, we could compare no music with rock music, no music with classical music, and, finally, rock music with classical music. Using this strategy, we would be performing three separate hypothesis tests. If we choose an alpha of .05 for each test, the alpha for the entire study will be greater than .05; in fact, it will be closer to .15. Thus, we have increased the alpha level by carrying out these multiple tests, and this increase in the alpha level is not acceptable for most research.

To think about why performing multiple hypothesis tests increases the probability of a Type I (or alpha) error, consider the difference between purchasing one lottery ticket and purchasing several lottery tickets. Your chances of winning the lottery are greater when you have several tickets in hand.

The One-way ANOVA controls the level of alpha.

An advantage of one-way ANOVA is that it controls for "familywise" error. That is, the comparisons among music conditions (no music with rock music, rock music with classical music, and no music with classical music) are considered a "family" of hypothesis tests. ANOVA controls the alpha at which we determine whether *any* mean difference exists. In so doing, ANOVA only examines differences and does not provide information about the direction of those differences.

The ANOVA test statistic is symbolized by F, in recognition of R. A. Fisher, who developed it.

One-way ANOVA is an expansion of the independent samples *t*-test. Remember *t*?

Reflect
To *t* or not to *t*, that is the question.

$$t = \frac{\text{Difference between Means}}{\text{Standard Error of the Mean Difference}}$$

In the one-way ANOVA formula, the numerator represents the variance between group means. The denominator represents the variability within each group.

$$F = \frac{\text{Variance between Groups}}{\text{Variance within Groups (Error)}}$$

Each statistic, *t* or *F*, is a ratio defined by differences between groups, compared with the variability within groups (the error). The *F* ratio partitions the variance in the dependent measure (the scores) into two components: that attributable to the independent variable (variance between groups) and that attributable to error (variance within groups).

Unlike *t*, the *F* ratio cannot be negative. Whereas the *t* statistic compares two means in the numerator, and the difference could be negative, the *F* ratio examines the variance between a set of means. Variability measures cannot be negative. If the means were identical, there would be zero variance and the *F* ratio would be zero. Any difference between means results in some amount of variability and a positive *F* ratio. This distinction will be an important one when we examine the critical value of *F*.

F Ratio

When we compute a one-way ANOVA, the statistic generated is the *F* ratio, defined as the ratio of between-groups variance to within-groups variance. Thus,

- The larger the differences between means, the larger is the *F* ratio;
- The smaller the variance within groups, the larger is the *F* ratio.

To explore this idea a bit more fully, let's consider some possible patterns of sample mean differences.

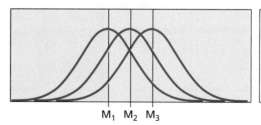

The differences between means are small, and the variances within groups are large. There is considerable overlap across distributions. This pattern in data is likely to result in a small (nonsignificant) F ratio.

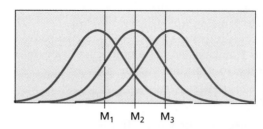

The differences between means are large, and the variances within groups are large. There is considerable overlap across distributions. This pattern in data is likely to result in a small (nonsignificant) *F* ratio.

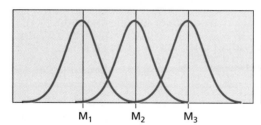

The differences between all three means are large, and the variances within groups are small. There is little overlap between distributions. This pattern in data is likely to result in a large (significant) *F* ratio.

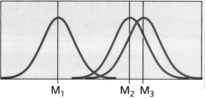

The differences between two of the three means is large, and the variance within each group is small. There is considerable overlap between two distributions, but little overlap with the third distribution. This pattern in data is likely to result in a large (significant) F ratio.

M_1 M_2 M_3

Components of Variance

In ANOVA, we try to break down the variability into smaller parts. The total variability breaks down into two components: **between-treatments variability** and **within-treatments variability.**

Between-treatments variability is the variability between the treatment conditions. In our example, we have the following treatment groups: Classical Music, Rock Music, and No Music. Differences between groups may be due to several sources, including the treatment effect (the music had an impact), individual differences (differences that existed in our participants prior to their participation in the study), and error (random variations).

> **Between-Treatment Variability** is the variability between treatment conditions.
>
> There are three sources of Between-Treatment Variability:
>
> > The Treatment Effect
> >
> > Individual Differences
> >
> > Error

Within-treatments variability is based on the differences *within* each treatment group. In our example, we have differences (variability) within each of the following groups: Classical Music, Rock Music, and No Music. Differences within groups may be due to two sources: individual differences (differences that existed in our participants prior to their participation in the study) and error (random variations).

> **Within-treatments variability** is the variability (differences) within each treatment group.
>
> There are two source of within-treatments variability:
>
> > Individual differences
> >
> > Error

> **Total variability** consists of
>
> Between-treatments variability
> Within-treatments variability

Note that it is not possible for a treatment effect to influence within-treatments variability, because such variability has to do with the variability within one group at a time and each group experiences only one treatment.

Now, to compare these two types of variability, we create a ratio:

$$F = \frac{\text{treatment effect } + \text{ individual differences } + \text{ error}}{\text{individual differences } + \text{ error}}$$

or

$$F = \frac{\text{Variance Between Treatments (between groups)}}{\text{Variance Within Treatments (within groups)}}$$

If there is no between-groups effect, we would expect the F ratio to equal 1. *But if the treatment made a difference,* the numerator will be larger than the denominator, and the F ratio will then be greater than 1.

Be Here Now

What F ratio would you expect in each of the following scenarios?

A. Differences between groups are large. Variability within each group is small.

B. Differences between groups are large. Variability within each group is large.

C. Differences between groups are small. Variability within each group is large.

Answers

A. A large F ratio, $F > 1$

B. A small F ratio, F near 1

C. A small F ratio, F near 1

Assumptions of ANOVA

As with the independent-samples t test, there are assumptions to consider in deciding to calculate an ANOVA:

- The variable being measured is normally distributed;
- The variances of the groups being assessed are equivalent (homogeneous variances); and
- Each sample (group) is independent.

Although these assumptions underlie the use of ANOVA, as with the independent-samples t-test, research suggests that violations of the assumptions are likely to have minimal impact on the statistical outcomes when the number of participants in each group is equal.

Stating Hypotheses: One-Way ANOVA

The steps we take in computing a one-way ANOVA are similar to those we follow in a t-test.

STEP 1   We begin by stating the null and alternative hypotheses.

In our music example, we express the null and alternative hypotheses as follows:

$$H_0:\ \mu_{\text{rock}} = \mu_{\text{classical}} = \mu_{\text{no music}}$$

or, more generally,

$$H_0:\ \mu_i = \mu_j$$
$$H_1:\ \mu_i \neq \mu_j,\ \text{for some i, j}$$

Notice the subscripts that we use—i and j—to denote the means. In the null hypothesis, we are indicating that there will be no difference between any of the means. The alternative hypothesis indicates that at least one mean is different.

STEP 2   Determine the critical value.

Next, we need to choose the alpha level we will use in deciding whether to reject the null hypothesis. As with t, we must select an alpha level that reflects the probability of making a Type I error that we are willing to accept. In this study, we will use an alpha of .05.

Again, as with the independent-samples t-test, we need to determine the number of degrees of freedom in order to identify the critical value of **F.**

Degrees of Freedom: One-Way ANOVA

To identify the critical value of **F**, we need to determine the number of degrees of freedom. When we calculated **df** for the independent-samples t-test, we computed only one number. That number reflected the **df** associated with the denominator of the independent-samples t-test $[(n_1 - 1) + (n_2 - 1)]$. We did not need to worry about the numerator, because, in the independent-samples t-test, we always compared two means; that is, the number of degrees of freedom associated with the numerator was always 1 (number of means -1). This is not the case with the one-way ANOVA, in which we might be comparing three means, as in our music example, or we might be comparing the effectiveness of five different doses of medication or the impact of six different room colors on mood. For one-way ANOVA, we must compute two values to represent the number of degrees of freedom: one value for the degrees of freedom associated with the numerator and one value for that associated with the denominator.

The degrees of freedom associated with the numerator (between-groups variance) of the F ratio are determined by taking the number of groups and subtracting 1. Often, this operation is symbolized by $k - 1 = df_b$. In our music example, there were three music conditions being compared (Rock Music, Classical Music, and No Music). Thus, the number of degrees of freedom for the numerator is $(k - 1) = (3 - 1) = 2$.

Once more, as with the independent-samples t-test, the number of degrees of freedom associated with the denominator (the within-groups variance) is the number in each group, minus 1, summed across the groups. In our example, we have music conditions; each type of music is played for seven babies. Thus, in our example, the number of degrees of freedom is 18. How did we determine that? We did the following:

$$df_w = (n_a - 1) + (n_b - 1) + (n_c - 1) = (7 - 1) + (7 - 1) + (7 - 1)$$
$$= 6 + 6 + 6 = 18.$$

When the numbers in each group are equal, we can compute the degrees of freedom by multiplying the number of groups by $n - 1$ $[df_w = k(n - 1) = 3(7 - 1) = 18]$ or subtracting the number of groups from the total number of participants $[df_w = n - k = 21 - 3 = 18]$.

Determining the Critical Value of F

As with the independent-samples t-test, we must determine the critical value that the observed F (based on our data) must exceed in order to conclude that at least one mean difference exists. To do this, we consult the table of F values in Appendix A3. Across the top row are the numbers of degrees of freedom associated with the numerator; down the first column are the numbers of degrees of freedom associated with the denominator. Within the table, numbers written in regular print represent critical values for an alpha of .05 and those in bold represent an alpha of .01. In our example, with 2 (numerator) and 18 (denominator) degrees of freedom, the critical value of F, given an alpha of .05, is 3.55.

We can now graph the critical region of F:

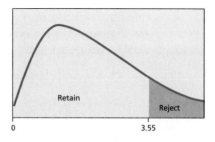

Notice that the graph begins at 0. Remember, because we are dealing with measures of variability, it is not possible to have a negative F ratio.

Please note, all possible numbers of df are not presented in the table of critical F values (Appendix A3). If your df does not appear in the table of F values, drop down to the next lower value of df.

Be Here Now

A researcher is interested in the effect of word type on recall. To examine this relationship, 12 participants are asked to study a list of nouns, 12 others are asked to study a list of verbs, 12 more are asked to study a list of adjectives, and a final 12 are asked to study a list of adverbs.

How many *df* are there? What is the critical value of *F* at $p < .05$?

Answer

$$df_b = (k - 1) = 4 - 1 = 3 \qquad df_w = k(n - 1) = 4(12 - 1) = 4(11) = 44$$

Critical $F(3, 44)\, p < .05 = 2.84$ (*Note*: Due to limited table values, 3,40 *df* were used)

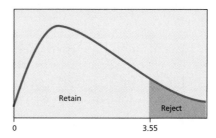

Relationship between *t* and *F* revisited

As has already been mentioned, there is a relationship between *t* and *F*. To see what that relationship is, note that it is possible to conduct an ANOVA, instead of an independent-samples *t*-test, to examine the difference between two means. In particular, as you look at the table of critical *F* values, observe that the first column is for one degree of freedom in the numerator. This situation occurs when you are comparing only two means $(k - 1 = 1)$. Notice also that the critical value for 1 and infinity (∞) *df* is 3.84. But this number is the square of the critical value of t for an infinite number of subjects. That is, $1.96^2 = 3.84$. Thus, $t^2 = F$, or $\sqrt{F} = t$. As has already been described, unlike *t*, *F* cannot assume a negative value.

Now that we have stated our hypotheses and calculated the number of degrees of freedom, we can proceed to the next step in hypothesis testing.

STEP 3   Compute the statistic.

Computing the *F* Ratio

We are now ready to calculate the appropriate test statistic: the *F* ratio. We will again use SPSS to accomplish this. If you would like to see how to compute the *F* ratio without the help of a computer, detailed formulas are shown in Appendix D.

Data entry for a one-way ANOVA is similar to that for the independent-samples *t*-test. It is important to remember that each row in the "Data View" page denotes a participant and each column denotes a variable. In our example, we must indicate the number of kicks made by participants for each music condition: No Music, Rock Music, and Classical Music, respectively.

We must click on the tab for "Variable View" and define one variable for our "Music" type and another variable to indicate the number of "Kicks." In this example, we will use "Music" and "Kick" as our variable names.

We can now click on the tab labeled "Data View." We must "code" our music information; we can use "1" to represent participants who had no music, "2" for those who listened to rock music, and "3" for those who listened to classical music. For each participant, we enter a code for music under the column headed "Music," and we enter the corresponding number of kicks under the column labeled "Kick."

Once data entry is complete, there should be 21 Music designations in the first column and 21 Kick scores in the second column. Each row presents the information on one participant. Each column presents the information for one variable. A sample of what the data look like in SPSS "Data View" is presented in Figure 1.

	Music	Kick
1	1.00	15
2	1.00	16
3	1.00	17
4	1.00	19
5	1.00	21
6	1.00	14
7	1.00	18
8	2.00	30
9	2.00	35
10	2.00	24
11	2.00	29
12	2.00	28
13	2.00	24
14	2.00	31
15	3.00	15
16	3.00	18
17	3.00	19
18	3.00	14
19	3.00	20
20	3.00	17
21	3.00	12

FIGURE 1 SPSS "Data View" sheet.

We are now ready for SPSS to calculate *F*.

There are several methods for computing a one-way ANOVA with SPSS. We are going to use the method listed under the General Linear Model menu choice.

To compute the one-way ANOVA, select

Analyze → General Linear Model → Univariate.

- "Kick" is the dependent variable. So move Kick to the box labeled "Dependent Variable." Recall from Chapter 1 that a dependent variable is the outcome that you are ultimately interested in. It is what results from the experimental manipulation. In our music study, we varied the type of music that each participant heard and recorded the number of kicks. We expect "Kick" to depend upon the type of music listened to.
- "Music" defines the group to which each participant was assigned. Type of Music is the independent variable—the factor that was experimentally manipulated. Move "Music" to the box labeled "Fixed Factor."
- Click on the "Post Hoc" button. Move "Music" to the box labeled "Post Hoc Tests for." Then check the boxes for "LSD" and "Scheffé." We will explain these choices when we describe the output.
- Click on the "Options" button. Highlight "Overall" and "Music," and use the arrow to move them to the box labeled "Display Means for." Check "Descriptive Statistics" and "Estimates of Effect Size."

There are several sections to the output ("Between-Subject Factors," "Descriptive Statistics," "Tests of Between-Subject Effects (ANOVA)," "Estimated Marginal Means," and "Multiple Comparisons"). We will not discuss the "Estimated Marginal Means"; much of the information contained in this section is already printed in the Descriptive Statistics section. We will describe the remaining sections in turn.

Output: Between-Subjects Factors

As presented in Figure 2, the between-subjects (or between-groups) factor was music.

First Column   The first column lists the levels of music. There were three levels: 1, 2, and 3. Remember that level 1 represents No Music, 2 Rock Music, and 3 Classical Music.

Between–Subjects Factors

Music	N
1.00	7
2.00	7
3.00	7

FIGURE 2 Between-subjects factors.

N   The number in each group is reported. There were 7 participants (babies) in each group.

Output: Descriptive Statistics

Figure 3 provides descriptive statistics for each group and for the overall sample. The first row gives the label for each column. The first column indicates the group being described. Remember, we used the code "1" to represent No Music, "2" to represent Rock Music, and "3" to represent Classical Music.

Descriptive statistics include the following:

First Column   In the first column, each "level," or type, of music is listed. We coded Music (1 = No Music, 2 = Rock, 3 = Classical).

Mean   The mean of each group is reported (group 1 = 17.1429, group 2 = 28.7143, and group 3 = 16.4286). Also reported is the mean across all 21 scores. This is the grand mean (20.7619).

Std. Deviation   The standard deviation of each group is reported (group 1 = 2.41030, group 2 = 3.90360, and group 3 = 2.87849). The standard deviation across all participants, regardless of group, is 6.48772.

N   There were 7 participants (babies) in each group. There were 21 participants in the study.

Output: ANOVA

Next is the ANOVA table (see Figure 4). As shown in the first column, the one-way ANOVA presents several estimates of variance. We will focus on three of these: between-groups variance, within-groups variance, and total variance (regardless of source). Each column lists a component of the ANOVA analysis.

Descriptive Statistics

Dependent Variable: Kicks

Music	Mean	Std. Deviation	N
1.00	17.1429	2.41030	7
2.00	28.7143	3.90360	7
3.00	16.4286	2.87849	7
Total	20.7619	6.48772	21

FIGURE 3   Descriptive statistics: kicks by music.

Tests of Between-Subjects Effects

Dependent Variable: Kicks

Source	Type III Sum of Squares	df	Mean Square	F	Sig.	Partial Eta Squared
Corrected Model	665.810[a]	2	332.905	34.047	.000	.791
Intercept	9052.190	1	9052.190	925.792	.000	.981
Music	665.810	2	332.905	34.047	.000	.791
Error	176.000	18	9.778			
Total	9894.000	21				
Corrected Total	841.810	20				

[a] R Squared = 0.791 (Adjusted R Squared = 0.768)

FIGURE 4 Anova Results: Kicks by Music.

The ANOVA table includes the following information:

Source   The three sources of variance we focus on are between-groups variance, which involves differences between groups; within-groups variance, which involves the variance within each group, also considered the error; and total variance, the variability across all participants regardless of group.

Type III Sum of Squares   For each source of variance, the sum of squares (SS) is computed. This is the sum of squared deviations from the mean. SS_b (between groups) is 665.810, while SS_w (within groups) is 176.000.

df   The df column lists the number of degrees of freedom associated with each source of variance. There are $2(k-1)$ degrees of freedom between groups and $18\,[(n_a-1)+(n_b-1)+(n_c-1)]$ degrees of freedom within groups. There are 20 df in total $(df_b + df_w)$.

Mean Square   For each source of variance, the mean square (MS) is computed by dividing SS by df. The MS_b (between groups) is 332.905 ($SS_b/df_b = 665/2$). The MS_w (within groups) is 9.778 ($SS_w/df_w = 176/18$).

F   The F ratio for music is 34.047, computed by dividing MS_b by MS_w (332.905/9.778).

Sig   The p at which F is significant is .000. Although .000 is printed, the actual alpha level is not 0;

rather, alpha is so small that, when rounded, it becomes 0. Since we used an alpha of .05, and .000 is less than .05, at least one mean is significantly different.

Partial Eta Squared The effect size for music, eta squared, is .791; we will discuss eta squared in more detail soon.

Figure 5 gives a summary of the elements in the ANOVA table. For the formulas for computing each SS, see Appendix D. As you examine the table, notice that SS_b and SS_w sum to SS_t. That is, the ANOVA divides the total variability into that attributed to differences between groups and that considered error within groups. The column for *df* is similarly divided.

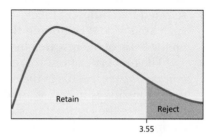

Based on the ANOVA output, we reject the null hypothesis: At least one mean is different from at least one other mean at alpha less than .05, our predetermined alpha level. We arrive at the same decision if we examine our critical value: Given our example with 2 and 18 degrees of freedom, the critical value of F at alpha less than .05 is 3.55.

We can now graph the critical region of F:

Notice that our observed F ratio of 34.047 exceeds the critical value of 3.55. Although we have rejected the null hypothesis and concluded that there is a mean difference, we are not finished with our analysis. We need to determine *where* the difference or differences exist. Was the difference between No Music and Rock Music significant, or was it Rock Music and Classical Music, or was it No Music and Classical Music, or were all three different from one another? We need to conduct a follow-up test.

Source	Sum of Squares	df	Mean Square	F
Between Groups	SS_b	$k - 1$	SS_b / df_b	MS_b / MS_w
Within Groups (Error)	SS_w	$k(n - 1)$ or $n_i - 1 ... + ... n_j - 1$	SS_w / df_w	
Total	$SS_b + SS_w = SS_t$	$df_b + df_w = df$		

FIGURE 5 General Configuration of the Anova Table.

Follow-up tests are necessary when you reject the null hypothesis and you have more than two groups. Had we not rejected the null hypothesis, we would be finished. No additional follow-up testing would be needed.

Multiple Comparisons among Means

When an ANOVA yields a significant value of F, **multiple comparisons** are used to identify where the significant difference (or differences) exists.

Generally, follow-up tests of a significant F ratio are termed multiple comparisons and may be either planned (or *a priori*—roughly, Latin for "before") or *post hoc* (roughly, Latin for "after the fact") tests. There are many possible strategies for following up a significant F ratio. Some follow-up tests are considered more conservative (less likely to reject the null hypothesis) than others. The actual test choices available in SPSS are many, and they are listed in the "Post Hoc" menu screen. In our music example, we have selected two different follow-up tests to give you experience in conducting such analyses. When doing the analysis, the researcher chooses a method on the basis of the research hypotheses.

For **planned comparisons**, the researcher may select a somewhat liberal follow-up procedure, such as the LSD.

The first type of test we requested is the LSD, which stands for *least significant difference*. This is the same as a *t*-test: The difference between each pair of means is tested at an alpha of .05. Generally, this follow-up test is used when the comparisons are planned, prior to data analysis; planned comparisons are called *a priori* hypotheses.

You may be wondering why we first compute a one-way ANOVA if, ultimately, we are only going to do a series of *t*-tests. Remember, however, that the ANOVA controls the familywise level of alpha. The significant F ratio provides assurance that there is, overall, at least one mean difference, at an alpha of $p < .05$.

For *post hoc* **comparisons**, the researcher may select a somewhat conservative follow-up procedure, such as the Scheffé test.

When the specific comparisons are not planned prior to data analysis, they are considered *post hoc*, and we take a more conservative approach to follow-up analyses. In this instance, the researcher may select the Scheffé follow-up test. The Scheffé adjusts the alpha in accordance with the number of follow-up tests (mean comparisons) to be performed. The more means being compared, the lower is the alpha level applied to each of the comparisons. In other words, the alpha level of any one comparison depends on the total number of comparisons (number of means); thus, the actual alpha of any one comparison is less than .05.

> *Reflect*
> A significant F ratio is the admission ticket necessary to enter the arena of comparative consciousness (multiple comparisons).

Figure 6 presents the results of the multiple-comparison tests requested. We have asked for both the LSD and Scheffé for demonstration purposes. Typically, a researcher will select the one method most in concert with the nature of her hypotheses (*a priori* or *post hoc*). The multiple-comparison output presents two sets of results. Across the top are the statistics computed. The first section of the printout presents the results for the Scheffé, while the second set depicts the LSD results.

Multiple Comparisons

Dependent Variable: Kicks

	(I) Music	(J) Music	Mean Difference (I-J)	Std. Error	Sig.	95% Confidence Interval	
						Lower Bound	Upper Bound
Scheffe	1.00	2.00	–11.5714*	1.67142	.000	–16.0279	–7.1149
		3.00	0.7143	1.67142	.913	–3.7422	5.1708
	2.00	1.00	11.5714*	1.67142	.000	7.1149	16.0279
		3.00	12.2857*	1.67142	.000	7.8292	16.7422
	3.00	1.00	–0.7143	1.67142	.913	–5.1708	3.7422
		2.00	–12.2857*	1.67142	.000	–16.7422	–7.8292
LSD	1.00	2.00	–11.5714*	1.67142	.000	–15.0830	–8.0599
		3.00	0.7143	1.67142	.674	–2.7972	4.2258
	2.00	1.00	11.5714*	1.67142	.000	8.0599	15.0830
		3.00	12.2857*	1.67142	.000	8.7742	15.7972
	3.00	1.00	–0.7143	1.67142	.674	–4.2258	2.7972
		2.00	–12.2857*	1.67142	.000	–15.7972	–8.7742

Based on observed means.

* The mean difference is significant at the 0.05 level.

FIGURE 6   Multiple-comparison test results: Music by Kicks

First Column (No Heading)	
Test Type	In the first column, the particular test being conducted is displayed. We instructed SPSS to compute the Scheffé and the LSD (*t*-test).
(i) music (j) music	*i* and *j* stand for the two means being compared. The first comparison is between means 1 (No Music) and 2 (Rock Music). Next is between 1 and 3. The list continues until every possible comparison is listed in every possible order. In effect, each comparison is listed twice. For example, "1 compared with 2" is listed as is "2 compared with 1"; the outcome is the same, only the sign (±) is different.
Mean Difference (i-j)	The difference between means is listed. As shown, the first difference is −11.57143, the result of subtracting the Rock Music mean from the No

Music mean. Notice that, alongside some differences, an asterisk (*) is printed. This denotes that the difference was significant at an alpha of .05. For example, the mean difference between 1 and 2 is followed by an asterisk (*); thus, the mean number of Kicks between No Music and Rock Music is significantly different ($p < .05$). However, the difference between mean 1 and mean 3 is not followed by an asterisk (*); thus, we failed to find a significant difference between the mean number of Kicks between No Music and Classical Music ($p > .05$, or ns).

Std. Error The standard error of the mean difference for each comparison is 1.182, a number based on the MS_w; thus, it is the same for all comparisons.

Sig. This is the alpha level at which the comparison is significant. For the comparison of means 1 and 2, the alpha printed is .000; this indicates that the alpha level was so small that, when rounded to three places, it approximated 0. Since an alpha that rounded to 0 is less than our preset level of alpha of .05, the difference is significant. Because we set alpha at .05, we will come to identical conclusions using the "Sig." column and letting the asterisk identify significant differences at the .05 level. Had we set alpha at a level other than .05, we would have used the "Sig." column to determine whether a difference was significant, and we would not depend upon the asterisks.

95% Confidence Interval Using the appropriate error term, we compute the 95 percent confidence interval around the mean difference. If the confidence interval includes 0, the difference between means is not significant. If the confidence interval does not include 0, the difference between means is significant. The confidence interval around the mean difference (-11.57143) between mean 1 and mean 2 ranges from -16.0279 to -7.1149. Since the interval does not include 0, the difference between mean 1 and 2 is significant. Thus, the "Confidence Interval" column provides the same information as the asterisk (or absence of the asterisk) does in the "Mean Difference" column.

Partial eta squared (η^2) is the proportion of the total score variance accounted for by the between-groups (independent) variable.

Effect Size

To further examine the effect of Music on Kicks, we asked SPSS to compute the effect size: **partial *eta* squared** (η^2). This statistic will give us a measure of the impact of music that is independent of the number of participants in the study. That is, it is not dependent on *df*. As noted in Figure 4, the partial *eta* squared (η^2) for Music is .791. Like Cohen's *d*, η^2 suggests that the type of music had a large effect on Kicks. As before, we examine the effect size only if the *F* ratio is significant.

Application: Hypothesis Testing with a One-Way ANOVA

To summarize briefly, the steps we followed in carrying out a one-way ANOVA are similar to those of earlier chapters:

STEP 1   State the hypotheses.

$$H_0: \mu_i = \mu_j$$
$$H_1: \mu_i \neq \mu_j, \text{ for some } i, j$$

STEP 2   Determine the critical value. With 2, 18 df, the critical value of *F* is 3.55 (see Appendix A3).

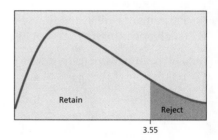

Retain

Reject

3.55

STEP 3   Compute the test statistic.

Tests of Between-Subjects Effects

Dependent Variable: Kicks

Source	Type III Sum of Squares	df	Mean Square	F	Sig.	Partial Eta Squared
Corrected Model	665.810[a]	2	332.905	34.047	.000	.791
Intercept	9052.190	1	9052.190	925.792	.000	.981
Music	665.810	2	332.905	34.047	.000	.791
Error	176.000	18	9.778			
Total	9894.000	21				
Corrected Total	841.810	20				

[a] R Squared = 0.791 (Adjusted R Squared = 0.768)

$$F(2, 18) = 34.047$$

Notice that the df for the numerator and denominator appear, respectively, in parentheses after the symbol F.

STEP 4 Evaluate the hypotheses.

$F(2, 18) = 34.047$, $p < .05$; there is a significant difference between at least two means.

Follow-up tests are necessary because F was significant. Since the examination between Music levels is a priori, we will focus on the LSD results: Rock Music resulted in significantly more Kicks than either No Music or Classical Music.

STEP 5 Determine the effect size.

The effect size $(\eta^2 = 79)$ is large.

Be Here Now

An industrial psychologist is interested in determining the impact of label color on product appeal to consumers. He has four groups of nine participants. Each group is assigned one label color: red, blue, yellow, or green. Participants are asked to rate the appeal on a scale of 1 (no appeal) to 9 (high appeal).

The ratings were as follows:

Red:	7	5	8	4	3	5	8	4	8
Blue:	7	5	8	9	6	7	8	8	5
Yellow:	2	3	4	1	5	2	4	3	1
Green:	8	9	7	6	5	8	9	4	4

Does the color of the label make a significant difference in product appeal?

Answer

STEP 1: Hypotheses: H_0: $\mu_i = \mu_j$ H_1: $\mu_i \neq \mu_j$, for some i, j

STEP 2: *Alpha* = .05, critical value of *F*, with 3 df_b and 32 df_w = 2.90.

STEP 3: Compute the statistic.

Descriptives: Means:

Red	5.7778
Blue	7.0000
Yellow	2.7778
Green	6.6667
Grand Mean (Overall)	5.5556

Be Here Now

ANOVA

APPEAL

	Sum of Squares	df	Mean Square	F	Sig.
Between Groups	99.778	3	33.259	11.190	.000
Within Groups	95.111	32	2.972		
Total	194.889	35			

STEP 4: Interpret the results.

$F(3,32) = 11.190$, $p < .05$. There is a significant difference. Follow-up LSD results suggest that yellow had significantly less appeal than any of the other colors examined.

STEP 5: Effect size:

The effect size ($\eta^2 = .51$) is moderate.

Reporting Results in APA Style

In APA style, you first report the descriptive statistics (means and standard deviations), followed by the inferential statistics (**F** statistics and follow-up tests). If there are more than two means, it is best to report the descriptive statistics in table form, being certain to highlight what is most important in the results. The following report is consistent with APA format:

As shown in Table 1, the mean number of kicks ranged from 16.43 ($SD = 2.88$) for Classical Music to 28.71 ($SD = 3.90$) for Rock Music.

A significant difference across Music levels was observed, $F(2,18) = 34.05$, $p < .05$. Follow-up tests (LSD) suggest that Rock Music resulted in more kicks than either No Music or Classical Music.

Table 1

Descriptive Statistics: Kicks by Music

Music	n	Mean	Standard Deviation
No Music	7	17.14	2.41
Rock	7	28.71	3.90
Classical	7	16.43	2.88

It's Out There... Adams II, Graves, and Adams (2006) examined the effectiveness of a conceptually based course on health-related knowledge. Six different groups were identified on the basis of when the subjects participated in the course. One-way ANOVA results of knowledge scores were significant $F(5,271) = 49.477$, $p < .05$. Among the LSD test results was a significant difference between those who recently completed the course and those who had not taken the course, as well as between those who recently completed the course and those who had completed the course in the past.

SUMMARY

One-way analysis of variance allows for the simultaneous examination of mean differences when there are two *or more* means. It controls the family-wise alpha level. The one-way ANOVA dissects the total variability into two sources: between groups and within groups. Significant *F* ratios for effects having three or more groups need to be followed up by a multiple-comparison test. The selection of the particular test depends on whether the hypothesis is *a priori* (planned prior to data collection) or *post hoc* (not planned prior to data collection). There are different choices of multiple-comparison tests, but we focused on the LSD and Scheffé tests. The effect size should be calculated as well, using η^2 when a significant value of *F* is obtained.

PRACTICE

F Ratio

1. What is the *F* ratio?
2. What is the relationship between *t* and *F*?
3. For each of the following, what *F* ratio would you expect?
   a. Small between-groups variance, large within-groups variance
   b. Small between-groups variance, small within-groups variance
   c. Large between-groups variance, small within-groups variance
   d. Large between-groups variance, large within-groups variance

Stating Hypotheses

4. For each of the following, state the null and alternative hypotheses:
   a. A psychologist is interested in measuring how well clients adjusted to one of five different therapy programs.
   b. A researcher wants to determine whether music affects one's mood. He randomly assigns participants to one of each of the following groups: Rap, classical, rock, and no music.

Computing df

5. For each of the following, compute the number of degrees of freedom associated with the numerator and denominator of the *F* ratio:

a. A researcher assigns 12 participants to each of four dosage levels:

0 milligrams, 12 participants
2 milligrams, 12 participants
4 milligrams, 12 participants
6 milligrams, 12 participants

b. A researcher had 30 participants available for her study. She randomly assigns 10 to Group 1, 10 to Group 2, and 10 to Group 3.

c. A clinical psychologist was interested in examining the effect of three therapies on anxieties. Seven clients participate in systematic desensitization, 7 participate in clinical diagnosis, and 8 do not participate in any therapy.

Critical Values of F

6. For each of the following, determine the critical value of F:

a. $df_b = 2$, $df_w = 27$, $p < .05$
b. $df_b = 3$, $df_w = 36$, $p < .01$
c. $df_b = 4$, $df_w = 20$, $p < .05$

Computing the F Ratio (For these problems, be sure to use the computer.)

7. A researcher is interested in the effect of presentation mode on learning. Children are presented nonsense words in one of three conditions: Words Only, Words Plus Picture, and Words Plus Video. The researcher records the number of words remembered correctly. Does presentation make a difference? Use alpha = .01. Her data are as follows:

Words Only	Words Plus Picture	Words Plus Video
1	5	1
2	4	3
3	5	2
5	8	4
4	6	1
2	8	1
1	7	2

8. A researcher is interested in examining the effect of context on persistence. Five participants are asked to work on an unsolvable task alone, 5 others are asked to work on the same task with a partner, another 5 are asked to work on the same problem within a small group, and a final 5 are asked to work on the same task in the company of a large group. The number of minutes each participant works on the problem is recorded. Does context make a difference? Use alpha = .05. The researcher's data are as follows:

Group: Alone	Partner	Small	Large
15	25	18	19
18	30	18	18
19	39	15	20
22	22	16	21
14	37	18	15

9. An educational psychologist wanted to test whether rewards influence the behavior of youngsters. She assigns 6 youngsters to a concrete reward group, 6 others to a verbal reward group, and another 6 to a no-reward group. The number of misbehaviors each youngster engages in is recorded. Does reward make a difference in behavior? Use alpha = .05. The psychologist's data are as follows:

Group: Concrete Reward	Verbal Reward	No Reward
12	15	18
13	12	16
19	15	13
18	15	20
21	20	13
16	19	18

Follow-up Tests (Multiple Comparisons)

10. Consider Problem 7. Should a follow-up be computed? If yes, what are the results?

11. Consider Problem 8. Should a follow-up be computed? If yes, what are the results?

12. Consider Problem 9. Should a follow-up be computed? If yes, what are the results?

Effect Size

13. Consider Problem 7. Should you compute the effect size? If yes, what are the results?

14. Consider Problem 8. Should you compute the effect size? If yes, what are the results?

15. Consider Problem 9. Should you compute the effect size? If yes, what are the results?

Reporting Results in APA Style

16. Report the results of problem 7 in APA format.

17. Report the results of problem 8 in APA format.

18. Report the results of problem 9 in APA format.

SOLUTIONS

F Ratio

1. $F\ Ratio = \dfrac{Mean\ Square\ Between}{Mean\ Square\ Within}$

2. $t^2 = F$, or $\sqrt{F} = t$

3. **a.** 1, small F value
   b. 1, small F value
   c. >1, large F value
   d. 1, small F value

Stating Hypotheses

4. **a.** Null hypothesis: There will be no significant difference between any pair of means.

   $$H_0: \mu_i = \mu_j$$

   Alternative hypothesis: The difference between at least one pair of means will be significant.

   $$H_1: \mu_i \neq \mu_j,\ \text{for some } i, j$$

   b. Null hypothesis: There will be no significant difference between any pair of means.

   $$H_0: \mu_i = \mu_j$$

   Alternative hypothesis: The difference between at least one pair of means will be significant.

   $$H_1: \mu_i \neq \mu_j\ \text{for some } i, j$$

Computing df

5. **a.** $df_b = (k - 1) = (4 - 1) = 3$

   $df_w = (n_a - 1) + (n_b - 1) + (n_c - 1) + (n_d - 1)$
   $\quad = (12 - 1) + (12 - 1) + (12 - 1) + (12 - 1) = 44$

   b. $df_b = (k - 1) = (3 - 1) = 2$

   $df_w = (n_a - 1) + (n_b - 1) + (n_c - 1)$
   $\quad = (10 - 1) + (10 - 1) + (10 - 1) = 27$

   c. $df_b = (k - 1) = (3 - 1) = 2$

   $df_w = (n_a - 1) + (n_b - 1) + (n_c - 1)$
   $\quad = (7 - 1) + (7 - 1) + (8 - 1) = 19$

Critical Values of F

6. **a.** 3.35
   b. 4.38
   c. 2.87

Computing the F Ratio

7. $F(2, 18) = 17.367, p < .01$

Tests of Between-Subjects Effects

Dependent Variable: recall

Source	Type III Sum of Squares	df	Mean Square	F	Sig.	Partial Eta Squared
Corrected Model	70.571[a]	2	35.286	17.367	.000	.659
Intercept	267.857	1	267.857	131.836	.000	.880
Group	70.571	2	35.286	17.367	.000	.659
Error	36.571	18	2.032			
Total	375.000	21				
Corrected Total	107.143	20				

[a] R Squared = 0.659 (Adjusted R Squared = 0.621)

8. The $F(3, 16) = 11.634, p < .05$

Tests of Between-Subjects Effects

Dependent Variable: minules

Source	Type III Sum of Squares	df	Mean Square	F	Sig.	Partial Eta Squared
Corrected Model	627.350[a]	3	209.117	11.634	.000	.686
Intercept	8778.050	1	8778.050	488.348	.000	.968
Group	627.350	3	209.117	11.634	.000	.686
Error	287.600	16	17.975			
Total	9693.000	20				
Corrected Total	914.950	19				

[a] R Squared = 0.686 (Adjusted R Squared = 0.627)

9. $F(2, 15) = 0.04, ns$

Tests of Between-Subjects Effects

Dependent Variable: misbehavior

Source	Type III Sum of Squares	df	Mean Square	F	Sig.	Partial Eta Squared
Corrected Model	0.778[a]	2	0.389	0.040	.961	.005
Intercept	4769.389	1	4769.389	487.225	.000	.970
Group	0.778	2	0.389	0.040	.961	.005
Error	146.833	15	9.789			
Total	4917.000	18				
Corrected Total	147.611	17				

[a] R Squared = 0.005 (Adjusted R Squared = –0.127)

Follow-up Tests (Multiple Comparisons)

10. Yes, *F* was significant ($F(2, 18) = 17.367, p < .01$). Both the Scheffé (*post hoc* strategy) and LSD (*a priori* strategy) have been requested. Scheffé results suggest that group 2 (Words Plus Picture) was different from either Group 1 (Words Only) or Group 3 (Words Plus Video). Similar conclusions follow from LSD results.

Multiple Comparisons

Dependent Variable: recall

	(I) group	(J) group	Mean Difference (I-J)	Std. Error	Sig.	95% Confidence Interval Lower Bound	Upper Bound
Scheffe	1.00	2.00	–3.57143(*)	.76190	.001	–5.6029	–1.5400
		3.00	0.57143	.76190	.758	–1.4600	2.6029
	2.00	1.00	3.57143(*)	.76190	.001	1.5400	5.6029
		3.00	4.14286(*)	.76190	.000	2.1114	6.1743
	3.00	1.00	–0.57143	.76190	.758	–2.6029	1.4600
		2.00	–4.14286(*)	.76190	.000	–6.1743	–2.1114
LSD	1.00	2.00	–3.57143(*)	.76190	.000	–5.1721	–1.9707
		3.00	0.57143	.76190	.463	–1.0293	2.1721
	2.00	1.00	3.57143(*)	.76190	.000	1.9707	5.1721
		3.00	4.14286(*)	.76190	.000	2.5422	5.7436
	3.00	1.00	–0.57143	.76190	.463	–2.1721	1.0293
		2.00	–4.14286(*)	.76190	.000	–5.7436	–2.5422

* The mean difference is significant at the 0.05 level.

11. Yes, F was significant ($F(3, 16) = 11.364$, $p < .05$). Both the Scheffé (*post hoc* strategy) and LSD (*a priori* strategy) have been requested. Both results suggest that group 2 (Partner) was different from each of the remaining groups (Alone, Small, and Large). Differences among the remaining groups were not significant.

Multiple Comparisons

Dependent Variable: minutes

	(I) Group	(J) Group	Mean Difference (I-J)	Std. Error	Sig.	95% Confidence Interval Lower Bound	Upper Bound
Scheffe	1.00	2.00	−13.00000(*)	2.68142	.002	−21.3584	−4.6416
		3.00	0.60000	2.68142	.997	−7.7584	8.9584
		4.00	−1.00000	2.68142	.986	−9.3584	7.3584
	2.00	1.00	13.00000(*)	2.68142	.002	4.6416	21.3584
		3.00	13.60000(*)	2.68142	.001	5.2416	21.9584
		4.00	12.00000(*)	2.68142	.004	3.6416	20.3584
	3.00	1.00	−0.60000	2.68142	.997	−8.9584	7.7584
		2.00	−13.60000(*)	2.68142	.001	−21.9584	−5.2416
		4.00	−1.60000	2.68142	.948	9.9584	6.7584
	4.00	1.00	1.00000	2.68142	.986	−7.3584	9.3584
		2.00	−12.00000(*)	2.68142	.004	−20.3584	−3.6416
		3.00	1.60000	2.68142	.948	−6.7584	9.9584
LSD	1.00	2.00	−13.00000(*)	2.68142	.000	−18.6844	−7.3156
		3.00	0.60000	2.68142	.826	−5.0844	6.2844
		4.00	−1.00000	2.68142	.714	−6.6844	4.6844
	2.00	1.00	13.00000(*)	2.68142	.000	7.3156	18.6844
		3.00	13.60000(*)	2.68142	.000	7.9156	19.2844
		4.00	12.00000(*)	2.68142	.000	6.3156	17.6844
	3.00	1.00	−0.60000	2.68142	.826	−6.2844	5.0844
		2.00	−13.60000(*)	2.68142	.000	−19.2844	−7.9156
		4.00	−1.60000	2.68142	.559	−7.2844	4.0844
	4.00	1.00	1.00000	2.68142	.714	−4.6844	6.6844
		2.00	−12.00000(*)	2.68142	.000	−17.6844	−6.3156
		3.00	1.60000	2.68142	.559	−4.0844	7.2844

* The mean difference is significant at the 0.05 level.

12. No, F was not significant ($F(2, 15) = 0.04$, *ns*).

Effect Size

13. Yes, F was significant ($F(2, 18) = 17.367$, $p < .01$).

$\eta^2 = .659$

14. Yes, F was significant ($F(3, 16) = 11.634$, $p < .05$).

($\eta^2 = .686$)

15. No, F was not significant ($F(2, 15) = 0.04$, *ns*).

Reporting Results in APA Style

16. As shown in Table 16, the mean number of words recalled ranged from 2.00 for Words Plus Video to 6.14 for Words Plus Picture

Table 16
Descriptive Statistics: Recall by Presentation Mode

Presentation Mode	n	Mean	Standard Deviation
Words Only	7	2.57	1.51
Words Plus Picture	7	6.14	1.57
Words Plus Video	7	2.00	1.15

A significant difference across presentation modes was observed: $F(2, 18) = 17.37$, $p < .01$. Follow-up tests (LSD) suggest that Words Plus Picture produced recall greater than either Words Only or Words Plus Video. Effect size was moderate ($\eta^2 = 66$).

17. As shown in Table 17, the mean number of minutes ranged from 17.00 for the Small Group condition to 30.60 for the Partner condition.

Table 17
Descriptive Statistics: Time by Group

Group	n	Mean	Standard Deviation
Alone	5	17.60	3.21
Partner	5	30.60	7.37
Small	5	17.00	1.41
Large	5	18.60	2.30

A significant difference across group conditions was observed: $F(3, 16) = 11.63$, $p < .05$. Follow-up tests (LSD) suggest that the Partner condition produced greater time than any of the other conditions. Effect size was moderate ($\eta^2 = 69$).

18. As shown in Table 18, the mean behavior scores ranged from 16.00 for the Verbal Reward Group to 16.50 for the Concrete Reward condition.

Table 18
Descriptive Statistics: Behavior by Reward

Group	n	Mean	Standard Deviation
Concrete Reward	6	16.50	3.51
Verbal Reward	6	16.00	2.97
No Reward	6	16.33	2.88

Differences between groups were not significant [$F(2, 15) = 0.04$, *ns*].

-11-

Two or More Between-Groups Variables: Factorial ANOVA

F O C U S

Factorial designs allow for the simultaneous examination of the impact of two or more variables on a dependent variable.

Often, we wish to compare the impact of two or more independent variables on a dependent variable. For example, we may want to examine the impact of the color of a room (yellow, red, or blue) and of noise on achievement. Although we *could* study each variable separately, we can examine the impact of a room's color and of noise on achievement *simultaneously* by using a **factorial design.** In addition, with a factorial design, we can determine whether the variables (Room Color and Noise) have a combined impact on achievement.

Factorial Designs

To study the impact of Room Color and Noise on achievement, we randomly assign 10 participants to a yellow room, 10 to a red room, and 10 to a blue room. Half of the participants in each room will study with noise in the background, while the other half will study in silence (no noise). Figure 1 depicts the assignment of participants to conditions. Notice that there are 5 participants in each "cell," or combination of conditions, presented in Figure 1. For example, although 10 participants will study in a red room, 5 of those participants will study with noise while 5 will study in silence (no noise).

		Noise Condition	
		Noise	No Noise
Room Color	Yellow	5	5
	Red	5	5
	Blue	5	5

FIGURE 1 Number of participants in each condition: Noise by Room Color.

Notice that we used random assignment in our example. This approach will allow us to determine whether one or more Room Colors, or Noise conditions, or some combination of Room Color and Noise, "caused" or is responsible for a change in achievement.

Figure 2 lists the achievement scores earned by our participants.

		Noise Condition	
		Noise	No Noise
Room Color	Yellow	79	75
		83	58
		79	70
		88	65
		91	55
	Red	55	85
		65	79
		45	65
		35	88
		45	79
	Blue	55	79
		45	85
		43	79
		55	69
		41	88

FIGURE 2 Scores of participants: Noise by Room Color.

Each variable is considered a **factor**.

Each condition within a factor is a **level**.

We often call the variables being manipulated **factors**.

Each of the conditions within a factor is considered a **level**. In our example, the factor *Room Color* has three levels (red, blue, and yellow) and the factor *Noise* has two levels (noise and no noise). We can use this information to provide a detailed description of our investigation and describe our study as a 3 by 2 between-groups factorial ANOVA (meaning that we have three room colors and two noise conditions).

To examine differences across levels of Room Color and Noise, we could conduct a series of *t*-tests. Using this strategy, we would perform 15 separate hypothesis tests. If we choose an alpha of .05 for each test, the alpha for the entire study will not be .05, but will be approximately .75. By doing these multiple tests, the alpha level increased, and this is not acceptable for most research. Using ANOVA, we avoid this problem. Also, by doing separate tests, we would not be able to examine whether the impact of one factor is influenced by that of another. This kind of influence is called an interaction, and by using ANOVA, we will have information about the interaction.

Factorial ANOVA designs can have as many independent factors as are needed for a study. Because, in life, little happens in isolation, examining an interaction between

Reflect
Life is complex. Few things happen in isolation.

factors is important. Although it is possible to have more than two independent variables, we will examine designs based on only two variables, or factors, in learning the basics of a factorial ANOVA.

Interactions

A **main effect** is the impact of one variable on the dependent variable.

An **interaction** occurs when the impact of one variable changes at one or more levels of another variable.

When we have a factorial design, we talk about **main effects** and **interactions**. Main effects describe a single variable that has an effect on the dependent variable. Interactions describe the combined effect of two (or more) variables on the dependent variable. Interactions occur when the impact of one variable changes at one or more levels of another variable. The following are some examples of how our data might reveal main effects and interactions.

In this first graph, there is a main effect for Color and a main effect for effect for Noise.

- Achievement changes at each level of Room Color.
- Achievement changes at each level of Noise.
- The difference between noise and no noise is the same at every level of Room Color.
- The lines are parallel, indicating that there is no interaction.

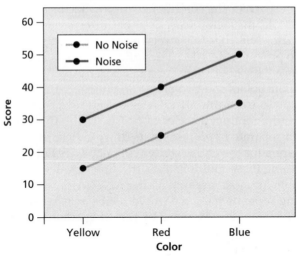

Below, there is an interaction of Room Color and Noise. The lines are not parallel. For example, in the yellow room, the no-noise condition produced lower scores than the noise condition. The reverse is true of blue-room participants. In the red-room condition, there appears not to be a difference between noise and no-noise conditions. When the lines of the graph cross, as in this example, we describe the interaction as a **disordinal interaction.**

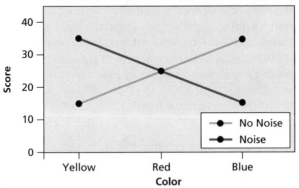

The next graph shows an interaction of Room Color and Noise. The lines are not parallel. In the yellow room, there appears to be little difference between the noise and no-noise conditions, but in the blue room there appears to be a large difference. When the lines in the figure do not cross, but are not parallel, we describe the pattern as an **ordinal interaction.**

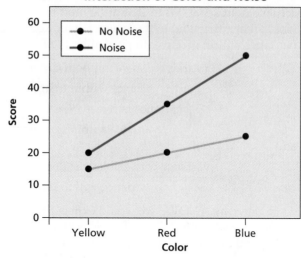

When there is a significant interaction, we focus on that interaction rather than the main effects. This is because the main effects don't tell us the whole story. When there is an interaction, main effects represent the influence of one factor on the dependent variable, *on average*. For example, let's say that you were doing market research on what type of movie is most appealing. You ask a group of males and a group of females which type of movie they'd rather see: an action adventure or a romantic comedy. You might find that, on average, the two types of movies are equally appealing; however, the females prefer the romantic comedies and the males prefer the action adventure movies. If we interpreted only the main effect (ignoring gender), we would lose important information.

In a factorial design, when there is an interaction, we interpret what are called the **simple effects.** Simple effects examine the effect of one variable at each level of another variable. For example, if we observed a significant interaction between Noise and Room Color, we would examine the effect of Noise (noise *vs.* no noise) at each level of Color (yellow, blue, and red).

Simple effects examine the impact of one variable at each level of another variable.

Reflect
The simple can reveal the complex.

F Ratios

Our 3 × 2 factorial ANOVA will generate three *F* ratios, each based on the portion of the variance attributed to either a main effect or the interaction. In our example, we generate an *F* ratio for

Room Color

Noise

Interaction of Room Color and Noise

Each **F** ratio will partition the variance in the dependent measure (the achievement scores) into that attributed to Room Color, that attributed to Noise, that Attributed to the Interaction of Room Color and Noise, and Error (variance within groups). In particular, the **F** ratios computed will be

$$F_{Color} = \frac{\text{variance between Room Color treatments (groups)}}{\text{variance within treatments (groups)}}$$

$$F_{Noise} = \frac{\text{variance between Noise treatments (groups)}}{\text{variance within treatments (groups)}}$$

and

$$F_{Interaction} = \frac{\text{variance between levels: Room Color and Noise (groups)}}{\text{variance within treatments (groups)}}$$

As in one-way ANOVA, we would expect the **F** ratio to equal 1 if there is no between-groups effect. *But if the treatment made a difference,* the numerator will be larger than the denominator, and the **F** ratio will then be greater than 1.

Assumptions of the Factorial ANOVA

As in a one-way ANOVA, there are assumptions to consider in deciding to calculate a factorial ANOVA:

- The variable being measured is normally distributed;
- The variances of the groups being assessed are equivalent (homogeneous variances); and
- Each sample (group) is independent.

Although these assumptions underlie the use of the factorial ANOVA, as with the independent-samples *t*-test and one-way ANOVA, research suggests that violations of the assumptions are likely to have a minimal impact on the statistical outcomes when the number of participants in each group is equal (Stevens, 1996).

Be Here Now

List each of the **F** ratios you would expect to obtain in each of the following scenarios.

A. A factorial design examining the impact of Presentation Order and Picture on recall.

B. A factorial design examining the impact of Medication and Counseling on anxiety.

C. An ANOVA examining the impact of different types of Video Clips on recall.

Answers

A. $F_{Presentation}$, $F_{Picture}$, and $F_{Interaction\ of\ Presentation\ and\ Picture}$

B. $F_{Medication}$, $F_{Counseling}$, and $F_{Interaction\ of\ Medication\ and\ Counseling}$

C. $F_{Video\ Clips}$

Stating Hypotheses: Factorial ANOVA

The steps we take in computing a factorial ANOVA are similar to those we use in computing a one-way ANOVA. However, rather than testing one hypothesis, we have three.

STEP 1   We begin by stating the null and alternative hypotheses. In our factorial design, we will have three sets of hypotheses. We can express the null and alternative hypotheses as follows:

Hypotheses for Room Color (main effect of Room Color):

$$H_0: \mu_{a1} = \mu_{a2} = \mu_{a3}$$
$$H_1: \mu_{ai} \neq \mu_{aj}, \text{ for some } i, j$$

In the null hypothesis, we indicate that there will be no difference between any of the means. The alternative hypothesis indicates that at least one mean is different. We use i and j to denote any pair of means.

Hypotheses for Noise (main effect of Noise):

$$H_0: \mu_{b1} = \mu_{b2}$$
$$H_1: \mu_{b1} \neq \mu_{b2}$$

Hypotheses for the interaction of Room Color and Noise:

$$H_0: \mu_{ab} = 0$$
$$H_1: \mu_{ab} \neq 0$$

STEP 2   Determine the critical value.

Next, we need to choose the alpha level that we will use in deciding whether to reject each null hypothesis. As with a one-way ANOVA and t, we must select an alpha level that reflects what we consider an acceptable probability of making a Type I error. In this study, we will use an alpha of .05.

As with a one-way ANOVA, we need to determine the degrees of freedom in order to determine the critical value of F. However, we must calculate three different sets of df and three critical values.

Degrees of Freedom: Factorial Designs

We need to determine three sets of degrees of freedom: one set for each main effect (in our example, Room Color and Noise) and one set for the interaction. For each, we will need to compute df for the numerator and df for the denominator. To do so, we need to know the number of participants at each level of each factor. Figure 1 is repeated for convenience; it contains the number of participants in each of the conditions of the study.

Numerator df: Room Color   The number of degrees of freedom associated with the numerator (between-groups variance) for the F ratio of Room Color is determined by taking the number of levels of Room Color and subtracting 1. Often, this is symbolized by $k - 1 = df_a$. In our example, there were three room colors being compared (yellow, red, and blue). Thus, the

		Noise Condition	
		Noise	No Noise
	Yellow	5	5
Room Color	Red	5	5
	Blue	5	5

FIGURE 1 Number of participants in each condition: Noise by Room Color.

number of degrees of freedom in the numerator for the main effect of Room Color is $(k - 1) = (3 - 1) = 2$.

Numerator df: Noise The number of degrees of freedom associated with the numerator (between-groups variance) for the F ratio of Noise is determined by taking the number of levels of Noise and subtracting 1. Often, this is symbolized by $k - 1 = df_b$. In our example, there were two levels of Noise being compared (noise and no noise). Thus, the number of degrees of freedom in the numerator for the main effect of Noise is $(k - 1) = (2 - 1) = 1$.

Numerator df: Interaction of Noise and Room Color The number of degrees of freedom associated with the numerator (between-groups variance) for the F ratio of the interaction of Room Color and Noise is determined by multiplying the df for Noise by the df for Room Color $[df_{ab} = (df_a)(df_b)]$. In our example, that number is $(2)(1) = 2$. Thus, $df_{ab} = 2$ in the numerator for the interaction of Room Color and Noise.

Denominator (Error) df The df_{error} associated with the denominator for each F ratio for a between-groups factorial ANOVA are the same. The number is computed by taking the number in each cell (see Figure 1) minus 1 and summing across cells. Here df_{error} for the denominator is $(5 - 1) + (5 - 1) + (5 - 1) + (5 - 1) + (5 - 1) + (5 - 1) = 24$. Because we have equal numbers of participants in each cell in our example, we can multiply the number of cells (6) by the number (n) in any one cell minus 1:

Number of cells multiplied by $(n - 1) = 6(n - 1) = 6(5 - 1) = (6)(4) = 24$. Thus, we obtain the same value: $24 = df_{error}$.

We can now indicate the number of degrees of freedom associated with each source of variance in our example:

The df_{color} (more generally labeled df_a) for F for the Room Color main effect are 2 and 24.

The df_{noise} (df_b) for F for the Noise main effect are 1 and 24.

The $df_{interaction}$ (df_{ab}) for F for the interaction of Room Color and Noise are 2 and 24.

Notice that, for each F ratio, we need two numbers for the degrees of freedom—one for the numerator and one for the denominator.

Determining the Critical Value

As with the independent-samples t-test and one-way ANOVA, we must determine the critical value that each observed F (based on our data) must exceed in order to conclude that at least one mean difference exists. To do this, we consult the table of critical values for F in Appendix A3. Across the top row are the degrees of freedom associated with the numerator; down the first column are the degrees of freedom associated with denominator. Within the table, numbers written in regular print represent critical values for an alpha of .05 and those in bold represent an alpha of .01. We must determine the critical value for each hypothesis we test.

Examining the main effect for Room Color, with 2 and 24 df_a we find that the critical value of F is 3.40.

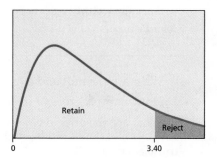

Examining the main effect for Noise, with 1 and 24 $\mathbf{df_b}$ we find that the critical value of F is 4.26.

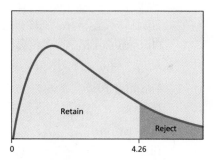

Examining the interaction of Noise and Room Color, with 2 and 24 $\boldsymbol{df}_{(ab)}$ we find that the critical value of F is 3.40.

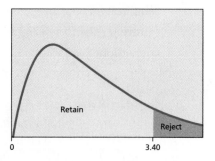

Notice that each of the graphs begins at 0. Remember, because we are dealing with measures of variability, it is not possible to have a negative F ratio.

STEP 3   Compute the statistic.

Be Here Now

A researcher was interested in examining the effect of word familiarity (familiar and unfamiliar) and picture cues (black-and-white picture, color picture, and no picture) on memory. The number of participants, by level of each variable, is presented in the following table:

	Familiarity	
Picture Condition	**Familiar**	**Unfamiliar**
Black/White Picture	8	8
Color Picture	8	8
No Picture	8	8

A. What are the null and alternative hypotheses?

B. How many degrees of freedom are there?

C. What are the critical values ($p < .05$)?

Answers

A. Hypotheses for Familiarity (main effect of Familiarity):

   $H_0: \mu_{a1} = \mu_{a2}$ $H_1: \mu_{a1} \neq \mu_{a2}$

   Hypotheses for Picture (main effect of Picture):

   $H_0: \mu_{b1} = \mu_{b2} = \mu_{b3}$ $H_1: \mu_{bi} \neq \mu_{bj}$, for some i, j

   Hypotheses for the Interaction of Familiarity and Picture:

   $H_0: \mu_{ab} = 0$ $H_1: \mu_{ab} \neq 0$

B. Familiarity: $k - 1 = 2 - 1 = 1$

   Picture: $k - 1 = 3 - 1 = 2$

   Interaction: $(2 - 1)(3 - 1) = 2$

   Error $(n - 1) + (n - 1) + (n - 1) + (n - 1) + (n - 1) + (n - 1) =$
   $(8 - 1) + (8 - 1) + (8 - 1) + (8 - 1) + (8 - 1) + (8 - 1) = 42$

C. Critical Values

   Familiarity: $F_a(1,42) = 4.08$ (because of F table limitations, 1,40 df used).

   Picture: $F_b(2,42) = 3.23$ (2,40 df used).

   Interaction: $F_{ab}(2,42) = 3.23$ (2,40 df used).

Application and Interpretation: Factorial ANOVA

We are ready to calculate the factorial ANOVA. We will again use SPSS to accomplish this task. To compute a factorial ANOVA without the help of a computer, see the Synthesis (Appendix D).

- Data entry for a factorial ANOVA is similar to that for a one-way ANOVA. It is important to remember that each row in the "Data View" page represents a participant and each column represents a variable. In our factorial design, we have two independent variables (Room Color and Noise) and one dependent variable (achievement score).

- We click on the tab for "Variable View" and define one variable to identify the Room Color, another to identify the Noise condition, and a third to identify the achievement scores. In this example, we will use "Color," "Noise," and "Score."

- We can now click on the tab labeled "Data View" and input an entry in the column labeled "Color" to indicate the room each participant was in. We will use 1 to indicate "Yellow," 2 to indicate "Red," and 3 to indicate "Blue." Next, we need to indicate the Noise condition each participant was in. We will use 1 to indicate "Noise" and 2 to indicate "No Noise." Last, we will enter the achievement score under the column heading "Score."

- Once data entry is complete, there should be 30 Room Color designations in the first column, 30 Noise designations in the second column, and 30 Scores in the third column. Each row presents the information on one participant. Each column presents the information for one variable. A sample of what the data look like in SPSS "Data View" is presented in Figure 3.

➡ **We are now ready for SPSS to calculate our *F* ratios**

- To compute the between-groups factorial ANOVA, select

  Analyze → General Linear Model → Univariate.

- "Score" is the dependent variable. So move "Score" to the box labeled "Dependent." (Recall from Chapter 1 that a dependent variable is the measured outcome that you are ultimately interested in; it is what depends upon the experimental manipulation.)

- "Color" and "Noise" each define a between-groups variable. Listed with each is the level of each group each participant was assigned. Move "Color" and "Noise" to the box labeled "Fixed Factor."

- Click on the *Post Hoc* button. Move "Color" into the box labeled "Post Hoc Tests for." Note that we are moving "Color" into this box because "Color" had three levels (yellow, red, and blue). If we get a significant main effect, we will want to check whether all three colors produced different results or whether only two results were different; in the latter case, we will be able to check which two are different. We did not move "Noise," because Noise has only two levels (noise and no noise). If we get a significant main effect for Noise, we will know that the Noise condition produced results different from those of No Noise; we can then look at the mean values and determine which condition produced the higher results.

- Click the *LSD* (Least Significant Difference) button.

	Color	Noise	Score
1	1	1	79
2	1	1	83
3	1	1	79
4	1	1	88
5	1	1	91
6	1	2	75
7	1	2	58
8	1	2	70
9	1	2	65
10	1	2	55
11	2	1	55
12	2	1	65
13	2	1	45
14	2	1	35
15	2	1	45
16	2	2	85
17	2	2	79
18	2	2	65
19	2	2	88
20	2	2	79
21	3	1	55
22	3	1	45
23	3	1	43
24	3	1	55
25	3	1	41
26	3	2	79
27	3	2	85
28	3	2	79
29	3	2	69
30	3	2	88

FIGURE 3 SPSS Data View sheet.

- Click on the *Options* button. Move "Overall," "Color," "Noise," and "Color*Noise" to the box labeled "Display Means for." Check "Descriptive Statistics" and "Effect Size."

The output has several sections ("Between-Subjects Factors," "Descriptive Statistics," "ANOVA: Tests of Between-Subjects Effects," and "Multiple Comparisons"). We will describe each section.

Interpretation: Between-Subjects Factors

As shown in Figure 4, the output begins with a summary of the number of participants in each condition. There are 10 participants in each level of Color and 15 in each level of Noise.

Interpretation: Descriptive Statistics

As presented in Figure 5, the descriptive statistics output box provides descriptive statistics for each cell of the design, each level of each variable, and the overall sample. The first row presents the label for each column. The first column indicates the cell or level being described.

Descriptive statistics include the following:

First Column   The first column is divided into two sections. The first section lists Color and its levels, and the second section lists Noise and its levels. Notice that, for each level of Color, each level of Noise, as well as Total, is in the second section. Total in the second section provides the overall statistics for each level of color. Total in the first section provides overall information for each level of Noise in the sample.

Mean   The mean of each cell, at each level of each variable and overall (Total), is reported. For example, the mean of those who participated in the yellow, noise condition was 84, while the mean of the yellow, no noise condition was 64.6. For the yellow level of color, regardless of noise, the mean was

Between-Subjects Factors

		Value Label	N
Color	1.00	yellow	10
	2.00	red	10
	3.00	blue	10
Noise	1.00	noise	15
	2.00	no noise	15

FIGURE 4 Between-subjects factors output: Number of participants per level of each factor.

Descriptive Statistics

Dependent Variable: score

Color	Noise	Mean	Std. Deviation	N
Yellow	Noise	84.0000	5.38516	5
	No noise	64.6000	8.26438	5
	Total	74.3000	12.15685	10
Red	Noise	49.0000	11.40175	5
	No noise	79.2000	8.84308	5
	Total	64.1000	18.59779	10
Blue	Noise	47.8000	6.72309	5
	No noise	80.0000	7.28011	5
	Total	63.9000	18.21141	10
Total	Noise	60.2667	18.98295	15
	No noise	74.6000	10.52073	15
	Total	67.4333	16.74903	30

FIGURE 5 Descriptive statistics: Score by Color and Noise.

74.3. Also reported is the mean across all 30 scores. This is the grand mean (67.4333).

Std. Deviation The standard deviation of each cell, at each level of each variable and overall, is reported. For example, the standard deviation of scores for those who participated in the blue, noise condition was 6.72309, while that for those who participated in the blue, no noise condition was 7.28011. Overall, the standard deviation for blue scores was 18.21141.

N Reports the number in each cell, as well as in each level of the study. Notice that each cell has 5 participants and each level of color has 10 participants, while 15 participants fall within each level of noise. Overall, there were 30 participants in the study.

Interpretation: Factorial ANOVA

The next section of the output is the ANOVA table, labeled "Tests of Between-Subjects Effects." This table provides the estimates of variance attributed to Color, Noise, and the Interaction of Color and Noise, our between-groups variables. The first column lists the sources of variance; the first row lists the statistics computed.

Let's focus on the sections of the output necessary to determine whether there are significant main effects for Color, and/or Noise, and/or an interaction. In particular, Figure 6 includes the following information:

Source	The first column lists the sources of variance. There are three estimates of between-groups variance (Color, Noise, and the Interaction of Color and Noise). Error reflects within-groups variance. Total describes variability across all participants, regardless of group.
Type III Sum of Squares	For each source of variance, the sum of squares (SS) is computed. A Type III sum of squares shows the sum of squares for an effect, adjusted for all other effects that do not contain it, and has the advantage of not being affected if there are different cell frequencies. This is the default setting for SPSS and is similar to the sum of squares of the one-way ANOVA (Chapter 10). It is the sum of squared deviations from the mean. SS_{Color} is 707.467, SS_{Noise} is 1540.833, $SS_{Color*Noise}$ (Interaction) is 4272.267, and SS_{Error} (the within-groups variance estimate, is 1614.800).
df	df lists the degrees of freedom associated with each source of variance. There are 2(or $3 - 1$) degrees of freedom associated with Color, 1(or $2 - 1$) degrees of freedom associated with Noise, and 2(or 2×1) degrees of freedom associated with the Interaction of Color and Noise. There are 24 (or $4 + 4 + 4 + 4 + 4 + 4$) degrees of freedom associated with Error.

Tests of Between-Subjects Effects

Dependent Variable: score

Source	Type III Sum of Squares	df	Mean Square	F	Sig.	Partial Eta Squared
Corrected Model	6520.567[a]	5	1304.113	19.382	.000	.802
Intercept	136417.633	1	136417.633	2027.510	.000	.988
Color	707.467	2	353.733	5.257	.013	.305
Noise	1540.833	1	1540.833	22.901	.000	.488
Color * Noise	4272.267	2	2136.133	31.748	.000	.726
Error	1614.800	24	67.283			
Total	144553.000	30				
Corrected Total	8135.367	29				

[a]R Squared = .802 (Adjusted R Squared = .760)

FIGURE 6 Factorial ANOVA results: Room color by noise.

Mean Square   For each source of variance, the mean square (MS) is computed by dividing SS by df. MS_{Color} is 353.733(SS_{Color}/df_{Color} = 707.407/2), MS_{Noise} is 1540.833(SS_{Noise}/df_{Noise} = 1540.833/1), $MS_{Interaction}$ is 2136.133($SS_{Interaction}/df_{Interaction}$ = 4272.267/2), MS_{Error}, the within-groups error, is 67.283(SS_{Error}/df_{Error} = 1614.80/24).

F   Three F ratios were computed.

F_{Color} = 5.257, computed by dividing MS_{Color} by MS_{Error} (353.733/67.283).

F_{Noise} = 22.901, computed by dividing MS_{Noise} by MS_{Error} (1540.833/67.283).

$F_{Interaction}$ = 31.748, computed by dividing $MS_{Interaction}$ by MS_{Error} (2136.133/67.283).

Sig (2-tailed)   Three p values are printed, one for each F ratio.

For F_{Color} p = .013.

For F_{Noise} p = .000.

For $F_{Interaction}$ p = .000

Thus, there appears to be a significant main effect for Color, a significant main effect for Noise, and a significant interaction. Also, recall that when .000 is printed, the actual alpha level is not 0; rather, alpha is so small that, when rounded, it approximates 0.

Partial Eta Squared   Partial eta squared (partial η^2) is the proportion of the total variability attributable to a factor, taken as if it were the only variable. Partial eta squared gives an estimate of the effect size, which in this case is .305 for color, .488 for noise, and .726 for the interaction.

Figure 7 presents a summary of the ANOVA table. To see the formulas for computing each SS, see the section titled "Synthesis" in Appendix D.

Source	Sum of Squares	df	Mean Square	F
Color	SS_{Color}	$k_{Color} - 1$	MS_{Color}	MS_{Color}/MS_{Error}
Noise	SS_{Noise}	$k_{Noise} - 1$	MS_{Noise}	MS_{Noise}/MS_{Error}
Interaction	$SS_{Color*Noise}$	$(df_{Color})(df_{Noise})$	$MS_{Color*Noise}$	$MS_{Color*Noise}/MS_{Error}$
Error	SS_{Error}	$(k_{Color})(k_{Noise})(n - 1)$	MS_{Error}	

FIGURE 7 General configuration of the ANOVA table.

Examining the main effect for Room Color, with 2 and 24 df_{Color}, we see that the critical value of F is 3.40. Our observed F_{Color} was 5.257. Because our observed F was greater than the critical F value, we conclude that the main effect for Color was significant.

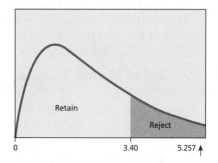

Examining the main effect for Noise, with 1 and 24 df_{Noise}, we see that the critical value of F is 4.26. Our observed F_{Noise} was 22.901. Because our observed F was greater than the critical F value, we conclude that the main effect for Noise was significant.

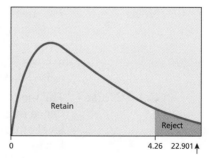

Examining the interaction of Noise and Room Color, with 2 and 24 $df_{Interaction}$, we see that the critical value of F is 3.40. Our observed $F_{Interaction}$ was 31.748. Because our observed F was greater than the critical F value, we conclude that there was a significant interaction of Color and Noise.

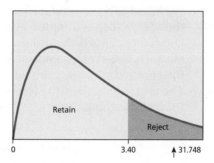

On the basis of the printed levels of alpha on the output, we reject each null hypothesis. Similarly, on the basis of the comparison of critical values with observed values of *F*, we reject each null hypothesis. However, the significant interaction will guide our interpretation of the results.

Interpretation: Estimated Marginal Means

Printed next is a series of output boxes that present the means. Much of the information in the section labeled "Estimated Marginal Means" is a repetition of the descriptive statistics already presented. These statistics for the estimated marginal means are presented in Figure 8.

Grand Mean   The mean, standard error, and 95 percent confidence interval are computed on the basis of all of the scores, regardless of variable or level, and are printed. The grand mean is 67.433. The 95 percent confidence interval is also printed (64.342 to 70.524).

Color   The mean, standard error, and 95 percent confidence intervals are computed for each level of Color. The mean for the red level is 64.100, the mean for the yellow level is 74.300, and the mean for blue is 63.900. The 95 percent confidence interval for yellow is from 68.946 to 79.654.

Noise   The mean, standard error, and 95 percent confidence intervals are computed for each level of Noise. The mean for the noise level is 60.267, while that for no noise is 74.600. The 95 percent confidence for the noise level ranged from 55.896 to 64.638.

Color*Noise   The mean, standard error, and 95 percent confidence intervals are computed for each cell (each level of noise at each level of color). For example, the mean for yellow and noise is 84.000, while that for yellow and no noise is 64.600. The 95 percent confidence interval for yellow and noise ranges from 76.429 to 91.571.

Multiple Comparisons

We asked SPSS to conduct a follow-up test of the main effect of Color, using the LSD test (see Chapter 10); recall that LSD stands for Least Significant Difference. This is the same as a *t*-test; the difference between each pair of means is tested at an alpha of .05. Generally, this follow-up test is used when the comparisons are planned (*a priori*). The results are printed in Figure 9. The first column lists the levels of color being compared. The top row labels the statistic computed.

(I) color (J) color   *i* and *j* stand for the two means being compared. The first comparison is between yellow and red. Next is yellow and blue. The list continues until every possible comparison is listed in every possible order.

Estimated Marginal Means

1. Grand Mean

Dependent Variable: score

Mean	Std. Error	95% Confidence Interval	
		Lower Bound	Upper Bound
67.433	1.498	64.342	70.524

2. Color

Dependent Variable: score

Color	Mean	Std. Error	95% Confidence Interval	
			Lower Bound	Upper Bound
Yellow	74.300	2.594	68.946	79.654
Red	64.100	2.594	58.746	69.454
Blue	63.900	2.594	58.546	69.254

3. Noise

Dependent Variable: score

Noise	Mean	Std. Error	95% Confidence Interval	
			Lower Bound	Upper Bound
Noise	60.267	2.118	55.896	64.638
No Noise	74.600	2.118	70.229	78.971

4. Color * Noise

Dependent Variable: score

Color	Noise	Mean	Std. Error	95% Confidence Interval	
				Lower Bound	Upper Bound
Yellow	Noise	84.000	3.668	76.429	91.571
	No noise	64.600	3.668	57.029	72.171
Red	Noise	49.000	3.668	41.429	56.571
	No noise	79.200	3.668	71.629	86.771
Red	Noise	47.800	3.668	40.229	55.371
	No noise	80.000	3.668	72.429	87.571

FIGURE 8 Descriptive statistics for Noise, Color, and the interaction of Noise and Color.

Multiple Comparisons

Dependent Variable: score
LSD

(I) color	(J) color	Mean difference (I-J)	Std. Error	Sig.	95% Confidence Interval	
					Lower bound	Upper bound
yellow	red	10.2000(*)	3.66833	.010	2.6289	17.7711
	blue	10.4000(*)	3.66833	.009	2.8289	17.9711
red	yellow	−10.2000(*)	3.66833	.010	−17.7711	−2.6289
	blue	.2000	3.66833	.957	−7.3711	7.7711
blue	yellow	−10.4000(*)	3.66833	.009	−17.9711	−2.8289
	red	−.2000	3.66833	.957	−7.7711	7.3711

Based on observed means.
* The mean difference is significant at the .05 level.

FIGURE 9 LSD results for Color.

In effect, each comparison is listed twice. That is, yellow is compared with red, as is red with yellow; the outcome is the same— only the sign (±) is different.

Mean Difference (I-J) The difference between each mean pair is listed. As shown, the first difference is 10.2000. This is the result of subtracting the mean of red from the mean of yellow. Notice that an asterisk (*) is printed alongside this difference. The asterisk denotes that the difference was significant at $p < .05$. The difference between red and yellow is significant ($p < .05$). However, the difference between red and blue is not (.2000, *ns*).

Std. Error The standard error of the mean difference for each comparison is 3.66833. Because it is based on MS_{error} it is the same for all comparisons.

Sig. This is the alpha level at which the comparison is significant. For the comparison of the yellow and red means, p is .010; because this is less than .05, it is considered significant. Had we set alpha at a level other than .05, we would have used the "Sig." column to determine whether a difference was significant, and not the asterisk, which flags all differences significant at $p < .05$.

95% Confidence Interval With the use of the appropriate error term, the 95 percent confidence interval around the mean difference is computed. The bounds of the confidence interval are determined from the critical value of the particular multiple-comparison test and the standard error of the mean difference (see Appendix D for the formulas used). If the confidence interval includes 0, the difference between means is not significant. If the confidence interval does not include 0, the difference between means is significant. The confidence interval around the mean difference between yellow and red (10.2000) ranges from 2.6289 to 17.7711. Because the interval does not include 0, the difference between the mean of yellow and the mean of red is significant.

SPSS uses an alpha equal to .05 as the default value. If you are using a different level of alpha, you will need to instruct SPSS accordingly.

Interpreting the Interaction

In our example, the interaction was significant. This fact has implications for how we interpret our results. Although we had two significant main effects (Room Color and Noise), we will not focus on either one. Rather, because the interaction was significant, we will interpret what are called *simple effects*—the impact of one variable at each level of the other variable. We will examine the effect of Noise at each level of Room Color.

We can accomplish this task by using SPSS to compute three *t*-tests (see Chapter 9). We will need to use the *Data* menu in order to separately analyze levels of Noise at each level of Room Color.

➡ **We are ready for SPSS to examine our simple effects.**

- To identify which level of Room Color to analyze, select

  Data → *Select Cases* → *check "If condition is satisfied"* → *check "If" box* → *highlight "Color" and move it to the workspace on the right* → *type=1* → *continue* → *OK.*

- The box on the right of the screen should look like this: Color = 1.

- This box instructs SPSS to analyze only those data points in the Yellow (1) level of Color.

- Now you can compute the *t*-test. Once the *t*-test is computed, repeat the process for each of the remaining levels of Color (2 and 3).
- If you computed the *t*-tests correctly, you will have obtained the results summarized in Table 1.

At each level of Color, there was a significant difference in the effect of Noise. In the yellow level of Room Color, the mean of noise was higher than the mean of no noise. This is unlike the red and blue levels of Room Color, in which the no-noise means were higher than the means of the noise condition. The simple-effects means are presented graphically in Figure 10 to depict the interaction. Often, when the interaction is significant, we share the results with a graph.

As Figure 10 shows, we can interpret this disordinal interaction as follows: Achievement scores are better under the no-noise condition when participants are tested in red and blue rooms, compared with yellow. The pattern is reversed for the noise condition, in which participants showed lower achievement scores in the red and blue rooms, compared with the yellow room.

TABLE 1
Simple Effects for the Interaction of Room Color and Noise

Color	Noise Level	Mean	*n*	*t*	*p*
Yellow	Noise	84.00	5	4.40	.002
	No Noise	64.60	5		
Red	Noise	49.00	5	4.68	.002
	No Noise	79.20	5		
Blue	Noise	47.80	5	7.27	<.001
	No Noise	80.00	5		

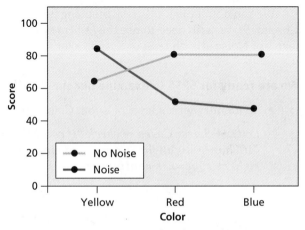

FIGURE 10 Interaction of Color and Noise.

Review of Steps Taken for the Factorial ANOVA

Let's review the steps we took in this example:

STEP 1   State the hypotheses.

Hypotheses for Room Color (main effect of Room Color):

$$H_0: \mu_{a1} = \mu_{a2} = \mu_{a3}$$
$$H_1: \mu_{ai} \neq \mu_{aj}, \text{ for some } i, j$$

Hypotheses for Noise (main effect of Noise):

$$H_0: \mu_{b1} = \mu_{b2}$$
$$H_1: \mu_{b1} \neq \mu_{b2}$$

Hypotheses for the Interaction of Room Color and Noise

$$H_0: \mu_{ab} = 0$$
$$H_1: \mu_{ab} \neq 0$$

STEP 2   Determine the critical values.

Examining the main effect for Room Color, with 2 and 24 $df_{\text{room color}}$ we find that the critical value of F is 3.40.

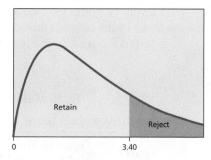

Examining the main effect for Noise, with 1 and 24 df_{noise} we find that the critical value of F is 4.26.

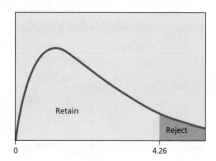

Examining the interaction of Noise and Room Color, with 2 and 24 $df_{\text{Interaction}}$ the critical value of F is 3.40.

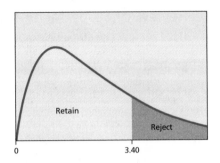

STEP 3   Compute the statistic.

Color: $F_a(2,24) = 5.257$

Noise: $F_b(1,24) = 22.901$

Interaction: $F_{ab}(2,24) = 31.748$

STEP 4   Evaluate the hypotheses.

From Figure 6 (presented again), we see that the main effects for Room Color and Noise were significant, as was the interaction of Room Color and Noise. Further interpretation of the interaction is thus necessary. The data presented in Figure 9 indicate that in the yellow level of Room Color the mean of noise was higher than the mean of no noise. This is unlike the red and blue levels of Room Color, in which the no-noise means were higher than the those of the noise condition. The interaction of room color and noise was depicted in Figure 10, which is duplicated here.

Tests of Between-Subjects Effects

Dependent Variable: score

Source	Type III Sum of Squares	df	Mean Square	F	Sig.	Partial Eta Squared
Corrected Model	6520.567ᵃ	5	1304.113	19.382	.000	.802
Intercept	136417.633	1	136417.633	2027.510	.000	.988
Color	707.467	2	353.733	5.257	.013	.305
Noise	1540.833	1	1540.833	22.901	.000	.488
Color * Noise	4272.267	2	2136.133	31.748	.000	.726
Error	1614.800	24	67.283			
Total	144553.000	30				
Corrected Total	8135.367	29				

ᵃR Squared = .802 (Adjusted R Squared = .760)

FIGURE 6 Factorial ANOVA results: Room Color by Noise (presented again).

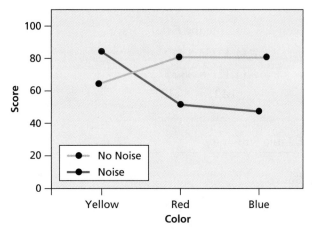

FIGURE 10 Interaction of Room Color and Noise (presented again).

STEP 5 Determine the effect size.

The partial-eta-squared estimates of effect size are moderate. The estimate of effect size for Room Color is .305 and for Noise is .488. The estimate for the interaction is .726.

Be Here Now

An educational psychologist wants to examine the effect of two methods of teaching reading on learning. She also wants to examine whether the effect of reading method varies by gender. Her data are as follows:

	Reading Method	
Gender	New	Traditional
Male	89, 79, 93, 94	65, 79, 66, 78
Female	90, 80, 89, 96	66, 66, 50, 49

Step 1: State Hypotheses

For Gender: H_0: $\mu_{a1} = \mu_{a2}$

H_1: $\mu_{a1} \neq \mu_{a2}$

For Reading Method: H_0: $\mu_{b1} = \mu_{b2}$

H_1: $\mu_{b1} \neq \mu_{b2}$

For the Interaction (Gender × Method): H_0: $\mu_{ab} = 0$

H_1: $\mu_{ab} \neq 0$

Step 2: Determine the critical values; we will use $p < .05$:

Gender: Critical $F_a(1,12) = 4.75$

Reading Method: Critical $F_b(1,12) = 4.75$

Interaction: Critical $F_{ab}(1,12) = 4.75$

(Continued)

Be Here Now

Step 3: Compute the F statistics:

$$F_a(1,12) = 3.412$$
$$F_b(1,12) = 38.307$$
$$F_{ab}(1,12) = 3.412$$

F Table Summary

Source	SS	df	MS	F	Sig	Partial *Eta* Squared
Gender	203.063	1	203.063	3.412	.090	.221
Reading	2280.063	1	2280.063	38.307	.000	.761
Gender $\times$ Rdg	203.063	1	203.063	3.412	.090	.221
Error	714.250	12	59.521			

Step 4: Evaluate hypotheses.
There was a main effect for Reading Method
($F(1,12) = 38.307, p < .05$). The mean for the new program of 88.750 was significantly higher than that of the traditional program (64.875). The main effect for Gender ($F(1,12) = 3.41$, *ns*) was not significant nor was the F for the interaction ($F(1,12) = 3.41$, *ns*).

Step 5: Examine Effect Size
The effect size for Reading Method of .76 is large.

Reporting Results in APA Style

Using our study examining the effect of Room Color and Noise on achievement, you first report the descriptive statistics (means and standard deviations) and then the inferential statistics (F ratios) and any follow-up tests (including tests for simple effects) that were done. You might write:

Table 2 presents the mean and standard deviations of achievement scores for each cell in the study of Room Color and Noise. The highest mean was obtained in the yellow room with noise ($M = 84.00, SD = 5.39$), and the lowest mean was obtained in the blue room with noise ($M = 47.80, SD = 6.72$).

The main effects for Room Color ($F_{Color}(2, 24) = 5.26, p < .05$) and Noise ($F_{Noise}(1, 24) = 22.90, p < .05$) were significant, as was the interaction of Room Color and Noise ($F_{Interaction}(2, 24) = 31.75, p < .05$). Further interpretation of the interaction (see Figure 10) indicated that in the yellow level of Room Color, the mean of noise was higher than the mean of no noise ($t(8) = 4.40, p < .05$). This is unlike the red ($t(8) = 4.68, p < .05$) or blue ($t(8) = 7.27, p < .05$) levels of Room Color, in which the means for no noise were higher than the means for the noise condition.

Table 2
Descriptive Statistics: Achievement Scores, Room Color by Noise

Room Color	Noise	n	Mean	Standard Deviation
Yellow	Noise	5	84.00	5.39
	No Noise	5	64.60	8.26
Red	Noise	5	49.00	11.40
	No Noise	5	79.20	8.84
Blue	Noise	5	47.80	6.72
	No Noise	5	80.00	7.28

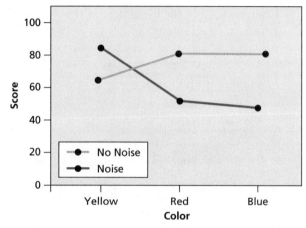

Figure 10. Interaction of Room Color and Noise (presented again).

It's Out There... Glindemann, Wiegand, and Geller (2007) examined situational variables related to alcohol consumption. Table 3 presents the blood alcohol levels of costumed and noncostumed males and females on Halloween. Glindemann and colleagues reported a significant main effect for costume $(F(1,87) = 5.94, p < .05)$. Those in costume had a higher blood alcohol level than those not in costume. No effect was found for gender or the interaction of gender and costume.

TABLE 3
Descriptive Statistics: Blood Alcohol by Gender and Costume

	Costume					
	In Costume			Not in Costume		
Gender	Mean	SD	N	Mean	SD	N
Male	.089	.06	34	.060	.05	19
Female	.090	.05	29	.055	.07	9

SUMMARY

The factorial ANOVA allows us to examine the effect of several independent variables in one analysis. Among the advantages of this statistical technique is the ability to examine the interaction between variables. An interaction occurs when the effect of one variable changes at different levels of a second variable. If the interaction is significant, the simple effects must be examined. If the interaction is not significant, the main effects may be interpreted.

PRACTICE

Interactions

1. Examine each of the three figures that follow. For each, does there appear to be an interaction? If so, what type of interaction?

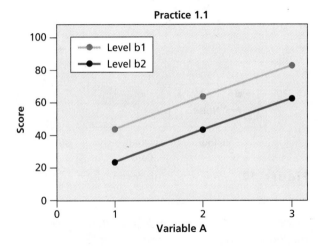

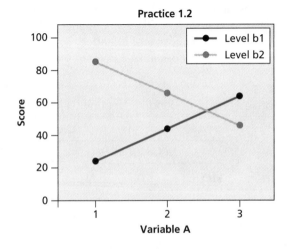

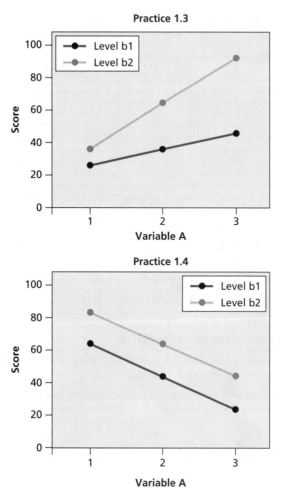

F Ratios

2. A researcher is interested in examining the impact of medication and distraction on postoperative pain. In particular, postoperative patients will be assigned to receive either medication or no medication. Half of those who receive medication will have a TV playing immediately after surgery (distraction), while the remaining half will be in a quiet environment (no distraction). Similarly, half of those who do not receive medication will be subjected to postoperative distraction, while half will not. The researcher's design is summarized as follows:

		Medication	
		Medication	No Medication
Distraction	TV Playing	5	5
	Quiet	5	5

What *F* ratios would you expect?

3. An educational psychologist is interested in examining the effect of gender and size of group on problem solving. Her design is summarized as follows:

		Gender	
		Male	**Female**
	Alone	8	8
Group	One Person Present	8	8
	Five Persons Present	8	8

What F ratios would you expect?

4. Use the data from Problem 2, and state how many degrees of freedom would be associated with each F ratio. At alpha = .05, what is the critical value of F?

5. Use the data from Problem 3, and state how many degrees of freedom would be associated with each F ratio. At alpha = .01, what is the critical value of F?

Computing Factorial ANOVA

6. A researcher was interested in the effect of type of car and time of day on the driving ability of senior citizens. She has 20 senior citizens drive each of four different types of car: subcompact, compact, sedan, and minivan. Half of the cars of each type of car are driven during daylight, and half are driven during evening hours. The researcher records the numbers of traffic cones knocked down on a test course by each driver. Conduct a factorial ANOVA of the following data the researcher obtained:

		Time of Day	
		Daytime	**Evening**
	Subcompact	2,1,3,2,2,1	2,3,2,1,2,3
Car Type	Compact	2,3,1,2,3,3	5,4,6,8,2,7
	Sedan	2,4,1,2,2,1	7,8,10,12,9,8
	Minivan	5,4,2,6,5,4	29,10,12,15,19,21

7. A psychologist is interested in comparing the exam scores of students who take a test anxiety reduction program. She believes that the effect will vary with gender. Ten male and 10 female students participate in the study. Half of the males and half of the females are given the anxiety reduction program. The other half of each group watch a movie. After the movie or anxiety program, the students are administered an achievement test. Conduct a factorial ANOVA on the following test scores the psychologist obtained:

		Gender	
		Male	**Female**
Test Anxiety	Participant (Program)	89,95,75,85,69	88,79,89,99,57
	Control (Movie)	86,91,86,88,75	95,84,79,69,77

Interpreting Factorial ANOVA Results and Reporting in APA Style

8. Interpret the output computed in Problem 6.

   Remember to examine each main effect and the interaction.

   Report your interpretation in APA style.

9. Interpret the output computed in Problem 7.

   Remember to examine each main effect and the interaction.

   Report your interpretation in APA style.

SOLUTIONS

Interactions

1. **1.1.** No interaction. The lines appear parallel
   1.2. Yes, there appears to be an interaction. The lines are not parallel. The lines cross. There appears to be a disordinal interaction.
   1.3. Yes, there appears to be an interaction. The lines are not parallel, but they do not cross. There appears to be an ordinal interaction.
   1.4. No interaction. The lines appear parallel

F Ratios

2. Three *F* ratios would be generated:

   Main Effect for Medication: $F_{\text{medication}}$ or F_a

   Main Effect for Distraction: $F_{\text{distraction}}$ or F_b

   Interaction of Medication and Distraction: $F_{\text{medication} \times \text{distraction}}$ or F_{ab}

3. Three *F* ratios would be generated:

   Main Effect for Group: F_{group} or F_a

   Main Effect for Gender: F_{gender}, or F_b

   Interaction of Group and Gender: $F_{\text{group} \times \text{gender}}$ or F_{ab}

4. Main Effect for Medication: $F_{\text{medication}}$: *df* = 1,16

   $\qquad$ *F* critical, *p* < .05 = 4.49

   Main Effect for Distraction: $F_{\text{distraction}}$: *df* = 1,16

   $\qquad$ *F* critical, *p* < .05 = 4.49

   Interaction of Medication and Distraction: $F_{\text{medication} \times \text{distraction}}$:

   $\qquad$ *df* = 1,16

   $\qquad$ *F* critical, *p* < .05 = 4.49

5. Main Effect for Group: F_{group}: $df = 2,42$
F critical, $p < .01 = 5.15$

Main Effect for Gender: F_{gender}: $df = 1,42$
F critical, $p < .01 = 7.27$

Interaction of Group and Gender: $F_{group \times gender}$:
$df = 2,42$
F critical, $p < .01 = 5.18$

Computing Factorial ANOVA

6. Univariate Analysis of Variance

Between-Subjects Factors

		N
Time	1.00	24
	2.00	24
Car_type	1.00	12
	2.00	12
	3.00	12
	4.00	12

Descriptive Statistics

Dependent Variable: Cones

Time	Car_type	Mean	Std. Deviation	N
1.00	1.00	1.8333	.75277	6
	2.00	2.3333	.81650	6
	3.00	2.0000	1.09545	6
	4.00	4.3333	1.36626	6
	Total	2.6250	1.40844	24
2.00	1.00	2.1667	.75277	6
	2.00	5.3333	2.16025	6
	3.00	9.0000	1.78885	6
	4.00	17.6667	6.91857	6
	Total	8.5417	6.87768	24
Total	1.00	2.0000	.73855	12
	2.00	3.8333	2.20880	12
	3.00	5.5000	3.91965	12
	4.00	11.0000	8.43154	12
	Total	5.5833	5.74950	48

Tests of Between-Subjects Effects

Dependent Variable: Cones

Source	Type III Sum of Squares	df	Mean Square	F	Sig.	Partial Eta Squared
Corrected Model	1250.667[a]	7	178.667	23.586	.000	.805
Intercept	1496.333	1	1496.333	197.536	.000	.832
Time	420.083	1	420.083	55.457	.000	.581
Car_type	543.000	3	181.000	23.894	.000	.642
Time* Car_type	287.583	3	95.861	12.655	.000	.487
Error	303.000	40	7.575			
Total	3050.000	48				
Corrected Total	1553.667	47				

[a]R Squared = 0.805 (Adjusted R Squared = 0.771)

1. Grand Mean

Dependent Variable: Cones

Mean	Std. Error	95% Confidence Interval	
		Lower Bound	Upper Bound
5.583	.397	4.780	6.386

2. Time

Dependent Variable: Cones

Time	Mean	Std. Error	95% Confidence Interval	
			Lower Bound	Upper Bound
1.00	2.625	.562	1.490	3.760
2.00	8.542	.562	7.406	9.677

3. Car_type

Dependent Variable: Cones

Car_type	Mean	Std. Error	95% Confidence Interval	
			Lower Bound	Upper Bound
1.00	2.000	.795	.394	3.606
2.00	3.833	.795	2.228	5.439
3.00	5.500	.795	3.894	7.106
4.00	11.000	.795	9.394	12.606

4. Time * Car_type

Dependent Variable: Cones

Time	Car_type	Mean	Std. Error	95% Confidence Interval	
				Lower Bound	Upper Bound
1.00	1.00	1.833	1.124	−.438	4.104
	2.00	2.333	1.124	.062	4.604
	3.00	2.000	1.124	−.271	4.271
	4.00	4.333	1.124	2.062	6.604
2.00	1.00	2.167	1.124	−.104	4.438
	2.00	5.333	1.124	3.062	7.604
	3.00	9.000	1.124	6.729	11.271
	4.00	17.667	1.124	15.396	19.938

7. Univariate Analysis of Variance

Between-Subjects Factors

		N
Program	1.00	10
	2.00	10
Gender	1.00	10
	2.00	10

Descriptive Statistics

Dependent Variable: score

Program	Gender	Mean	Std. Deviation	N
1.00	1.00	82.6000	10.52616	5
	2.00	82.4000	15.86821	5
	Total	82.5000	12.69514	10
2.00	1.00	85.2000	6.05805	5
	2.00	80.8000	9.60208	5
	Total	83.0000	7.91623	10
Total	1.00	83.9000	8.21178	10
	2.00	81.6000	12.39355	10
	Total	82.7500	10.30010	20

Tests of Between-Subjects Effects

Dependent Variable: score

Source	Type III Sum of Squares	df	Mean Square	F	Sig.	Partial Eta Squared
Corrected Model	49.750[a]	3	16.583	.135	.938	.025
Intercept	136951.250	1	136951.250	1114.557	.000	.986
program	1.250	1	1.250	.010	.921	.001
gender	26.450	1	26.450	.215	.649	.013
program * gender	22.050	1	22.050	.179	.677	.011
Error	1966.000	16	122.875			
Total	138967.000	20				
Corrected Total	2015.750	19				

[a]R Squared = 0.025 (adjusted R Squared = –0.158)

Estimated Marginal Means

1. Grand Mean

Dependent Variable: score

Mean	Std. Error	95% Confidence Interval	
		Lower Bound	Upper Bound
82.750	2.479	77.495	88.005

Interpreting Factorial ANOVA Results and Reporting in APA Style

8. Table 4 presents the descriptive statistics. The main effect for car was significant ($F(3, 40) = 23.89, p < .01$), as was the main effect for Time of Day ($F(1, 40) = 55.46, p < .05$). The interaction of Type of Car and Time of Day was also significant ($F(3, 40) = 12.66, p < .05$). Partial *eta* squared for the interaction was .487, which is moderate. The interaction is depicted in Figure 11.

Table 4

Descriptive Statistics: Time of Day by Type of Car

Type of Car	Time of Day					
	Daytime			Evening		
	M	*SD*	*n*	*M*	*SD*	*n*
Subcompact	1.83	0.75	6	2.17	0.75	6
Compact	2.33	0.82	6	5.33	2.16	6
Sedan	2.00	1.10	6	9.00	1.79	6
Minivan	4.33	1.37	6	17.67	6.92	6

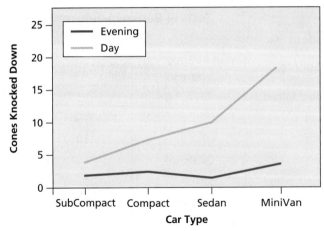

Figure 11. Interaction of Type of Car and Time of Day.

9. Table 5 presents the descriptive statistics. None of the F ratios were significant at $p < .05$. In particular, $F (1, 16)$ for Gender was .22, $F (1, 16)$ for Program was .01, and $F (1, 16)$ for the interaction of Program and Gender was .18.

Table 5
Descriptive Statistics: Gender by Program

	Test Anxiety					
	Program			No Program (Movie)		
Gender	*M*	*SD*	*n*	*M*	*SD*	*n*
Male	82.6	10.53	5	85.20	6.06	5
Female	82.4	15.87	5	80.80	9.60	5

-12-

The Related-Samples *t*-Test: A Within-Groups Design

In This Chapter

Within-Groups vs. Between-Groups Designs

The Related-Samples *t*-Test
- Hypotheses about Change or Related Data
- Assumptions about the Related-Samples *t*-Test
- Conducting the Related-Samples *t*-Test
- Degrees of Freedom for the Related-Samples *t*-Test

Application: The Related-Samples *t*-Test

Reporting the Results in APA Style

Directional Tests Using the Related-Samples *t*-Test

Matched Samples

FOCUS

Related-samples or dependent-samples *t*-tests examine change between pairs of measurements on the same sample.

We have already learned that the *t*-statistic is used to test for differences between a sample mean and a population mean and to test for differences between the means of two independent samples. When a research design involves independent samples, it is referred to as a **between-groups design.** Under different circumstances, we may take multiple measurements using the *same* sample. When measuring *change* in the *same* sample, we use another *t*-test called the **related-samples *t*-test** or the **dependent-samples *t*-test.** With the related-samples *t*-test, we examine the change in *pairs* of measurements. For example, we may be interested in the increase or decrease in two measurements over time. Any time we repeatedly measure the same participants, we are employing a **within-groups**, or **repeated-measures**, design. When we have two measures per participant, SPSS refers to this type of *t*-test as a **paired-samples *t*-test.**

In one variation of the related-samples *t*-test, we might be interested in differences between *pairs* of measurements from samples that have been *matched* on a variable that we think is important

Reflect
All roads lead to t.

to our study. For example, we may match pairs of participants on their IQ scores and compare different reading programs. In a somewhat different example, we might consider comparing volunteer activities of married couples. In each pair of measurements, there are actually two participants; however, because the pairs of participants have been matched on one or more variables of interest, we consider the resulting pairs of scores a repeated measure. This application is called a **matched-samples *t*-test.**

Matched-samples *t*-tests examine differences in matched pairs of measurements.

There are many potential questions about differences between pairs of measurements. We can evaluate the potential benefit of an antidepressant medication, we can evaluate whether a new memory technique improves test performance, or whether a new therapy reduces shyness in children. Clearly, the questions about changes in measurements are numerous. We can also answer important questions by examining differences between matched samples.

Reflect
Change is not merely a difference.

Within-Groups vs. Between-Groups Designs

Each time we decide how to conduct our research, we are also making a decision as to which statistic to compute. Imagine that we want to test the impact of a new cell phone policy in a local high school. We could divide the students into two groups, one that would be allowed to bring phones to school and another that would not. We would record the students' monthly cell phone usage in minutes and compare the two groups. This would be a **between-groups design**, and we would test whether a significant difference existed between the group with the cell phones and the group without. Because we have two independent groups, we would use an independent samples *t*-test to analyze these data.

Between-groups designs compare measures across different groups of participants.

Alternatively, using a different research strategy, we could measure cell phone usage in the same group of students, change the school policy, and then measure cell phone usage again. We would use the related-samples *t*-test to analyze these data. This is an example of a **within-groups**, or **repeated-measures, design.**

Within-groups designs repeatedly measure characteristics of the same participants.

Which is the better study? There is no one answer to this question. Each approach has its benefits.

Ideally, in a between-groups design, all students assigned to one group should have the same cell phone usage; after all, they were in the same group. In reality, however, this does not occur. Some students in the study use the phone a great deal, while others rarely use it, regardless of the group to which they were assigned. The fluctuations that naturally occur in any group of high school students (or any group of humans or animals, for that matter) are called *individual differences*. When we analyze the set of data in a between-groups study, these individual differences remain in the error term (denominator) of the between-groups (independent-samples) *t*-test. Anything that makes the denominator larger will make the overall *t*-statistic smaller. In particular, the independent-samples *t*-test is computed using the formula

$$t = \frac{\text{Difference between Means}}{\text{Standard Error of the Mean Difference}}$$

Recall from Chapter 9 that the standard error of the mean difference of the independent-samples *t*-test has two components:

Standard Error of the Mean Difference = Error Due to Sampling + Error Due to Individual Differences

In contrast, the related-samples *t*-test is a within-groups design: We are examining differences within one group. When we compute the standard error of the mean difference, we are able to remove the variability contributed by individual differences because we repeatedly measure the same participants. We would expect that a student who uses the cell phone frequently will continue to use the cell phone frequently even after the school policy changes. Of course, the policy change may affect his or her overall cell phone use—there may be an increase or decrease in usage—but we would expect that a "high-usage" student would continue to show relatively high usage. Because we measure the characteristics of participants more than once, we can statistically minimize the effect of individual differences. Also, because each participant serves as his or her own control, individual differences can be removed from the error term of the related-samples *t*-test. Anything that makes the denominator smaller makes the overall *t*-statistic larger, and we have a greater likelihood of rejecting the null hypothesis.

An **advantage** of the related sample *t*-test: Individual differences are removed from the error term.

Thus, an **advantage** of the related-samples *t*-test is that the standard error of the mean difference does not include individual differences and tends to be smaller.

A **disadvantage** of the related sample *t*-test: *df* are smaller and the critical value is larger.

However, there is a **disadvantage**: a loss of degrees of freedom. We will go into greater detail later in the chapter about this point, but it is important to note here that if we lose degrees of freedom, the critical value we use to evaluate our hypotheses will be larger. As a consequence, it will be more difficult to reject the null hypothesis.

Each design has advantages and disadvantages. As a researcher, you will make the decision to use a between-groups design or a repeated-measures design on the basis of many factors, such as the number of participants available and the nature of the variable being measured.

Reflect
The path is chosen with the first step.

The Related-Samples *t*-Test

Imagine that we invent a happiness rating scale and ask individuals to rate how happy they are at a given moment. We use a rating scale of 1 to 20, where 1 represents minimal happiness and 20 represents extreme happiness. Imagine all the questions that we might resolve with this happiness rating scale. We could determine whether happiness changes significantly after a major life event such as marriage, the birth of a child, or finding a job. Or we could investigate whether there is a significant change in happiness after a major expenditure, such as purchasing a car or a house. Hypotheses relative to each of these questions are tested by using a **related-samples *t*-test** that compares happiness ratings before and after the event of interest.

For example, consider the following happiness scores obtained from incoming students who were asked to rate their general feelings before the beginning of the semester and again at midterm:

Happiness Scores

Student	Midterm	Beginning
A	16	11
B	17	13
C	8	9
D	13	12
E	15	10
F	9	7
G	5	2
H	11	15
I	18	18
J	10	8
K	15	14
L	8	6
M	20	18
N	17	18

Was there a significant change in happiness during the first half of the semester?

Hypotheses about Change or Related Data

Let's begin by formulating a null and an alternative hypothesis:

H_0: There is no significant change in happiness from the beginning of the semester to the midterm of the semester.

H_1: There is a significant change in happiness from the beginning of the semester to the midterm of the semester.

Our question about change in happiness is nondirectional, or two tailed. In other words, in this case, our question is not whether there is an increase in happiness (or whether there is a decrease in happiness); rather, our question is whether a change in *any* direction occurred in this sample between the beginning of the semester and midterm. Our hypotheses could also be stated to reflect the difference between means over time:

$$H_0: \mu_{\text{midterm}} - \mu_{\text{beginning}} = 0$$
$$H_1: \mu_{\text{midterm}} - \mu_{\text{beginning}} \neq 0$$

To evaluate whether there is a change in happiness, we will look at the difference (**D**) between happiness ratings at the beginning of the semester and happiness ratings at midterm, for each student who participated in our study. The average difference is symbolized as *D* bar: $\overline{D}$.

Our null and alternative hypotheses can be written by referring to the mean population difference, μ_D:

$$H_0: \mu_D = 0$$
$$H_1: \mu_D \neq 0$$

Consistent with the other hypothesis testing that we have done, we will use the average change in our sample to make inferences about changes in happiness in the population. We will reject our null hypothesis (that there is no change in happiness) if we observe an average difference in happiness ratings that is sufficiently larger than we would expect to find by chance.

The question of interest is: How much of a difference between happiness ratings is necessary for us to conclude that there is a significant change? To answer that question, we use the related-samples *t*-test.

> **Reflect**
> Apparent change is not necessarily real change.

Assumptions of the Related-Samples *t*-Test

There are several things we assume to be true when we use the related-samples *t*-test:

- The population distribution of difference scores is normally distributed.
- The participants are a random sample of the population.
- The variance of the measures assessed are equal.

Although these assumptions underlie the use of the related-samples *t*-test, research suggests that violations of the assumptions are likely to have minimal effect on the statistical outcomes. Specifically, violations of the assumption of a normal distribution have been found to have minimal effects on the outcomes (Boneau, 1960).

Conducting the Related-Samples *t*-Test

Let's again consider the inferential statistical tests that we have already presented:

To compare a mean with a population value when the population variance (and the standard error of the mean) is known, we use a z-statistic:

$$z = \frac{M - \mu}{\sigma_\mu}$$

To compare a mean with a population value when the population variance (and the standard error of the mean) is not known, we use a single-sample t-statistic:

$$t = \frac{M - \mu}{S_M}$$

where

$$S_M = \frac{S}{\sqrt{n}}$$

To compare two independent means, we use an independent-samples t-statistic:

$$t = \frac{M_1 - M_2}{S_{M_1 - M_2}}$$

Remember the strong resemblance among the formulas: In each formula, the numerator consists of the difference between means and the denominator is an estimate of error.

The definitional formula for the related-samples t-test can be stated as

$$t = \frac{\text{Mean Difference (or Mean Change)}}{\text{Standard Error of the Mean Difference}} = \frac{\overline{D}}{S_{\overline{D}}}$$

where

$$S_{\overline{D}} = \frac{S_D}{\sqrt{n}}$$

Degrees of Freedom for the Related-Samples t-Test

The number of degrees of freedom for the related-samples t-test is equal to the number of pairs, minus 1. Thus, for the related-samples t-test,

$$df = n_{\text{pairs}} - 1$$

In our example,

$$df = 14 - 1$$
$$df = 13$$

Now we are ready to tackle our example of the related-samples t-test to see whether the difference from the beginning of the semester to the midterm reflects a significant change.

Application: The Related-Samples *t*-Test

Let's look again at our pairs of happiness scores:

Happiness Scores

Student	Midterm	Beginning
A	16	11
B	17	13
C	8	9
D	13	12
E	15	10
F	9	7
G	5	2
H	11	15
I	18	18
J	10	8
K	15	14
L	8	6
M	20	18
N	17	18

STEP 1   State the null and the alternative hypotheses.

We have already stated our null and alternative hypotheses:

$$H_0: \mu_D = 0$$
$$H_1: \mu_D \neq 0$$

In words, the null hypothesis states that there is no change in happiness from the beginning of the semester to the midterm. The alternative hypothesis suggests that there is a change. Notice again that this is a nondirectional test: We are not specifying whether happiness increases or decreases.

STEP 2   Choose the alpha level and determine the critical value.

For this example, we will use an alpha level of .05. We have already calculated our *df*(*df* = 13). Next, we identify the critical value of *t* in the table of *t* values (see Appendix A2). We look under the column labeled .05 for two tails, look across the degrees of freedom row labeled 13, and find that the critical value of *t* is 2.160.

We draw the hypothetical distribution of values of *t* when the null hypothesis is true:

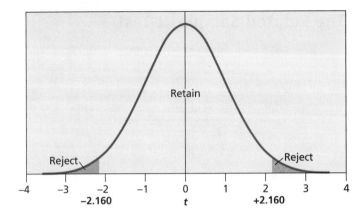

When the null hypothesis is true, and when there is no sampling error, we would expect no difference between beginning and midterm happiness.

Because our test is two-tailed, we have two critical regions, one in each tail of the *t* distribution. If our value for *t* falls into either tail region, we will reject the null hypothesis and conclude that there was a significant change in happiness. To reject the null hypothesis, our calculated value of *t* must fall beyond +2.160 or −2.160.

STEP 3   Calculate the appropriate test statistic.

Remember, when entering the data for the related-samples *t*-test, we must keep the pairs of happiness ratings intact. In using SPSS, it is important to recall each row in the data entry page (the spreadsheet) represents a participant and each column represents a variable.

In our example, we must indicate which scores were from the midterm and which were from the beginning of the semester. To do this,

Click on the tab for "Variable View," and define one variable for our midterm measurements and another variable to identify the beginning measurement. In this example, we will use "Midterm" and "Beginning" as our variable names.

Note that how you organize the "Data View" page (which variable you enter first) impacts the direction (positive or negative) of the *t*-statistic computed.

Click on the tab labeled "Data View," and enter each happiness rating at midterm under the column heading "Midterm" in the SPSS spreadsheet and each beginning happiness rating under the column heading "Beginning" in the spreadsheet. We entered "Midterm" first so that, when SPSS compares scores, if there was increased happiness, the difference ("Midterm"–"Beginning") would be positive.

Once we have entered all of the data, our spreadsheet consists of 14 pairs of happiness ratings. Each row presents the information on one participant.

	Midterm	Beginning
1	16	11
2	17	13
3	8	9
4	13	12
5	15	10
6	9	7
7	5	2
8	11	15
9	18	18
10	10	8
11	15	14
12	8	6
13	20	18
14	17	18

FIGURE 1 SPSS spreadsheet for entering data to test for a change in happiness rating.

Each column presents the test data collected at a particular time (Beginning and Midterm). Once entered in SPSS, your data should look like that shown in Figure 1.

We are now ready to calculate our related-samples *t*-statistic with SPSS.

→ **To conduct a related-samples *t*-test, choose**

Analyze → Compare Means → Paired-Samples T-Test.

- We must identify the pair of variables under study. Click on "Beginning" and "Midterm," the two variables that we want to compare. Notice that the pair appears in the section labeled "Current Selection." Use the arrow to move the pair to the "Paired Variables" box so that SPSS calculates the difference scores and the related-samples *t*.

Figure 2 presents the SPSS output for the results of the related-samples *t*-test for happiness ratings.

In the first box of output, SPSS presents the descriptive statistics for each pair of variables. The top row labels the statistics computed; below each label are the values for the Midterm and Beginning happiness ratings.

T-Test

Paired Samples Statistics

		Mean	N	Std. Deviation	Std. Error Mean
Pair 1	Midterm	13.0000	14	4.52344	1.20894
	Beginning	11.5000	14	4.87931	1.30405

Paired Samples Correlations

		N	Correlation	Sig.
Pair 1	Midterm & Beginning	14	.864	.000

Paired Samples Test

		Paired Differences					t	df	Sig. (2-tailed)
		Mean	Std. Deviation	Std. Error Mean	95% Confidence Interval of the Difference				
					Lower	Upper			
Pair 1	Midterm - Beginning	1.50000	2.47293	.66092	.07217	2.92783	2.270	13	.041

FIGURE 2 SPSS output for the paired-samples *t*-test for change in happiness ratings.

Paired Samples Statistics

Pair 1
Consists of the Midterm and Beginning happiness ratings.
There is only one pair of variables in our analysis.

Midterm
One of the measures for Pair 1 is the Midterm happiness ratings.

Beginning
The other measure for Pair 1 is the Beginning happiness ratings.

Mean
The mean happiness rating at Midterm was 13.00, and the mean Beginning happiness rating was 11.50.

N
There were 14 happiness ratings at Midterm and 14 happiness ratings at the Beginning of the semester.

Std. Deviation
The standard deviation at Midterm was 4.52344, while the standard deviation at the Beginning was 4.87931.

Std Error Mean   The estimate of the standard error of the mean for the Midterm ratings was 1.20894; the estimate of the standard error of the mean for the Beginning happiness ratings was 1.30405.

This information will be used later for reporting in APA format.

The second output box, labeled "Paired Samples Correlations," contains information about the relationship between the Midterm and Beginning ratings. We will wait until a future chapter (Chapter 14) to fully explain the information that is provided here about the relationship between the Midterm and Beginning happiness ratings. We anticipate a strong relationship, insofar as we are repeatedly measuring the same sample.

The third output box contains the results of the *t*-test. The top row labels the statistics computed; below each label are the values calculated by SPSS.

Paired Samples Test

Pair 1 Midterm-Beginning   SPSS provides information about the differences between Midterm and Beginning happiness ratings.

Paired Differences

Mean   The average difference ($\overline{D}$) between the pairs of happiness ratings is 1.50.

Std. Deviation   The standard deviation of the differences between pairs of happiness ratings is 2.47293.

Std. Error Mean   The estimate of the standard error of the mean difference ($S_{\overline{D}}$) is .66092.

95% Confidence Interval of the Difference   As we have previously seen, we can determine whether there is a significant difference between the happiness ratings by computing confidence intervals, or bands, around the observed difference. If 0 falls within the interval, we do not have a significant difference; if the interval does not include 0, the difference is significant.

Lower   The lower point of the interval is .07217.

Upper   The upper point of the interval is 2.29783.

The difference is significant.

Since 0 does not fall between .07217 and 2.29783, the difference is significant.

t   The *t* value is 2.270.

df   There are 13 degrees of freedom (*df*).

Sig (2-tailed) The p at which the t is significant is .041. Since we use $p < .05$, and .041 is less than .05, the difference between happiness ratings is significant.

There are several ways to evaluate our hypotheses with SPSS. We can use the value of p in the column labeled "Sig (2-tailed)" and reject the null hypothesis if $p < .05$. Or we can use the confidence interval and reject the null hypothesis when it does not include 0. Finally, we can compare our calculated value of t with the critical value of t.

We see that the calculated value of t based upon our sample t is 2.270. The df are reported as 13. On the basis of a critical value of $t = \pm 2.160$, we reject the null hypothesis. Our calculated value of t exceeds the critical value of t; the value we obtained falls into a critical region. Thus, we found a significant change in happiness ratings between the beginning of the semester and midterm. Now we look at the mean happiness ratings. We find that the midterm happiness ratings were larger than the happiness ratings at the beginning of the semester. Therefore, we conclude that the happiness ratings increased significantly in this sample.

Because the results indicate a significant difference, we compute the effect size with Cohen's d:

$$\text{Cohen's } d = \frac{\overline{D}}{S_D}$$

$$\text{Cohen's } d = \frac{1.50}{2.47}$$

$$\text{Cohen's } d = .61$$

Thus, in our example, the effect size is a moderate .61.

Remember the interpretation guidelines of Cohen's d (Cohen, 1988): .2 is considered a small effect, .5 a moderate effect, and .8 a large effect.

Reporting the Results in APA Format

We have already learned how to report the results of a t-test for single-sample and independent-samples t-statistics. Here we review the format and consider the various statements we might make when we reject the null hypothesis for a related-samples t-statistic. For our previous example, we might write any one of the following:

In this sample, there was a significant change in happiness during the first half of the semester, $t(13) = 2.27$, $p = .041$. The average change in happiness from the beginning of the semester to the midterm was 1.50($SD = 2.47$). The effect size was moderate (Cohen's $d = .61$).

In this sample, the average change in happiness from the beginning of the semester to the midterm was 1.50($SD = 2.47$). This constitutes a statistically significant change, $t(13) = 2.270$, $p = .041$. The effect size was moderate (Cohen's $d = .61$).

Be Here Now

An industrial psychologist wishes to examine the change in employee morale after the implementation of a new training program. An employee morale inventory was administered prior to training. Three months after the training program, the inventory was administered again.

Posttraining	Pretraining
43	32
37	29
50	39
44	40
22	24
71	66
65	57
59	50
83	79
34	31

Is there a significant change in employee morale after training? Test at $\alpha = .05$.

Answer

In words, the null hypothesis is: "There is no significant change in employee morale after training." The alternative hypothesis is: "There is a significant change in employee morale after training."

Stated symbolically, the null and alternative hypotheses would read as follows:

$$H_0: \mu_D = 0$$
$$H_1: \mu_D \neq 0$$

Notice that this is a two-tailed, or nondirectional, test.

$$df = 9 \; [df = 10 - 1 = 9]$$

Given an alpha level of .05, the critical value of t is ± 2.262.

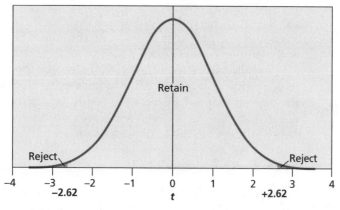

(Continued)

Be Here Now

Enter the data into SPSS, and make certain that we name two variables (Post and Pre).

The result of $t = 4.742$ exceeds the critical value of t. We note that Sig (2-tailed) = .001, which is less than our p value of .05. Thus, we conclude that the change in employee morale between pretraining ($M = 44.70$, $SD = 17.91$) and posttraining ($M = 50.8$, $SD = 18.63$) is significant, $t(9) = 4.74$, $p = .001$.

Because the test result is significant, we will calculate Cohen's d. The mean difference in employee morale was 6.10($SD = 4.07$). Calculating Cohen's d yields

$$\text{Cohen's } d = \frac{\overline{D}}{S_D} = \frac{6.10}{4.07} = 1.50$$

Thus, the effect size is large.

Directional Tests Using the Related-Samples t-Test

Sometimes there is sufficient information to predict an increase or a decrease in a measured variable and thus use a directional (or one-tailed) test. Let's illustrate with an example.

A cognitive psychologist is interested in examining potential **deficits** in cognitive abilities after cardiac bypass surgery. A short form of a cognitive skills test is administered to patients when they report to the hospital for presurgery blood tests and chest X-rays. The same cognitive skills test is administered when patients visit their cardiologists for a postsurgery follow-up examination. The measure is scored so that the higher the score, the higher is the cognitive functioning. The data are on the next page.

STEP 1 State the null and alternative hypotheses.

$$H_0: \mu_D \geq 0$$
$$H_1: \mu_D < 0$$

In words, the null hypothesis states that there is no significant decrease in cognitive capabilities (there is no difference between the two test scores or the difference is positive).

The alternative hypothesis stipulates that there is a significant decrease in cognitive capabilities after bypass surgery. Notice that we organized the data so that any decrease in cognitive capability will be reflected in a negative difference between test scores. For this reason, we put the variable "After Surgery" in the leftmost column.

STEP 2 Choose the alpha level and draw the distribution.

For this example, we will use an alpha level of .01. The sample is based on 20 patients, so $df = 20 - 1 = 19$. We find the critical value of t from the table of

Cognitive Skills Scores

Patient	After Surgery	Before Surgery
1	47	53
2	39	40
3	43	44
4	51	49
5	33	38
6	52	52
7	41	47
8	65	64
9	46	49
10	53	57
11	39	38
12	44	39
13	58	56
14	35	33
15	36	37
16	49	51
17	63	66
18	28	28
19	47	48
20	52	54

t values. Notice that we are predicting a specific direction for the outcome, a **decrease** in cognitive abilities over time, so this is a one-tailed (or directional) test. We look under the column labeled .01 for one tail, look across the degrees of freedom row labeled 19, and find that the critical value of *t* is 2.539 in the table.

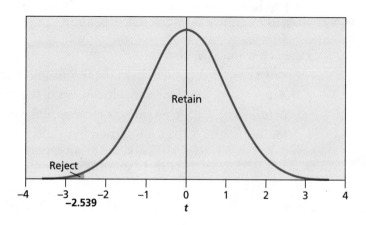

Next, we draw the hypothetical distribution of values of t when the null hypothesis is true. In a perfect world, when the null hypothesis is true, we expect no change in cognitive test scores or there will be an improvement in test scores postsurgery. In that case, we would expect a value of t equal to or greater than 0. With a decrease in cognitive test scores after surgery, we expect that t will be less than 0.

Because this is a one-tailed test, we have one critical region in the lower tail of the distribution, bounded by our critical value of $t(t = -2.539)$. To reject the null hypothesis, our calculated value of t must be a negative number larger than -2.539.

STEP 3 Calculate the appropriate test statistic.

Remember that each row in the data entry page in SPSS represents a participant and each column represents a variable. In our example, we must indicate which scores were obtained before surgery and which were obtained after surgery.

Click on the tab for "Variable View," and define one variable, "After," for cognitive test scores obtained after bypass surgery and another variable, "Before," to identify the cognitive test scores obtained prior to surgery. By organizing our data in this way, we will obtain a negative difference when there is a decrease in test scores.

When we now click on the tab labeled "Data View," we can enter each pair of scores in the SPSS spreadsheet. Once we have entered all of the data, our spreadsheet consists of 20 pairs of scores.

We are now ready for SPSS to calculate t.

To conduct a related-samples *t*-test, choose

Analyze → Compare Means → Paired-Samples T-Test.

- We will highlight "After" and "Before" in the left column to identify the pair of variables for the related-samples t-test. Then we will move the pair to the "Paired Variable" box on the right.

Figure 3 presents the SPSS output for the results of the related-samples t-test for cognitive skills test scores before and after surgery.

Paired Samples Statistics

Pair 1	Consists of the Before and After surgery scores.
After	One variable consists of the After surgery scores.
Before	One variable consists of the Before surgery scores.
Mean	The mean cognitive skills score After surgery was 46.05, and the mean cognitive skills score Before surgery was 47.15.

T-Test

Paired Samples Statistics

		Mean	N	Std. Deviation	Std. Error Mean
Pair 1	After	46.0500	20	9.73585	2.17700
	Before	47.1500	20	10.01722	2.23992

Paired Samples Correlations

		N	Correlation	Sig.
Pair 1	After & Before	20	.958	.000

Paired Samples Test

		Paired Differences					t	df	Sig. (2-tailed)
		Mean	Std. Deviation	Std. Error Mean	95% Confidence Interval of the Difference				
					Lower	Upper			
Pair 1	After-Before	−1.10000	2.88189	.64441	−2.44876	.24876	−1.707	19	.104

FIGURE 3 SPSS results for the related-samples test of changes in cognitive skills test scores after surgery. The first output box, labeled Paired Samples Statistics, contains information about each time of measurement.

N	Twenty cognitive skills scores were obtained After surgery, and 20 cognitive skills scores were obtained Before surgery.
Std. Deviation	The standard deviation After surgery was 9.73585, while the standard deviation Before surgery was 10.01722.
Std Error Mean	The estimate of the standard error of the mean for the After surgery scores was 2.177; the estimate of the standard error of the mean for the After surgery scores was 2.23992.

The second output box, labeled "Paired Samples Correlations," contains information about the relationship between the cognitive skills scores obtained After and Before surgery. We would expect that these values would be strongly related, because each pair of scores comes from the same individual. We discuss correlation in detail in Chapter 14.

The third output box contains the results of the t-test. The top row labels the statistics computed; below each label are the values calculated by SPSS.

Paired Samples Test

Pair 1 After–Before	SPSS provides information about the differences between scores After and Before surgery.
Paired Differences	
Mean	The average difference ($\overline{D}$) between the pairs of cognitive skills scores is −1.10.
Std. Deviation	The standard deviation of the differences between pairs of cognitive skills scores is 2.88189.
Std. Error Mean	The estimate of the standard error of the mean difference is .64441.
95% Confidence Interval of the Difference	With nondirectional tests, we can determine whether there is a significant difference between the cognitive skills scores by computing confidence intervals around the observed difference. If 0 falls within the interval, we do not have a significant difference; if the interval does not include 0, the difference is significant. Remember, though, that here we are conducting a directional test.
Lower	The lower point of the interval is −2.44876.
Upper	The upper point of the interval is .24876.
t	The t value is −1.707.
df	There are 19 degrees of freedom (df).
Sig (2-tailed)	The value of p at which t is significant is .104. SPSS calculates the probability of making a Type I error that is based on a two-tailed test. Because we are conducting a one-tailed test, this value should be divided by 2 if we wish to calculate the actual value of p in a directional test. In this case, $p = .052$, and this value is greater than our alpha level of .01.

As we have previously demonstrated, there are at least two ways to evaluate our hypotheses with SPSS: We can modify the value of *p* in the column labeled "Sig (2-tailed)" to reflect our probability in a one-tailed test and reject the null hypothesis if $p < .01$. Or we can compare our calculated value of *t* with the critical value of *t*.

The calculated value of *t* based upon our sample is $t = -1.707$. There are 19 *df*. Based upon a critical value of $t = -2.539$, we fail to reject the null hypothesis, because our calculated value of *t* fails to exceed the critical value of *t*. That is, -1.707 is not as large a negative number as -2.539. The value of *t* that we obtained does not fall into a critical region. Thus, we found no significant decrease in cognitive skills test scores after cardiac bypass surgery.

SPSS computes the confidence bands based on a two-tailed test. Because we have conducted a one-tailed test, we will not be able to use the confidence bands provided.

> **Reflect**
> All things change. Yet, do all things change significantly?

Using APA format, we might phrase our conclusions in the following way:

The average change observed in this sample on cognitive skills test scores before and after cardiac bypass surgery was $-1.10(SD = 2.88)$. This difference was not significant, $t(19) = -1.707$, $p = .052$.

It's Out There...

Larson (2007) examined the pre- and post–adventure camp self-esteem scores of 31 adolescents with behavioral problems. The mean difference between the pre and post test of $5.58(SD = 33.69)$ was significant, $t(30) = 2.66$, $p < .01$.

Matched Samples

Sometimes we wish to compare matched samples. For example, suppose we wish to compare achievement motivation in fraternal twins when one of the twins participates in organized sports. Clearly, the samples of twins are not independent; rather, they are related. In this case, the related-samples *t*-test is the most appropriate choice of statistical test to use to determine whether the twins differ in achievement motivation. We call this a **matched-samples *t*-test**.

Let's explore the use of the matched-samples *t*-test with hypothetical twins' data. Suppose that we assess 20 sets of fraternal twins, using an achievement motivation test in which higher scores indicate higher achievement motivation. The paired, or matched, data we obtain follow:

We use the same basic hypothesis-testing strategy as in the related-samples *t*-test.

Be Here Now

A local high school prohibits cell phones on school premises. A group of parents petition the school board, arguing that cell phones will enhance safety and assuring the board that students will not use their cell phones for social purposes during school hours. A pilot test is conducted with seniors; they are allowed to bring their cell phones if they participate in a study to determine whether cell phone use increases during school hours. Information about cell phone usage (monthly minutes) is obtained the month prior to the implementation of the pilot test and a month after the test is in place. The following data are collected from a sample of participating seniors:

	After	Before
A	232	199
B	129	108
C	217	256
D	311	301
E	502	499
F	108	111
G	293	313
H	491	489
I	106	101
J	111	99
K	433	496
L	329	324
M	537	499
N	242	268
O	96	109

Will there be a significant increase in the time students spend on the cell phone after the change in school policy? Test at $\alpha = .05$.

Answer

In words, the null hypothesis is: "There is no significant increase in cell phone usage during school hours." The alternative hypothesis is: "There is a significant increase in cell phone usage during school hours."

Stated symbolically, the null and alternative hypotheses would read as follows:

$$H_0: \mu_D \leq 0$$
$$H_1: \mu_D > 0$$

Be Here Now

Notice that this is a one-tailed, or directional, test.

$$df = 14 \quad df = 15 - 1 = 14$$

Given an alpha level of .05, the critical value of *t* is +1.761.

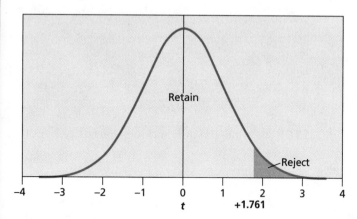

Enter the data into SPSS, and make certain to name the two variables (After and Before).

The *t*-test result of −.339 fails to exceed the critical value of *t*. Thus, we conclude that there is no significant increase in cell phone usage after a change in school policy, $t(14) = -.339$, *ns*.

Because the test result is not significant, we do not calculate Cohen's *d*.

Achievement Motivation Scores

Twin Set	Participant	Nonparticipant
1	91	76
2	13	17
3	79	88
4	26	13
5	80	55
6	71	69
7	24	10
8	35	41
9	38	48
10	13	20
11	31	15
12	48	8
13	68	60

(Continued)

Achievement Motivation Scores

Twin Set	Participant	Nonparticipant
14	54	17
15	25	33
16	84	69
17	11	22
18	76	71
19	91	78
20	63	47

STEP 1   State the null and the alternative hypotheses.

$$H_0: \mu_D = 0$$
$$H_1: \mu_D \neq 0$$

In words, the null hypothesis states that there is no difference in the achievement motivation of fraternal twins when one of the twins participates in organized sports. The alternative hypothesis suggests that there is a significant difference in achievement motivation. Notice that this is a nondirectional test, because we are not specifying a direction of difference in achievement motivation.

STEP 2   Choose the alpha level and draw the distribution.

For this example, we will use an alpha level of .05. In the matched-samples t-test, the calculation of the number of degrees of freedom is the same as in the related-samples t-test: $df = n - 1 (df = 20 - 1 = 19)$. Next, we identify the critical value of t in the table of t values (see Appendix A2). We look under the column labeled .05 for two tails, look across the degrees of freedom row labeled 19, and find the critical value of t is 2.093.

We draw the hypothetical distribution of values of t when the null hypothesis is true.

If there is no sampling error, and the null hypothesis is true, we would expect no difference between the twins' achievement motivation scores.

Because this is a two-tailed test, we have two critical regions, one in each tail. If our value for t falls into either tail region, we will reject the null hypothesis and conclude that there is a significant difference in achievement motivation (positive or negative). To reject the null hypothesis, our calculated value of t must be greater than $+2.093$ or less than -2.093.

STEP 3   Calculate the appropriate test statistic.

Remember, when entering the data for the related-samples t-test, we must keep the pairs of twins together. In using SPSS, it is important to recall that

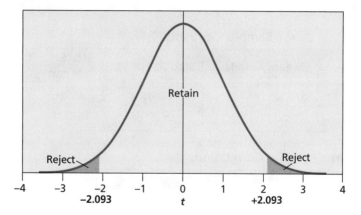

each row in the data entry page (the spreadsheet) represents a participant and each column represents a variable. In our example, each row represents a pair of twins.

Click on the tab for "Variable View," and define one variable for the participating twin and another variable for the nonparticipating twin. In this example, we will use "Participant" and "Nonparticipant" as our variable names.

When we now click on the tab labeled "Data View," we can enter the achievement motivation scores under the appropriate column headings in the SPSS spreadsheet. Again, we must remember to keep the sets of twins intact.

Once we have entered all of the data, our spreadsheet consists of 20 pairs of achievement motivation scores. Each row presents the information on one set of twins. One column presents the scores for the twin who participates in sports (Participant), and the other column presents the scores for the twin who does not participate (Nonparticipant).

We are now ready to calculate our matched-samples *t*-statistic with SPSS.

To conduct a matched-samples *t*-test, choose

Analyze →Compare Means → Paired-Samples T-Test.

- We must identify the pair of variables under study. Click on "Participant" and "Nonparticipant," the two variables that we want to compare. Notice that the pair appears in the section labeled "Current Selection." Use the arrow to move the pair to the "Paired Variables" box so that SPSS calculates the difference scores and the matched-samples *t*.

Figure 4 presents the SPSS output for the results of the matched-samples *t*-test.

T-Test

Paired Samples Statistics

		Mean	N	Std. Deviation	Std. Error Mean
Pair 1	Participant	51.0500	20	27.84634	6.22663
	Nonparticipant	42.8500	20	26.52958	5.93219

Paired Samples Correlations

		N	Correlation	Sig.
Pair 1	Participant & Nonparticipant	20	.848	.000

Paired Samples Test

		Paired Differences					t	df	Sig. (2-tailed)
		Mean	Std. Deviation	Std. Error Mean	95% Confidence Interval of the Difference				
					Lower	Upper			
Pair 1	Participant - Nonparticipant	8.20000	15.05289	3.36593	1.15503	15.24497	2.436	19	.025

FIGURE 4 SPSS output for paired-samples t-test for differences in achievement motivation of fraternal twins.

In the first box of output, SPSS presents the descriptive statistics for each pair of variables. The top row labels the statistics computed; below each label are the values for the descriptive statistics for achievement motivation.

Paired Samples Statistics

Pair 1   Consists of the achievement motivation scores for Participants and Nonparticipants.

Participant   Consists of the achievement motivation scores for the twin who participates in organized sports.

Nonparticipant   Consists of the achievement motivation scores for the twin who does not participate in organized sports.

Mean   The mean achievement motivation score for twins categorized as Participant was 51.05, and the mean achievement motivation score for twins categorized as Nonparticipant was 42.85.

N	There were 20 achievement motivation scores for twins who participate in sports and 20 achievement motivation scores for twins who do not participate.
Std. Deviation	The standard deviation for Participants was 27.84634 and the standard deviation for Nonparticipants was 26.52958.
Std Error Mean	The estimate of the standard error of the mean for Participants was 6.22663; the estimate of the standard error of the mean for Nonparticipants was 5.93219.

The second output box, labeled "Paired Samples Correlations," contains information about the relationship between Participant and Nonparticipant scores. We anticipate a strong relationship, because we are analyzing data from fraternal twins. Correlation will be discussed in Chapter 14.

The third output box contains the results of the *t*-test. The top row labels the statistics computed; below each label are the values calculated by SPSS.

Paired Samples Test

Pair 1	Participant–Nonparticipant
	SPSS provides information about the differences between the achievement motivation in twins who participate in organized sports and in those who do not.

Paired Differences

Mean	The average difference ($\overline{D}$) between the achievement motivation scores of the twins is 8.20.
Std. Deviation	The standard deviation of the differences between the achievement motivation scores of the twins is 15.05289.
Std. Error Mean	The estimate of the standard error of the mean difference is 3.36593
95% Confidence Interval of the Difference	We can determine whether there is a significant difference between the achievement motivation scores by computing confidence intervals around the observed difference. If 0 falls within the interval, we do not have a significant difference; if the interval does not include 0, the difference is significant.
Lower	The lower point of the interval is 1.15503

Upper	The upper point of the interval is 15.24497.
	Since 0 does not fall between 1.15503 and 15.24497, the difference is significant.
t	The *t* value is 2.436.
df	There are 19 degrees of freedom (*df*).
Sig (2-tailed)	The value of *p* at which *t* is significant is .025. We used *p* < .05, and because .025 is less than .05, the difference between the achievement motivation scores of twins who participate in sports and who do not participate in sports is significant.

There are several ways to evaluate our hypotheses with SPSS. We can use the value of *p* in the column labeled "Sig (2-tailed)" and reject the null hypothesis if *p* < .05. Alternatively, we can use the confidence interval and reject the null hypothesis when it does not include 0. Finally, we can compare our calculated value of *t* with the critical value of *t*.

We see that the calculated value of *t* based upon our sample *t* is 2.436 and the *df* are reported as 19. On the basis of a critical value of *t* = ± 2.093, we reject the null hypothesis. Our calculated value of *t* exceeds the critical value of *t*; the value we obtained falls into a critical region. Thus, we found a significant difference in achievement motivation between twins who participate in organized sports and those who do not. In examining the mean achievement motivation scores, we learned that Participants were higher in achievement motivation than Nonparticipants.

Because the results indicate a significant difference, we compute the effect size with Cohen's *d*:

$$\text{Cohen's } d = \frac{\overline{D}}{S_D}$$

$$\text{Cohen's } d = \frac{8.20}{15.05}$$

$$\text{Cohen's } d = .54$$

Thus, in our example, the effect size is .54, which is moderate.

In presenting the results in APA format, we might write the following:

For this sample, there was a significant difference between the achievement motivation of twins who participated in organized sports and twins who did not participate, $t(19) = 2.44$, $p = .025$. The mean of fraternal twins who participated in organized sports was $M = 51.05$, $(SD = 27.85)$ and that of those who did not participate in sports was $M = 42.85(SD = 26.53)$. The effect size was moderate (Cohen's $d = .54$).

For this sample of fraternal twins, those who participated in organized sports ($M = 51.05$, $SD = 27.85$) showed a significantly higher achievement motivation than those who did not participate in sports ($M = 42.85$, $SD = 26.53$), $t(19) = 2.44$, $p = .025$. The effect size was moderate (Cohen's $d = .54$).

It's Out There... Raina, Rogers, and Holm (2007) compared the effect of the environment on activity performance in matched samples of women over the age of 70 who had had heart failure. Paired-samples *t*-tests indicated several differences between women performing activities in a clinic and their matched participants performing activities at home. For example, women in the clinic had significantly higher scores for using the stairs ($M = 2.92, SD = 0.43$) than did women at home ($M = 2.43, SD = 1.18$), $t(54) = 3.05, p = .004$. However, women at home ($M = 2.99, SD = 0.04$) had significantly higher scores for telephone use than did women in the clinic ($M = 2.88, SD = 0.15$), $t(54) = -5.26, p < .001$.

SUMMARY

When we repeatedly measure the same participants, we are employing a **within-groups**, or **repeated measures**, design. This approach differs from a **between-groups** design, in which we compare independent groups of participants. Each design has advantages and disadvantages. We use the **related-samples *t*-test** to examine differences when we measure the same sample twice. Just as with the single-sample *t*-test and the independent-samples *t*-test, we can use Cohen's *d* to measure the effect size. In one variation of the related-samples *t*-test, called the **matched-samples *t*-test**, we use *pairs* of measurements from samples that have been *matched* on a variable.

PRACTICE

The Related-Samples t-Statistic

1. In words, what is the formula for the related-samples *t*-statistic?
2. What is the definitional formula of the related-samples *t*-statistic?

Within-Groups and Between-Groups Designs

3. For each of the following, identify the design as either a within-groups or between-groups design:
    a. Comparing groups of patients receiving different forms of psychotherapy.
    b. Comparing blood test results before and after chemotherapy.
    c. Comparing television viewing time during the school year and during summer vacation.
    d. Comparing television viewing time for children who attend an inner city school with children who attend a suburban school.
4. Individual differences are removed from what portion of the related-samples *t*-statistic?

Stating Hypotheses

5. A researcher is interested in determining whether attending an after-school program changes students' test scores. Students are tested before they begin the program and again after four weeks of participation. What are the null and alternative hypotheses?

6. A researcher wants to test whether participants in a weight reduction program lose weight. Participants are weighed prior to the program and again after 10 weeks of participation. What are the null and alternative hypotheses?

df and Critical Values

7. The researcher in Problem 5 had 15 students in the after-school program. What are the *df*? What is the critical value at alpha=.05?

8. The researcher in Problem 6 used the pre- and postparticipation weights of 11 participants. What are the *df*? What is the critical value at alpha=.01?

For each of the items that follow, be sure to

Nondirectional Tests Using the Related-Samples t-test

State the null and alternative hypotheses.
Chose the alpha and critical value.
Compute the statistic.
Evaluate the hypothesis on the basis of the test results.
When appropriate, compute Cohen's *d*.

9. A psychologist is interested in the benefits of group grief counseling. An invitation to participate in group grief counseling is circulated in local churches and hospice centers. At the first group meeting, the psychologist administers a bereavement inventory that provides an index of the negative components of the grief process (e.g., anger, guilt, denial, etc.), in which higher scores reflect greater difficulties in coping with grief. The following list gives the scores of the participants at the first group meeting and after eight weeks of grief counseling.

Bereavement Inventory Scores

Week 1	Week 8
24	10
17	12
31	30
4	4
15	9
25	22
26	30
19	7
20	17
16	15

Does group grief counseling have an effect on the bereavement process? Use a nondirectional test and alpha=.05.

10. A cognitive psychologist is interested in the effect of a general collaborative experience on test-taking ability. A group of students is given 10 minutes of a collaborative exercise in which students are asked to solve anagrams (anagrams are words in which the letters are presented in a jumbled order; for example, "bif" is an anagram for the word, "fib"). The time it takes each student to complete a brief quiz in class is measured both before and after the unrelated collaborative exercise. The data obtained are as follows:

Quiz Time (in minutes)

Before Collaboration	After Collaboration
6	10
10	13
14	14
11	10
9	12
13	15
16	19
24	23
18	21

Does an unrelated collaborative exercise have an impact on quiz time? Use a nondirectional test and alpha=.05.

Directional Tests Using the Related-Samples t-test

11. In investigating the enhancement of self-esteem with a commitment to a regimen of physical exercise, a sports psychologist administered a self-esteem survey to individuals touring a new health and wellness center. Higher survey scores indicate greater self-esteem. If the individual chose to become a member of the center, another self-esteem survey was administered. The sports psychologist reported the survey results as follows:

Self-Esteem Scores

Before Membership	After Membership
37	33
45	48
20	21
48	47
17	12
13	13

(Continued)

Self-Esteem Scores

Before Membership	After Membership
6	8
22	18
34	37
44	40
29	30
13	10
38	36
24	32
10	14
6	9
13	19
17	18

Does making a commitment to a regimen of physical exercise by joining a health and wellness center enhance self-esteem? Use a directional test and alpha=.05.

Matched-Samples t-test

12. A marriage and family therapist examined whether there is a difference in the amount of time husbands spend socializing with friends and acquaintances, compared with the amount of time wives do so. Over a seven-day period, the number of replies to personal emails were counted and used as an index of social interaction. The therapist reported the following data:

Number of Emails

Wife	Husband
20	71
51	32
103	20
31	19
13	2
4	39
44	18
12	34
73	22
29	21
65	24
19	5
77	16
11	1

Is there a significant difference in the electronic socializing of husbands and wives? Use a nondirectional test and alpha=.05.

APA Format

13. Write your answer to Problem 10 in APA format.

SOLUTIONS

The Related-Samples t-statistic

1. It is the ratio of the mean difference to the standard error of the mean difference.

2. The definitional formula for the related-samples *t*-test can be stated as

$$t = \frac{\text{Mean Difference (or Mean Change)}}{\text{Standard Error of the Mean Difference}} = \frac{\overline{D}}{S_{\overline{D}}}$$

where

$$S_{\overline{D}} = \frac{SD}{\sqrt{n}}$$

Within-Groups and Between-Groups Designs

3. a. Between groups
   b. Within groups/Repeated measures
   c. Within groups/Repeated measures
   d. Between groups

4. The denominator

Stating Hypotheses

5. $H_0: \mu_D = 0$ No difference in achievement exists (Pre–Post).
   $H_1: \mu_D \neq 0$ A difference is achievement exists (Pre–Post).

6. $H_0: \mu_D \leq 0$ Weight increases or remains the same over time (Pre–Post)
   $H_1: \mu_D > 0$ Weight decreases over time (Pre–Post)

df and Critical Values

7. $df = n - 1 = 15 - 1 = 14$
   $t(14)$ critical value, at $p < .05 = 2.145$ (two-tailed)

8. $df = n - 1 = 11 - 1 = 10$
   $t(10)$ critical value, at $p < .01 = 2.764$ (one-tailed)

Nondirectional Tests Using the Related-Samples t-test

9. $H_0: \mu_D = 0$ Group grief counseling has no effect on the bereavement process.

   $H_1: \mu_D \neq 0$ Group grief counseling has an effect on the bereavement process.

T-Test

Paired Samples Statistics

		Mean	N	Std. Deviation	Std. Error Mean
Pair 1	week1	19.7000	10	7.48406	2.36667
	week8	15.6000	10	9.15545	2.89521

Paired Samples Correlations

		N	Correlation	Sig.
Pair 1	week1 & week8	10	.802	.005

Paired Samples Test

	Paired Differences					t	df	Sig. (2-tailed)
	Mean	Std. Deviation	Std. Error Mean	95% Confidence Interval of the Difference				
				Lower	Upper			
Pair 1 week1-week8	4.10000	5.46606	1.72852	.18982	8.01018	2.372	9	.042

With $df = n - 1 = 10 - 1 = 9$, the critical t value at $\alpha = .05$ is 2.262.

Using SPSS, and naming one variable Week 1 and the other Week 8, we enter the bereavement scores. The SPSS output is shown above:

On the basis of these findings, we conclude that, for this sample, eight weeks of group grief counseling had a significant effect on the bereavement process, $t(9) = 2.37$, $p < .05$.

With a significant finding, we need to calculate the effect size:

$$\text{Cohen's } d = \frac{\overline{D}}{S_D} = \frac{4.10}{5.47} = .75$$

This is a large effect size.

10. H_0: $\mu_D = 0$ A collaborative experience has no effect on students' performance on the quiz.

   H_1: $\mu_D \neq 0$ A collaborative experience has a significant effect on students' performance on the quiz.

With $df = n - 1 = 9 - 1 = 8$, the critical value for $\alpha = .05$ is 2.306.

Using SPSS, and naming one variable After and the other Before, we enter the quiz times. The SPSS output looks like this:

T-Test

Paired Samples Statistics

		Mean	N	Std. Deviation	Std. Error Mean
Pair 1	before	13.4444	9	5.38774	1.79591
	after	15.2222	9	4.73756	1.57919

Paired Samples Correlations

		N	Correlation	Sig.
Pair 1	before & after	9	.936	.000

Paired Samples Test

		Paired Differences					t	df	Sig. (2-tailed)
		Mean	Std. Deviation	Std. Error Mean	95% Confidence Interval of the Difference				
					Lower	Upper			
Pair 1	before-after	−1.77778	1.92209	.64070	−3.25523	−.30033	−2.775	8	.024

These results indicate that there was a significant change in test-taking ability as a consequence of the intervening collaborative experience; $t(8) = -2.78$, $p < .05$.

With a significant finding, we need to calculate the effect size:

$$\text{Cohen's } d = \frac{\overline{D}}{S_D} = \frac{-1.78}{1.92} = -.93$$

This is a large effect size.

Directional Tests Using the Related-Samples t-test

11. $H_0: \mu_D \leq 0$ There is no significant increase in self-esteem after making a commitment to a regimen of physical exercise.

    $H_1: \mu_D > 0$ There is a significant increase in self-esteem after making a commitment to a regimen of physical exercise.

    With $df = n - 1 = 18 - 1 = 17$, the critical value for $\alpha = .05$, one tail, is 1.740.

    Using SPSS, and naming one variable After and the other Before, we enter the self-esteem scores. The SPSS output looks like this:

T-Test

Paired Samples Correlations

		N	Correlation	Sig.
Pair 1	after & before	18	.963	.000

Paired Samples Statistics

		Mean	N	Std. Deviation	Std. Error Mean
Pair 1	after	24.2222	18	13.78642	3.24949
	before	24.7222	18	13.27007	3.12779

Paired Samples Test

		Paired Differences					t	df	Sig. (2-tailed)
		Mean	Std. Deviation	Std. Error Mean	95% Confidence Interval of the Difference				
					Lower	Upper			
Pair 1	after-before	−.50000	3.69817	.87167	−2.33906	1.33906	−.574	17	.574

The average self-esteem score before making a commitment to a physical exercise regimen was 24.72(SD = 13.27); the average score after making the commitment by joining a health and wellness center was 24.22(SD = 13.79). The increase in self-esteem was not significant, $t(17)$ = 0.57, *ns*.

Matched-samples t-test

12. $H_0:\mu_D = 0$ There is no difference in the sociability of husbands and wives.

$H_1:\mu_D \neq 0$ There is a significant difference in the sociability of husbands and wives.

With $df = n - 1 = 14 - 1 = 13$, the critical value for α = .05 is 2.160.

Using SPSS, and naming one variable Wife and the other Husband, we enter the cumulative number of email replies. The SPSS output looks like this:

T-Test

Paired Samples Statistics

		Mean	N	Std. Deviation	Std. Error Mean
Pair 1	Wife	39.4286	14	30.22662	8.07840
	Husband	23.1429	14	17.79307	4.75540

Paired Samples Correlations

		N	Correlation	Sig.
Pair 1	Wife & Husband	14	−.083	.778

Paired Samples Test

		Paired Differences					t	df	Sig. (2-tailed)
		Mean	Std. Deviation	Std. Error Mean	95% Confidence Interval of the Difference				
					Lower	Upper			
Pair 1	Wife-Husband	16.28571	36.32212	9.70750	−4.68606	37.25748	1.678	13	.117

On the basis of this sample of married couples, no difference in sociability was found between husbands and wives, $t(13) = 1.68$, *ns*.

13. In this sample, there was a significant change in test-taking ability as a consequence of the intervening collaborative experience; $t(8) = -2.78$, $p < .05$. The effect size was large (Cohen's $d = -.93$).

-13-

Repeated-Measures Analysis of Variance (ANOVA)

F O C U S | The repeated-measures analysis of variance (ANOVA) expands the design of the related-samples *t*-test. In the related-samples *t*-test, we have two measurements for each participant, and we are interested in whether there is a difference from the first measurement to the second. In the repeated-measures ANOVA, we are not limited to two measurements per participant; we can have three, or four, or a dozen or more measurements on each participant. In this chapter, we will examine the basic concepts of the repeated-measures ANOVA, which is also called a within-groups or within-subjects ANOVA.

Repeated-Measures ANOVA

Background and Assumptions

The repeated-measures ANOVA permits us to explore data related to questions that can be quite complex. At its most basic level, a repeated-measures ANOVA comprises multiple measurements of a variable for each participant in a study. For example, we may have grades for a group of students for each of four marking periods over a school year. In this chapter, we will present the basic form of the repeated-measures ANOVA.

In a repeated-measures ANOVA, we use F to represent the test statistic. As you look at the formula defining the repeated-measures ANOVA, you may notice the similarity to the related-samples t:

$$F = \frac{\text{Mean Difference (or Mean Change)}}{\text{Standard Error of the Mean Difference (Individual differences removed)}}$$

The assumptions for the repeated-measures ANOVA are similar to that of the related-samples t-test:

- The population distribution of difference scores is normally distributed.
- The participants are a random sample of the population.
- The variance of the measures assessed are equal.

These assumptions are fairly robust (see Stevens, 1996). In other words, violations of the assumptions are not likely to have serious effects on the ANOVA statistic.

SPSS will automatically test the assumption of equal variances and will display the results in what is known as a test for sphericity. SPSS will also show multivariate results for the repeated-measures ANOVA in the output. Interpreting the multivariate results is appropriate for more advanced statistical applications, and you will be directed past those analyses in this chapter.

Stating Hypotheses

The steps for a repeated-measures ANOVA are similar to those of other hypothesis tests. We begin by stating our null and alternative hypotheses.

Let's start with an example based on the idea that a major goal for first grade is to learn how to read. Suppose that a first-grade teacher is interested in learning whether the scores for her class changed significantly over the course of the school year. Let's say that the following are the scores for each marking period for her students:

Be Here Now

For each of the following, state the appropriate statistical test to use and identify the variables for each study:

1. To determine whether a new medication reduces high blood pressure, the new medication is given to a sample of participants with high blood pressure. Blood pressure readings are taken once a week for a period of three months.

2. A survey measuring interest in astronomy is given on the first day of an astronomy class and on the last day of the semester.

Answers

1. Repeated-measures ANOVA, repeated measure = weekly blood pressure readings, dependent variable = blood pressure readings.

2. Related-samples *t*-test, repeated measure = pre- and postsurvey, dependent variable = interest in astronomy (note that we could also use a repeated-measures ANOVA).

Student	1st Marking Period	2nd Marking Period	3rd Marking Period	4th Marking Period
Charley	50	60	70	72
Hamish	55	62	67	74
Ian	49	58	69	81
Tekanu	59	65	71	72
Phil	51	61	71	77
Keryn	42	56	60	65
Juli	47	58	72	74
Nicola	40	55	60	64
Lynette	44	61	73	75
Charissa	44	56	68	70

Notice that, for each student in the class, there are four grades, one for each marking period. The question is whether the scores for the class changed significantly over the school year.

The null hypothesis would state that there was no change in reading scores over the school year. The alternative hypothesis would state that there had been a change. Our hypotheses would look like this:

$$H_0: \mu_1 = \mu_2 = \mu_3 = \mu_4 \text{ or } \mu_1 = \mu_2 = \ldots = \mu_j$$
$$H_1: \mu_i \neq \mu_j, \text{ for some } i, j$$

The null hypothesis indicates that there was no change across the four marking periods; the alternative hypothesis indicates that a significant difference exists between at least two of the marking periods.

Degrees of Freedom

The number of degrees of freedom (*df*) for the numerator of the repeated-measures ANOVA is equal to the number of measurements on each participant minus 1. We use the letter *k* to denote how many measurements we have. In our example, because we have scores for four marking periods, our degrees of freedom are as follows:

$$df_{numerator} = k - 1$$
$$df_{numerator} = 4 - 1$$
$$df_{numerator} = 3$$

The number of degrees of freedom for the within-subjects factor—the numerator—is 3.

We also need to know the number of degrees of freedom associated with the error term—the denominator. We calculate that number by taking the number of participants minus 1 and multiplying it by the number of measurements minus 1:

$$df_{denominator} = (n - 1)(k - 1)$$
$$df_{denominator} = (10 - 1)(4 - 1)$$
$$df_{denominator} = (9)(3)$$
$$df_{denominator} = 27$$

So, in our example, the number of degrees of freedom associated with the denominator is 27. Now we can determine the critical value at which to test our hypothesis.

Determining the Critical Value

We next need to choose the alpha level to use in deciding whether to reject the null hypothesis. As before, we select an alpha level that reflects the probability of making a Type I error that we are willing to accept. In this example, we will use an alpha of .05.

Now we have all the information we need to determine the critical value that the observed *F* must exceed before we can reject the null hypothesis and conclude that there is a significant difference between at least two of the means.

To find the critical value, we return to the ANOVA table in Appendix A3. We look across the top row to find the number of degrees of freedom associated with the numerator. We look down the column on the left to find the number of degrees of freedom associated with denominator. Within the table, numbers written in regular print represent critical values for an alpha of .05 and those in bold represent an alpha of .01. Given our example with $df_{numerator} = 3$ and $df_{denominator} = 27$, the critical value of *F* with an alpha of .05 is 2.96.

So, for our measurements over the school year, if we obtain an *F* statistic that exceeds 2.96, we will reject the null hypothesis and conclude that there was a significant difference between at least two of the marking periods.

Reflect

Repetition can lead to change...but does it lead to growth?

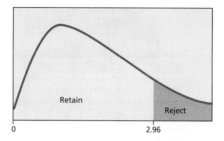

Be Here Now

A researcher wants to determine the difference between pre- and post-participation in a new mathematics program for second graders. Twelve second graders are tested before they participate in the program, after they participate in the program, and at the end of third grade. At alpha .05, what is the critical value of F?

Answer

$df_{numerator} = k - 1 = 3 - 1 = 2$

$df_{denominator} = (n - 1)(k - 1) = (12 - 1)(3 - 1) = (11)(2) = 22$

The critical value of $F(2,22)$ at $p = .05$ is 3.44.

Application: The Repeated-Measures ANOVA

Computing the Repeated-Measures ANOVA

To compute the repeated-measures ANOVA for our example, we will use SPSS.

Data entry for a repeated-measures ANOVA is similar to that for the related-samples t-test. It is important to remember that each row in the "Data View" page represents a participant and each column represents one of the measurements, in this case, reading scores. We must indicate which reading scores were given for Marking Period 1, Marking Period 2, Marking Period 3, and Marking Period 4 for each participant. Click on the tab for "Variable View," and define one variable for each of the marking periods. We will use "mp1" for Marking Period 1, and so forth.

Now click on the tab labeled "Data View," and enter each score under its appropriate column, being careful to keep each child's scores together and in order across each row. That is, the data for each row must represent the scores for the same child.

Once data entry is complete, there should be four columns with 10 scores in each column. Figure 1 shows how the data should look.

We are now ready for SPSS to calculate F for the repeated-measures ANOVA.

➡ **To compute the repeated-measures ANOVA, select**

Analyze → General Linear Model → Repeated Measures.

	mp1	mp2	mp3	mp4
1	50.00	60.00	70.00	72.00
2	55.00	62.00	67.00	74.00
3	49.00	58.00	69.00	81.00
4	59.00	65.00	71.00	72.00
5	51.00	61.00	71.00	77.00
6	42.00	56.00	60.00	65.00
7	47.00	58.00	72.00	74.00
8	40.00	55.00	60.00	64.00
9	44.00	61.00	73.00	75.00
10	44.00	56.00	68.00	70.00

FIGURE 1 SPSS "Data View" sheet.

- Where there is a box for *Within-Subjects Factor Name:*, the word "factor" appears. You may replace the word "factor" with the word "reading," which is what we are measuring in this example.
- *Number of Levels*: asks how many times this variable was measured. For our example, we will enter the number "4."
- Next, click *Add*, and you will see "reading (4)" appear in the large box.
- Because we are focusing on a simple repeated-measures ANOVA in our example, you can skip the rest of the boxes on the screen and click "Define."
- You will be shown a new screen with the four marking-period variables in a box on the left, and a box on the right that contains something that looks like this:

$$\_\_?\_\_ (1)$$
$$\_\_?\_\_ (2)$$
$$\_\_?\_\_ (3)$$
$$\_\_?\_\_ (4)$$

- If we had had three measurements, there would have been 3 of these lines; if we had had a dozen measurements, there would have been 12. Highlight the four variables in the box on the left side, and use the arrow to move the variables to the box on the right side.
- Before clicking on "OK," it's a good idea to click on "Options" and select "Descriptive Statistics" and "Estimates of Effect Size." Click "Continue" to return to the previous screen, and then click "OK."

The output will now come into view. There are several sections to the output. Some sections will be important to understand, while others will be skipped in this basic introduction to the repeated-measures ANOVA.

Interpretation: Descriptive Statistics

There are several parts to the SPSS output for a repeated-measures ANOVA. The first box, shown in Figure 2, lists the within-subjects factors and their variable names. These correspond to the repeated measurements of the participants. Next, as shown in Figure 3, descriptive statistics are provided for each of the measurements.

Within-Subjects Factors

Measure: MEASURE 1

Reading	Dependent Variable
1	mp1
2	mp2
3	mp3
4	mp4

FIGURE 2 SPSS output showing within-subjects factor (repeated measurements).

Descriptive Statistics

	Mean	Std. Deviation	N
mp1	48.1000	5.93390	10
mp2	59.2000	3.15524	10
mp3	68.1000	4.62961	10
mp4	72.4000	5.14674	10

FIGURE 3 SPSS output showing descriptive statistics.

Descriptive statistics include the following:

First Column	The first column lists the variable names for each of the repeated measurements.
Mean	The next column shows the mean for each of the variables (Marking Period 1 = 48.1000, Marking Period 2 = 59.2000, Marking Period 3 = 68.1000, and Marking Period 4 = 72.4000).

Std. Deviation	The standard deviation of each set of measurements is reported (Marking Period 1 = 5.93390, Marking Period 2 = 3.15524, Marking Period 3 = 4.62961, and Marking Period 4 = 5.14674).
N	The number of participants contributing data for each measurement is reported. Here, the number is the same for all measurements, namely, 10.

The next output box present results for "Multivariate Tests." This output box will not be described here; it is beyond the scope of this book.

Printed next is "Mauchly's Test of Sphericity" (see Figure 4). The sphericity test is the repeated-measures equivalent of the test of equal variances described in Chapter 10. The test is also called Mauchly's W. If Machly's W is not significant, we can assume that we did not violate the assumption. If we have violated the assumption, the analysis gets more complicated and is beyond the scope of this text. We will assume that we have met the assumptions of the analysis. For more information on the sphericity assumption, see Leik, 1997.

Mauchly's Test of Sphericity[b]

Measure: MEASURE 1

Within Subjects Effect	Mauchly's W	Approx. Chi-Square	df	Sig.	Epsilon[a]		
					Greenhouse-Giesser	Huynh-Feldt	Lower-bound
factor1	.439	6.351	5	.277	.667	.858	.333

Tests the null hypothesis that the error covariance matrix of the orthonormalized transformed dependent variables is proportional to an identity matrix.

[a] May be used to adjust the degrees of freedom for the averaged tests of significance. Corrected tests are displayed in the Tests of Within-Subjects Effects table.

[b] Design: Intercept
Within Subjects Design: factor1

FIGURE 4 SPSS output of Mauchly's *W*.

The Mauchly *W* output includes a number of statistics that are beyond the scope of the text and will not be discussed here.

Within Subjects Effect

Factor 1	The test is being conducted for Factor 1, our marking periods.
Mauchly's W	The value of Mauchly's *W* is .439.
Sig.	The alpha at which the data are significant is .277. Because we are using an alpha of .05,

we have no evidence to conclude that we have violated the sphericity assumption.

Interpretation: Repeated-Measures ANOVA

The output box labeled "Tests of Within-Subjects Effects"(see Figure 5) shows the test statistics for the repeated-measures ANOVA. Several statistics are presented. We will focus on the first row of each section, "reading Sphericity Assumed." As noted, we have no evidence to suggest that we violated the sphericity assumption (see Figure 4).

The SPSS output shows the following:

Tests of Within-Subjects Effects

Measure: MEASURE 1

Source		Type III Sum of Squares	df	Mean Square	F	Sig.	Partial Eta Squared
reading	Sphericity Assumed	3464.100	3	1154.700	118.814	.000	.930
	Greenhouse-Geisser	3464.100	2.001	1731.380	118.814	.000	.930
	Huynh-Feldt	3464.100	2.573	1346.422	118.814	.000	.930
	Lower-bound	3464.100	1.000	3464.100	118.814	.000	.930
Error (reading)	Sphericity Assumed	262.400	27	9.719			
	Greenhouse-Geisser	262.400	18.007	14.572			
	Huynh-Feldt	262.400	23.155	11.332			
	Lower-bound	262.400	9.000	29.156			

FIGURE 5 SPSS output showing results for test statistics for within-subjects effects.

Source	The first column is labeled "Source," for the source of variance. There are two sources in this example: "reading" and "Error (reading)."
Second Column	This column does not have a heading; it shows the various test statistics computed for the

repeated-measures ANOVA. Because we are assuming that we have not violated any assumptions, we will be using the row labelled "Sphericity Assumed."

Type III Sum of Squares   This column presents the Type III sum of squares, the default setting for SPSS. A Type III sum of squares shows the sum of squares for an effect, adjusted for all other effects that do not contain it, and has the advantage of not being affected if there are different cell frequencies. This statistic is similar to the sum of squares of the one-way ANOVA (Chapter 10).

df   The *df* column provides the degrees of freedom associated with each source of variance. In our example, the number of degrees of freedom associated with the numerator (the four measurements taken) equals 3. *df* associated with the denominator (error) is 27.

Mean Square   The mean square is shown for each of the test statistics displayed. The mean square is computed by dividing the Type III sum of squares by *df*. The Type III sum of squares (3464.100) divided by the number of degrees of freedom (3) is equal to the mean square (1154.700). Similarly, the error mean square is computed by dividing the Type III sum of squares by *df*. The error mean square is 9.719(262.400/27 = 9.719).

F   The *F* ratio is 118.814. It is computed by dividing the mean square for reading by the mean square for error, which in this case is 1154.7000/9.719 = 118.814.

Sig   The value of *p* at which *F* is significant is shown as .000. Although .000 is printed, the actual alpha level is not 0; rather, alpha is so small that it rounds to 0. Because we used an alpha of .05, and .000 is less than .05, we can conclude that at least one mean is significantly different from at least one other mean.

On the basis of the ANOVA output, we reject the null hypothesis. That is, at least one mean is different from at least one other mean at alpha less than .05, our predetermined alpha level. We arrive at the same decision if we examine our critical value. Given our example with 3 and 27 degrees of freedom, the critical value of *F* at $p < .05$ is 2.96. Our *F* of 118.814 exceeds the critical value.

Partial Eta Squared (partial η^2)

This is the proportion of the total variability attributable to a factor, taken as if it were the only variable. Partial *eta* squared gives an estimate of the effect size. Here, its value is .930. In other words, 93% of the variability in the reading scores can be explained by the marking period in which the scores were measured. Thus, a large portion of the variance in scores is accounted for by the marking period. Note that we use partial *eta* squared as the estimate of effect size because we are comparing multiple means. Cohen's *d*, which compares two means at a time, could not be used here.

The final two boxes, "Tests of Within-Subjects Contrasts" and "Tests of Between-Subjects Effects," are more advanced topics and will not be discussed here.

Post Hoc Tests

SPSS does not have an option for conducting *post hoc* tests for a simple repeated-measures ANOVA. If you are interested in determining whether any particular marking period is significantly different from any other marking period, one approach would be to compute all possible pairs of related-samples *t*-tests (for details, see Chapter 12). We will use this approach to follow up our *a priori* hypothesis that there will be a difference across marking periods. The related-samples *t*-test is similar to the least significant difference (LSD) test used to follow up a significant one-way ANOVA. However, as with the one-way ANOVA, this method inflates the alpha level that was set.

Table 1 summarizes the outcomes of the set of related-samples *t*-tests. As shown in the table, the related-samples *t*-test results indicate that all means were significantly different from each other.

TABLE 1

Results of Related-Samples *t*-tests

Means Compared	Related *t* value	*p*
mp1 and mp2	−10.09	<.01
mp1 and mp3	−11.91	<.01
mp1 and mp4	−13.77	<.01
mp2 and mp3	−7.88	<.01
mp2 and mp4	−9.09	<.01
mp3 and mp4	−1.04	<.01

Advantages and Disadvantages of the Repeated-Measures ANOVA

Longitudinal studies repeatedly measure characteristics of the same participants over time.

Because the participants act as their own controls in a repeated-measures ANOVA, this is a technique that works particularly well with a small sample size. Also, the repeated-measures ANOVA is useful when we are interested in following one group of participants over time, a design known as a **longitudinal study.** For example, we might be interested in studying language development in young children. To accomplish this, we might conduct a longitudinal study and measure the vocabulary of the same children when they are one, two, three, four, and five years old.

A **carry-over effect** may occur if effects of one condition affect conditions that follow.

The repeated-measures ANOVA also has less error variance and greater power for within-subjects effects than between-groups ANOVA designs. As with a related-samples *t*-test, this is because errors due to individual differences are removed from the error (denominator) term.

A **practice effect** occurs when participants get better as they move through conditions.

The repeated-measures ANOVA also has some limitations. Because participants are measured under all conditions in a repeated-measures ANOVA, there are some issues that you should keep in mind. **Carry-over effects** occur when the effects experienced in one condition affect the performance of—or carry over to—the conditions that follow. There also may be a **practice effect**, in which participants get better at whatever is being measured as they move through conditions. Another potential problem in a repeated-measures ANOVA is a **fatigue effect**, wherein participants get worse as they move through conditions, due to factors such as becoming tired, bored, or hungry.

A **fatigue effect** occurs when participants get worse as they move through conditions because of factors such as becoming tired.

A **counterbalancing** refers to varying the presentation of conditions in a repeated-measures design.

In some investigations, these potential limitations can be addressed by means of a technique called **counterbalancing,** in which the conditions to be experienced by participants are presented in a varied order.

Be Here Now

A researcher plans to have her participants complete push-ups, sit-ups, and jumping jacks in that order. What would you advise her to do?

Answer

She should counterbalance her conditions and watch for fatigue effects.

Review of Steps Taken in the Repeated-Measures ANOVA

To summarize, using the example we just completed, here are the steps in conducting a repeated-measures analysis of variance:

STEP 1   State the hypotheses.

$$H_0: \mu_1 = \mu_2 \ldots = \ldots \mu_j$$
$$H_1: \mu_i \neq \mu_j, \text{for some } i, j$$

STEP 2   Determine the critical value.

The number of degrees of freedom (the repeated factor, or the numerator) for the repeated-measures ANOVA is $df_{numerator} = k - 1$. Here, $df_{numerator} = 3$.
The degrees of freedom error (the denominator) is $df_{denominator} = (n - 1)(k - 1)$. Here, $df_{denominator} = 27$.
Given our example with $df_{numerator} = 3$ and $df_{denominator} = 27$, the critical value of F with an alpha of .05 is 2.96.

STEP 3   Compute the statistic.
Using SPSS, we find that $F(3, 27) = 118.814$, $p < .001$.

STEP 4   Evaluate the test statistic.
We reject the null hypothesis and conclude that at least one mean is different from at least one other mean at alpha less than .05, our predetermined alpha level. Conduct follow-up related-samples t-tests to interpret a significant F ratio.

STEP 5   Determine the effect size.
Partial *eta* squared is .930. This is a large effect.

Reporting Results in APA Style

Reporting the results of a repeated-measures ANOVA follows the same format as reporting the results of other ANOVA tests. We report the descriptive statistics (means and standard deviations) first, highlighting what is most important in the results. Then, we report the F statistic information and the effect size. For significant results, we discuss findings from *post hoc* tests.

Notice that only two decimal places are reported. From our reading example, you might state the following:
Descriptive statistics are shown in Table 2. Note that the average scores increased steadily over the course of the school year. The result for the repeated-measures ANOVA was significant, $F(3, 27) = 118.81$, $p < .001$. Partial *eta* squared was .93.

Post hoc analyses using the related samples t-test indicated that each marking period was significantly different from any other (see Table 3).

Table 2
Descriptive Statistics by Marking Period

Variable	Mean	Standard Deviation	n
Marking Period 1	48.10	5.93	10
Marking Period 2	59.20	3.16	10
Marking Period 3	68.10	4.63	10
Marking Period 4	72.40	5.15	10

Table 3
Results of LSD Post Hoc Tests

Means Compared	Related t value	p
mp1 and mp2	−10.09	< .01
mp1 and mp3	−11.91	< .01
mp1 and mp4	−13.77	< .01
mp2 and mp3	−7.88	< .01
mp2 and mp4	−9.09	< .01
mp3 and mp4	−1.04	< .01

It's Out There...

Christ, Schanding, Jr., and Thomas (2007) conducted a repeated-measures ANOVA to examine changes in math assessments repeated in Novel, Neutral, and Reward conditions. The repeated-measures **F** ratio was significant, $F(2,462) = 44.5$, $p < .05$. Follow-up tests indicated significant differences between Novel ($M = 8.7$) and Reward ($M = 12.0$), as well as between Novel and Neutral ($M = 11.6$), conditions.

Let's try another example, following the steps in hypothesis testing. A researcher is interested in the artistic perceptions of university students. He asks six students to observe works of art in three different settings: shown on slides, viewed on a computer, and seen in person in a museum. He measures how long the participants spend viewing the art in each setting. He counterbalances the three conditions. The data obtained are as follows:

Student	Slides	Computer	Museum
A	46	48	48
B	40	42	36
C	40	36	42
D	46	36	48
E	42	48	48
F	36	48	42

STEP 1 State the hypotheses.

$$H_0: \mu_1 = \mu_2 = \mu_3$$
$$H_1: \mu_i \neq \mu_j, \text{ for some } i, j$$

STEP 2 Determine critical value.

df for the numerator is 2.

df for the denominator is 10.

Using alpha = .05 with 2 and 10 *df*, we find that the critical value of $F = 2.92$.

Descriptive Statistics

	Mean	Std. Deviation	N
Slides	41.6667	3.88158	6
Computer	43.0000	5.89915	6
Museum	44.0000	4.89898	6

Mauchly's Test of Sphericity[b]

Measure: MEASURE 1

Within Subjects Effect	Mauchly's W	Approx. Chi-Square	df	Sig.	Epsilon[a]		
					Greenhouse-Giesser	Huynh-Feldt	Lower-bound
factor1	.537	2.488	2	.288	.683	.853	.500

Tests the null hypothesis that the error covariance matrix of the orthonormalized transformed dependent variables is proportional to an identity matrix.

[a] May be used to adjust the degrees of freedom for the averaged tests of significance. Corrected tests are displayed in the Tests of Within-Subjects Effects table.

[b] Design: Intercept
Within Subjects Design: factor1

FIGURE 6 Selected output boxes: repeated-measures ANOVA, time spent viewing art.

STEP 3 Compute the test statistic.

Figure 6 presents selected output boxes of the repeated-measures ANOVA for the art-viewing data. We see that $F(2,10) = .407$, *ns*.

STEP 4 Evaluate the hypotheses.
Fail to reject the null hypothesis.

STEP 5 Determine the effect size.
Printed is the partial *eta* squared of .075; however, because the F ratio is not significant, we will not interpret it.

Other ANOVA Designs

A **mixed-factors design** combines repeated (within-groups) and independent (between-groups) factors in the same analysis.

Building on the repeated-measures ANOVA design, we may want to compare different, independent groups within the repeated measures. For example, remember the first example in this chapter: examining the change in reading level over four marking periods. We may want to compare males and females for each of the four marking periods. We would then have a repeated-measures component (the four measurements over the year) and a between-groups component (gender). This approach is sometimes called a **mixed-factors design.**

A physical therapist wants to test the change in range of motion over two months for a sample of clients with rotator cuff injuries. He makes an initial evaluation and gives each client a score of 1 (low) to 5 (high). He then reevaluates each client after 1 month of physical therapy and after 2 months of physical therapy. Using SPSS and an alpha = .05, test whether there is a significant difference among the means. The data are as follows:

Participant	Initial Rating	At One Month	At Two Months
1	2	3	3
2	3	3	4
3	2	3	4
4	2	3	4
5	2	3	4
6	3	4	3
7	2	4	3
8	2	3	5
9	1	2	3
10	1	3	4

Answer

The result of the repeated-measures analysis was significant, $F(2,18) = 20.69$, $p < .001$. Partial *eta* squared was .70.

It's nice to know that ANOVAs can be quite flexible in helping us to design studies and examine data for questions we have about life! However, the statistics and the interpretations of many of the mixed-factors designs can become rather complex and are best left to a more advanced statistics or research design course.

SUMMARY

The repeated-measures ANOVA is a powerful technique for examining change over time or when there are two or more repeated measurements for each participant. The steps for hypothesis testing follow those of other ANOVA procedures; the effect size is determined by means of partial *eta* squared.

Reflect
In statistics, the more things change, the more they are likely to be significant. (with apologies to Alphonse Karr)

PRACTICE

Background

1. Identify the appropriate test statistic for each of the following:

   a. A sixth-grade teacher pretests his class on math skills in September. In June, he tests his class again to determine how much the students gained in their math skills.

   b. A gardener plants tomatoes in her yard. She sections her garden into three areas and uses a different type of fertilizer in each area. She wants to determine which type of fertilizer will grow the best tomatoes.

   c. A university asks its male basketball players to wear four different brands of sneakers for three weeks per brand. The students rate the performance of the sneakers on a scale of 1 (awful) to 5 (outstanding).

   d. A university compares the GPAs of all men and women majoring in psychology.

2. A researcher is planning to investigate how much vocabulary increases between the ages of two and five. She was planning to conduct a one-way ANOVA with a group of 25 two-year-olds, 25 three-year-olds, 25 four-year-olds, and 25 five-year-olds. How might she change this study so that it becomes a repeated-measures ANOVA?

Computing the Repeated-Measures ANOVA

3. A researcher is interested in rating noncola soft drinks. He asks a sample of five participants to rate three popular noncola soft drinks on a scale of 1 (awful) to 10 (delicious). Using the following data with alpha = .05, test the null hypothesis that there is no difference among the drinks:

Participant	Drink 1 Rating	Drink 2 Rating	Drink 3 Rating
1	7	8	7
2	5	10	8
3	9	6	9
4	4	7	9
5	6	9	7

4. Rowena is interested in factors that affect how people encode information for short-term memory. A sample of 10 participants is asked to memorize a short rhyming poem, a nonrhyming poem of a similar length, and a poem of the same length containing nonsense words. Rowena counterbalances the conditions and times of each participant in each condition. The data obtained, reported in seconds, are as follows:

Participant	Rhyming Poem	Nonrhyming Poem	Nonsense Poem
1	27	34	41
2	32	39	44
3	26	35	40
4	29	33	42
5	34	36	46
6	39	44	46
7	29	36	49
8	25	37	44
9	30	38	45
10	29	40	49

Using alpha = .05, test whether there is a significant difference in the times required to memorize the three types of poems.

5. Stephano administers an IQ measure to six students who have been classified as perceptually impaired. Scores on four of the subtests are as follows:

Participant	Picture Completion	Block Design	Object Assembly	Visual-Spatial Knowledge
1	11	7	12	11
2	9	12	11	13
3	5	7	13	9
4	8	11	9	14
5	13	8	14	12
6	9	9	8	11

Using alpha = .05, test whether there is a significant difference among the means.

6. A researcher is interested in the importance of various traits in long-term relationships. Participants who have been in a relationship for at least 10 years are asked to rate their partners on four traits, using a scale of 1 (not at all important) to 5 (vitally important). The results obtained are as follows:

Participant	Intelligence	Sense of Humor	Honesty	Physical Appearance
1	4	5	5	2
2	3	4	4	3
3	5	5	5	1
4	4	5	4	1
5	3	4	5	2

Using alpha = .05, test whether there is a significant difference among the means. Use LSD to calculate any necessary *post hoc* analyses.

7. A researcher is interested in the effects of color on reaction time. Reaction time is measured by pressing the space bar on a computer keyboard in response to large blocks of color presented on a computer screen. Three colors are used: green, yellow, and red. The data (in hundredths of a second) for 10 participants are as follows:

Participant	Green	Yellow	]Red
1	21	22	21
2	22	23	20
3	21	22	20
4	20	22	21
5	20	21	21
6	22	23	22
7	21	22	20
8	22	23	22
9	20	22	20
10	21	23	20

Using alpha = .05, test whether there is a significant difference among the means.

Advantages and Disadvantages of the Repeated-Measures ANOVA

8. A researcher wonders whether the type of cheese used as a reward affects how quickly rats will learn to run through a maze. She can obtain only three rats, so she decides to use a repeated-measures ANOVA design to make the most of her small sample size. She times each rat going through the maze for 30 trials. For the first 10 trials, she uses cheddar cheese as a reward. For the second 10 trials, she uses gorgonzola cheese, and for the final 10 trials, she uses blue cheese. What would you tell her about the way she set up this experiment?

9. A student is working on a project that requires each participant to complete a set of three lengthy questionnaires, administered in the same order to all participants. He finds that many of the participants' answers to the third questionnaire are incomplete. What explanation might you give him for this finding?

Reporting Results in APA Format

10. Write the results you obtained in Problem 6 in APA style.

SOLUTIONS

Background

1. **a.** Related-samples *t*-test
   b. One-way ANOVA
   c. Repeated-measures ANOVA
   d. Independent-samples *t*-test

2. The researcher could initially measure the vocabulary of a group of two-year-old children and then measure the same children at ages three, four, and five. This approach would give her four measurements per child. Thus, she could conduct a longitudinal study.

Computing the Repeated-Measures ANOVA

3. Step 1: State the hypotheses.

$$H_0: \mu_1 = \mu_2 = \mu_3$$
$$H_1: \mu_i \neq \mu_j, \text{for some } i, j$$

Step 2: Determine the critical value.

Using alpha = .05 with 2, 8 *df*, we find that the critical value of F is 4.46.

Step 3: Compute the test statistic.

$$F(2, 8) = 1.67, \ p = .248$$

Descriptive Statistics	Mean	Std. Deviation	N
Drink 1	6.2000	1.92354	5
Drink 2	8.0000	1.58114	5
Drink 3	8.0000	1.00000	5

Step 4: Evaluate the hypotheses.

Fail to reject the null hypothesis; conclude that there are no significant differences among the drinks.

4. Step 1: State the hypotheses.

$$H_0: \mu_1 = \mu_2 = \mu_3$$
$$H_1: \mu_i \neq \mu_j, \text{for some } i, j$$

Step 2: Determine the critical value.

Using alpha = .05 with 2, 18 *df*, we find that the critical value of F is 3.55.

Step 3: Compute the test statistic.

Descriptive Statistics	Mean	Std. Deviation	N
Rhyming Poem	30.0000	4.13656	10
Nonrhyming Poem	37.2000	3.22490	10
Nonsense Poem	44.6000	3.06232	10

$$F(2,18) = 89.82, p < .001$$

Step 4: Evaluate the hypotheses.

Reject the null hypothesis; conclude that at least one of the means is significantly different from at least one of the other means.

Step 5: Determine the effect size.

Partial *eta* squared is .91

5. Step 1: State the hypotheses.

$$H_0: \mu_1 = \mu_2 = \mu_3 = \mu_4$$
$$H_1: \mu_i \neq \mu_j, \text{ for some } i, j$$

Step 2: Determine the critical value.

Using alpha $= .05$ with 3, 15 *df*, we find that the critical value of *F* is 3.29.

Step 3: Compute the test statistic.

Descriptive Statistics	Mean	Std. Deviation	N
Picture Completion	9.1667	2.71416	6
Block Design	9.0000	2.09762	6
Object Assembly	11.1667	2.31661	6
Visual–Spatial Assembly	11.6667	1.75119	6

$$F(3,15) = 2.34, p = .114$$

Step 4: Evaluate the hypotheses.

Fail to reject the null hypothesis; conclude that there are no significant differences among the means.

6. Step 1: State the hypotheses.

$$H_0: \mu_1 = \mu_2 = \mu_3 = \mu_4$$
$$H_1: \mu_i \neq \mu_j, \text{ for some } i, j$$

Step 2: Determine critical value.

Using alpha $= .05$ with 3, 12 *df*, we find that the critical value of *F* is 3.49.

Step 3: Compute the test statistic.

Descriptive Statistics	Mean	Std. Deviation	N
Intelligence	3.8000	3.83666	5
Sense of Humor	4.6000	4.54772	5
Honesty	4.6000	4.54772	5
Physical Appearance	1.8000	1.83666	5

$$F(3,12) = 15.41, p < .001$$

Step 4: Evaluate the hypotheses.

Reject the null hypothesis; conclude that at least one of the means is significantly different from at least one of the other means.

Step 5: Determine the effect size.

Partial *eta* squared is .79.

Post hoc analyses:

Results of Related-Samples t-tests

Means Compared	Related *t* value	*p*
intelligence–humor	−4.00	<.05
intelligence–honesty	−2.14	*ns*
intelligence–physical	2.83	<.05
humor–honesty	0.00	*ns*
humor–physical	4.80	<.01
honesty–physical	5.72	<.01

7. Step 1: State the hypotheses.

$H_0: \mu_1 = \mu_2 = \mu_3$

$H_1: \mu_i \neq \mu_j$, for some i, j

Step 2: Determine the critical value.

Using alpha = .05 with 2, 18 *df*, we find that the critical value of *F* is 3.55.

Step 3: Compute the test statistic.

$F(2,18) = 21.00, p < .001$

Descriptive Statistics	Mean	Std. Deviation	N
green	21.0000	3.81650	10
yellow	22.3000	3.67495	10
red	20.7000	3.82327	10

Step 4: Evaluate the hypotheses.

Reject the null hypothesis; conclude that at least one of the means is significantly different from at least one of the other means.

Step 5: Determine the effect size.

Partial *eta* squared is .70.

Advantages and Disadvantages of the Repeated-Measures ANOVA

8. The researcher should have counterbalanced the presentations of the types of cheeses. The results she obtains are most likely going to be

due to practice effects, not the type of cheese used! She might have been better off trying to obtain more rats and using a one-way ANOVA design.

9. He is obtaining a fatigue effect. He should have counterbalanced the order of the questionnaires and given participants a break with some refreshments while completing the questionnaires!

Reporting Results in APA Format

10. Descriptive statistics are shown in Table 1. Note that the average scores for sense of humor and honesty indicate that these traits were valued the most by the participants and that, on average, physical appearance was valued the least. The results for the repeated-measures analysis were significant, $F(3,12) = 15.41$, $p < .001$. Partial *eta* squared was .79. *Post hoc* analyses using related-samples *t*-tests yielded four significant pairwise comparisons (see Table 2).

Table 1
Descriptive Statistics for Traits

Trait	Mean	Standard Deviation	N
Intelligence	3.80	.84	5
Sense of Humor	4.60	.55	5
Honesty	4.60	.55	5
Physical Appearance	1.80	.84	5

Table 2
Results of Related-Samples t-tests

Means Compared	Related *t* value	*p*
intelligence–humor	−4.00	<.05
intelligence–honesty	−2.14	*ns*
intelligence–physical	2.83	<.05
humor–honesty	0.00	*ns*
humor–physical	4.80	<.01
honesty–physical	5.72	<.01

-14-

Correlation and Regression

In This Chapter

Correlation
- What is Correlation?
- Scatterplots
- Characteristics of Correlation
- Calculating the Pearson Product-Moment Correlation
- Testing Hypotheses about Correlations
- Interpreting Correlations
  The Coefficient of Determination
  Restriction of Range
  Outliers
- Application: Hypothesis Testing with Correlation

- Reporting Correlation Results in APA Style
- The Correlation Matrix
- Summary of Correlation

Regression
- What is Regression?
- Calculating Simple Regression
- Application: Regression and Multiple Regression
  Regression Method
  Regression Output
- Reporting Regression Results in APA Style
- Summary of Regression

F O C U S | This chapter introduces two closely related techniques used to examine relationships among variables: correlation and regression. We use the information from correlation and regression to examine relationships among variables and to make predictions about how these variables behave in life.

Correlation

What Is Correlation?

Correlation is one of the major approaches to analyzing data in the social sciences. **Correlation** measures the relationship between two variables. Usually, but not always, the variables we examine are variables we do not manipulate. So, for example, we might be interested in the relationship between height and weight or weight and blood pressure. We can't randomly assign a person to a height, a weight, or a level of blood pressure; these are naturally occurring variables. Examples of other variables that are frequently used in correlations are IQ scores, SAT or GRE scores, measures of personality, and grade point average (GPA).

A key to interpreting correlation is remembering that we are interested in the *relationship* between two variables; we are *not* trying to show cause and effect. That is, we are not trying to demonstrate that variable **X** caused variable **Y**—only that they are associated.

The statistic we obtain that describes the relationship between the variables of interest is referred to as the **correlation coefficient**. The symbol for the correlation coefficient is a lowercase *r*. The value of *r* can range from −1.00 to + 1.00. If you calculate *r* and get a number greater than + 1.00 or less than −1.00, you've done something wrong!

There are two important guidelines having to do with correlation. First, to calculate a correlation coefficient, you must have *two measures, or scores, for each participant*. Second, it is important to *keep the two scores for each participant together*. For example, if you are interested in the relationship between height and weight, you don't want to pair Pauline's height with Robert's weight! That would not yield a correct correlation coefficient and might lead to some inaccurate conclusions.

There are several types of correlations; which type you use depends on the scale of measurement associated with the variables. When the data are continuous, we use a **Pearson product-moment correlation**. That is the most common type of correlation, and it is what we will focus on here. Pearson is the default correlation of SPSS. There are other types of correlations (e.g., Spearman rho and phi) that you may encounter, but they are all interpreted in the same manner as the Pearson.

> **Reflect**
> Relationships work best when pairs are kept together.

Scatterplots

When you calculate a correlation coefficient, it is always a good idea to plot your data in the form of a **scatterplot,** or **scatter diagram** (or have SPSS plot it for you). A scatterplot is a graph of the data (**X** and **Y** data points) collected from each participant. It gives you an indication of the strength

Correlation measures the relationship between two variables.

The **correlation coefficient**, or *r*, is a value ranging between −1.00 and + 1.00; it describes the relationship between two variables.

A **scatterplot**, or **scatter diagram**, is a graph of the paired coordinates associated with **X** and **Y** points.

of the correlation, as well as the direction of the relationship (positive or negative).

To create a scatterplot by hand, choose one of your variables for the x-axis of a graph and the other for the y-axis. Next, place an appropriate range of values for the data along each axis, so that the graph will clearly depict all of the data points. For example, if the actual values of the variable range from 5 to 20, use 0 to 25 as the values on the axis. Finally, plot the data for each participant by moving along the x-axis to the value that represents that person's score on the **X** variable, then moving vertically to the value that represents the same person's score on the **Y** variable, and placing a dot or data point at the intersection.

For example, let's consider age and height measurements for a sample of $n = 10$ children ranging in age from 5 to 10 years. The data are as follows:

Child	Age (in years)	Height (in inches)
Carolyn	5	36
Thomas	6	38
Kevin	7	40
Margaret	8	42
Elizabeth	9	45
Gregory	6	41
Timothy	8	45
Kaelin	9	48
Denny	10	54
Pauline	5	37

The variable *age* will be placed on the x-axis, and the variable *height* on the y-axis, of the scatterplot. Notice that on both axes, the values on the lines extend just below the number associated with the lowest data point and just above the number associated with the highest data point. Let's look at how we would plot the data, using the first pair of values for Carolyn. To plot the data, we go to the value 5 on the x-axis and on up to 36 on the y-axis, and place a data point. For Thomas, we go to 6 on the x-axis and 38 on the y-axis, and place a data point. Notice that we keep the values together for each child. We continue doing this for each pair of scores in the data set. When all the data have been plotted, we will have something that looks like this:

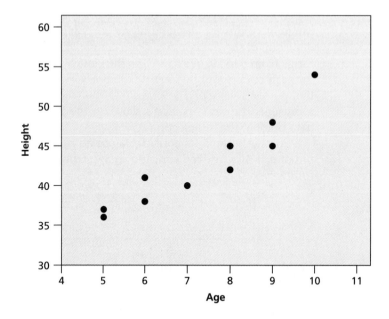

In looking at this scatterplot, you can see that as age increases, so does height.

Using these data, create and interpret a scatterplot.

SAT	GPA
520	2.10
430	1.80
660	3.20
340	1.50
500	2.20
710	3.80
220	1.20
560	2.80

Answer

As SAT goes up, GPA goes up.

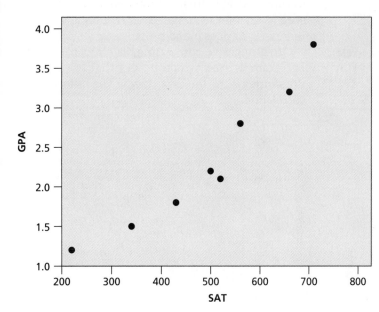

Characteristics of Correlation

Correlation tells us about the strength, direction, and shape of the relationship between two variables.

We'll use scatterplots to illustrate each of these characteristics of correlation coefficients.

In a **positive correlation**, as one variable increases, the other variable increases.

The correlation coefficient tells us whether a relationship between two variables is positive or negative. In a **positive** relationship, as the values of one variable increase, values of the other variable increase. The following is a scatterplot illustrating data collected regarding the positive relationship between Blood Pressure readings and Anxiety levels:

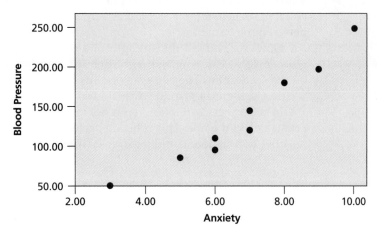

In looking at this relationship between level of anxiety and blood pressure, if the plot were based on real data, you would predict that as anxiety increases, so does blood pressure.

In a **negative correlation**, as one variable increases, the other *decreases*. For example, you would expect that the more cigarettes smoked per day, the shorter is the life expectancy. The following scatterplot shows a negative correlation:

In a **negative correlation**, as one variable increases, the other variable decreases.

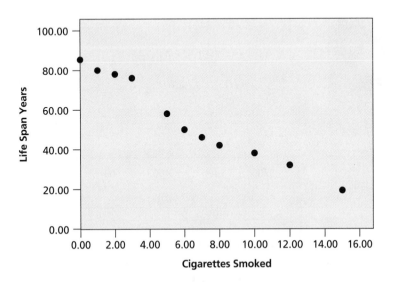

In looking at this relationship between life span and number of cigarettes smoked per day, if the plot were based on real data, you would predict that the more one smoked, the shorter that person's life span would be.

Next, let's consider the strength of the relationship. Remember, *r* can range from −1.00 to +1.00. The closer you get to −1.00 or +1.00, the stronger the relationship. A correlation of +1.00 refers to a **perfect positive correlation**; a correlation of −1.00 denotes a **perfect negative correlation**. An *r* of +.82 is equal *in strength* to an *r* of −.82. A zero correlation indicates no relationship between the variables.

A **perfect positive correlation** would yield an *r* of +1.00.

A **perfect negative correlation** would yield an *r* of −1.00.

How can you tell the strength of a relationship by looking at a scatterplot? Imagine drawing a straight line through the center of the data points on the scatterplot. The closer the points are to the line, the stronger the correlation is, regardless of the direction.

In a perfect correlation, whether positive or negative, all of the data points fall on the line. This is how a perfect positive correlation and a perfect negative correlation would look on scatterplots:

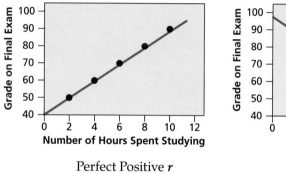

Perfect Positive *r*

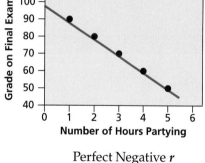

Perfect Negative *r*

The following scatterplot depicts the relationship between achievement and motivation scores:

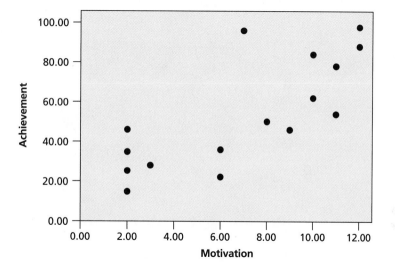

The correlation indicated in this plot is +.75. The data are scattered or spread out, rather than falling along a straight line. Notice, though, that the relationship is positive: The data points are scattered in such a way that they suggest a positive—but not perfect—relationship (correlation).

With a correlation of −.60, the data are more scattered. Notice how the pattern of the data (age and wellness ratings) suggests a negative trend.

With a correlation of +.30 (SAT score and GPA), the data are even more scattered.

As the correlation approaches zero, the data become highly scattered.

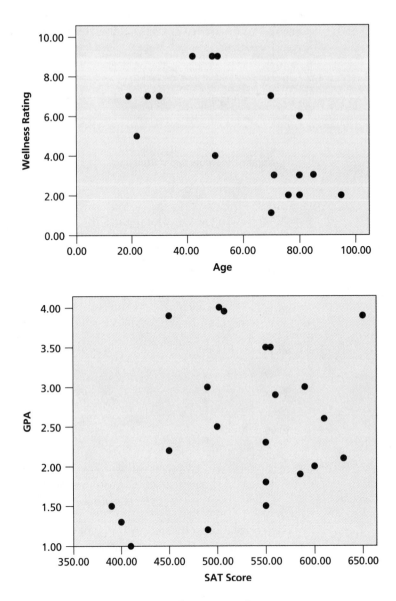

When there is no correlation, $r = 0$. The data points are completely scattered; neither a positive nor a negative direction of relationship is suggested. This situation is illustrated in the following scatterplot, which depicts the relationship between shoe size and IQ.

The last piece of information we can gather from a scatterplot is the shape or form of the relationship: Are the data linear (lying on a straight line) or curvilinear (curved)? The scatterplot that depicted the relationship between grade on a final exam and number of hours spent studying was a strong, positive linear relationship. We know this because (1) the data points are clustered

Reflect
Life is rarely perfect.
Scatterplots illustrate this.

tightly; (2) as one variable goes up, the other goes up; and (3) the data are arranged in a straight line.

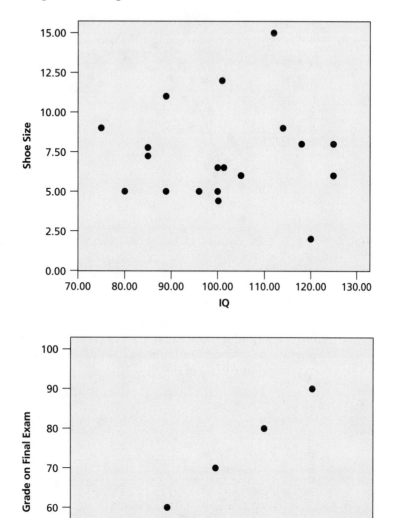

What type of relationship might be curvilinear? Let's return to anxiety. Research over the past six decades has demonstrated that a moderate amount of anxiety is related to a good performance, whether in athletics, in music recitals, or on exams. Too much or too little anxiety is related to lower performance. The correlation between the amount of anxiety a person is experiencing and his or her performance forms an inverted-U shape on a scatterplot:

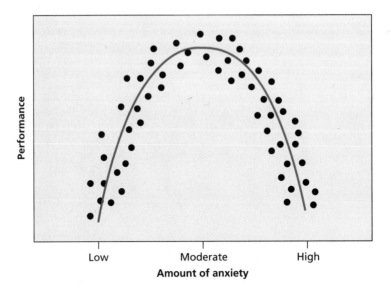

Amount of anxiety

A correlation between age and the need for physical assistance also would produce a U-shaped scatterplot, but not an inverted one. The following scatterplot indicates that at birth and in old age, there is a high need for physical assistance, but far less physical assistance is typically needed during the middle years of life:

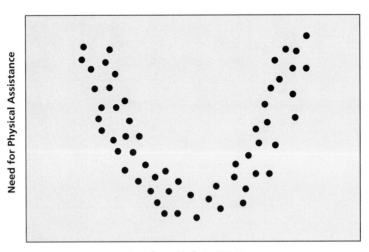

Age from Birth to 100 Years

Curvilinear relationships can level off. For example, look at the scatterplot for number of cookies you have eaten and a rating of whether you'd like to eat another cookie.

It would be difficult to draw a *straight line* through the center of the data points for any of these curvilinear relationships. In fact, if we computed the

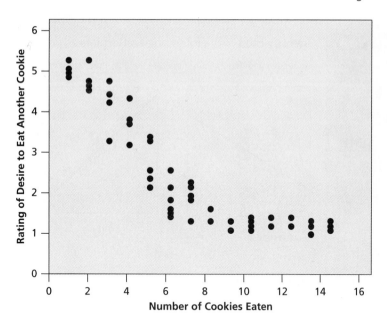

Pearson product-moment correlation, it would be near zero, because these relationships are not linear. When data are curvilinear, another type of correlation coefficient is needed (see Glass & Hopkins, 1995, for more information). Thus, it is important to examine a scatterplot of your data to help determine whether the data are linear or curvilinear and therefore to decide whether the Pearson product-moment correlation is the appropriate type of correlation to compute.

From this point forward, we'll concentrate on the Pearson product-moment correlation coefficient, which we'll now refer to simply as a correlation.

Calculating the Pearson Product-Moment Correlation

To calculate the Pearson product-moment correlation, we begin with the formula for the correlation, using words rather than symbols:

$$r = \frac{\text{the degree to which } X \text{ and } Y \text{ vary together}}{\text{the degree to which } X \text{ and } Y \text{ vary separately}}$$

The numerator—the degree to which **X** and **Y** vary together—is called the **covariance**. We use **"cov"** to denote the covariance. The denominator reflects the standard deviation of *X* and the standard deviation of *Y*.

Covariance indicates the degree to which two variables vary together.

Substituting symbols for words, we obtain the following formula:

$$r = \frac{\text{cov}}{S_x S_y}$$

For the formula for computing the correlation coefficient without the aid of a computer, see Appendix D.

Be Here Now

Identify each of the following scatterplots as positive, negative, or curvilinear:

1.

2.

3.

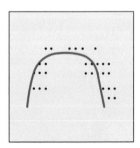

Answers

1. Positive
2. Negative
3. Curvilinear

Testing Hypotheses about Correlations

So far, we have described relationships. Although scatterplots help us understand the nature of a relationship, we must compute a statistic to determine whether there is evidence of a linear relationship or whether the appearance of a linear pattern is likely to be due to chance. To do so, we conduct a hypothesis test to learn whether our correlation is significantly different from zero.

Suppose that we have tested five children on two measures: the number of nonsense words each can remember after 10 minutes and the number of meaningful words each can remember after 10 minutes. We want to know whether there is a relationship between the number of nonsense words and the number of meaningful words a child can recall. The data we obtain are as follows:

	Nonsense	Meaningful
Tanisha	3	10
Ian	2	3
Sophie	4	8
Helen	0	4
Brett	1	5

We will use the same steps we used to test hypotheses (involving z, t, and F) in previous chapters.

STEP 1 State the Null and Alternative Hypotheses.

In our example, we want to test if there is a relationship between the number of nonsense words and the number of meaningful words a child can recall.

Rho (ρ) is the symbol for the correlation coefficient of a population.

The null hypothesis for correlation assumes no relationship ($r = 0$). If we collect data, we can test whether a relationship exists, which is expressed by the alternative hypothesis ($r \neq 0$). The symbol for the population correlation coefficient is ρ, or rho, the letter in the Greek alphabet that corresponds to r in the English alphabet. It is pronounced like the English word "row."

Thus, our hypotheses would look like this:

$$H_0\!: \rho = 0$$
$$H_1\!: \rho \neq 0$$

STEP 2   Determine the critical value.

In this example, we will set alpha to .05. To determine the critical value of r, we need to compute the degrees of freedom.

For the **correlation coefficient**, $df = n - 2$, where n is the number of pairs of scores.

For the **correlation coefficient**, we lose two degrees of freedom (we are measuring two variables), so $df = n - 2$. In our example, we have five children; therefore,

$$df = 5 - 2 = 3$$

Notice that n denotes the pairs of scores, not the number of individual scores, in the data set.

We now can look up the critical value of r. Critical values of the Pearson correlation coefficient are listed in Appendix A4. We are conducting a two-tailed test, so we go to the row in the column headings that is labeled "two tailed." Next, we choose the column that corresponds to an alpha of .05. We go down that column until we reach the value that corresponds to the row for 3 df. The df are listed in the first column. The critical value of r with 3 df is .878 for a two-tailed test with alpha $<$.05. This value is labeled in the following figure, which also identifies the critical regions in which we reject the null hypothesis:

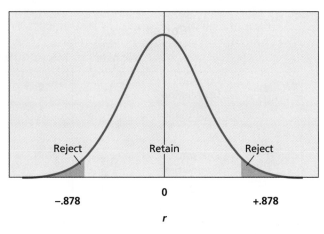

STEP 3   Compute r, the appropriate test statistic.

To have SPSS compute r, we must create a data file. Remember, we know two things about each participant: the recall score for nonsense words and the score for meaningful words. Each row in our data file represents the data on one participant. Figure 1 presents the data file, once it is entered into SPSS.

Once the data are entered into an SPSS file, you are ready to have SPSS compute *r*.

To perform a test for correlation, you need to select

Analyze → *Correlate* → *Bivariate.*

- Move the variables "Nonsense" and "Meaningful" to the "Variables" Box.

As we previously mentioned, SPSS calculates the Pearson product-moment correlation coefficient by default. SPSS will also provide information about a two-tailed test by default. However, you may change either of these SPSS default settings.

Once completed, your output should look like that presented in Figure 2.

SPSS provides one output box labeled "Correlations." Across the top row and down the first column are the variable labels.

	Nonsense	Meaningful
1	3	10
2	2	3
3	4	8
4	0	4
5	1	5

FIGURE 1 Sample data file: Recall of nonsense and meaningful words.

Correlations

		Nonsense	Meaningful
Nonsense	Pearson Correlation	1	.705
	Sig. (2-tailed)		.184
	N	5	5
Meaningful	Pearson Correlation	.705	1
	Sig. (2-tailed)	.184	
	N	5	5

FIGURE 2 SPSS output: Recall of nonsense and meaningful words.

Each cell in the "Correlations" output box that has the correlation between Meaningful and Nonsense Words contains the following three pieces of information:

Pearson Correlation	Within each cell, the top number is the correlation coefficient *r*.
Sig. (2-tailed)	In each cell, beneath *r*, is the value of *p* at which the correlation is significant.
N	The number of participants (data points) for which *r* was computed.

Now let's explore the particular values in the cells for our example.

First, look at the output cell where "Nonsense" is the column heading as well as the row heading.

Pearson correlation	$r = 1$. The correlation of any variable with itself is 1.
Sig. (2-tailed)	No significance level is computed.
N	The correlation is based on five pairs of recall scores.

Next, look at the cell for which "Nonsense" is the column heading and "Meaningful" is the row heading.

Pearson correlation	$r = .705$. The correlation between the recall of Nonsense and Meaningful words is $+ .705$.
Sig. (2-tailed)	The level of significance is .184. Since this is greater than .05, the correlation is not significant.
N	The correlation is based on five pairs of recall scores.

Notice that these results are printed twice. SPSS prints out the correlation between Nonsense and Meaningful, as well as the correlation between Meaningful and Nonsense; these results will be identical because the data are identical—only the order of the variables differs. Numbers printed on the diagonal will always be made up of ones, because these represent the variable correlated with itself—a perfect correlation every time!

STEP 4   Evaluate the hypothesis.

With 3 *df*, and for $p < .05$, the critical value of *r* is .878. The observed value of *r* was .705. Therefore, we fail to reject the null hypothesis. In other words, we failed to find evidence of a significant linear relationship between the number of nonsense words and the number of meaningful words recalled.

As a final exercise, we will plot our data points on a scatterplot to see if there is anything about the pattern of the data points that might warrant additional attention.

➡   **To have SPSS create a scatterplot, select**

> *Graph →Legacy Dialogs →Scatter/Dot →Simple Scatter →
> Define.*

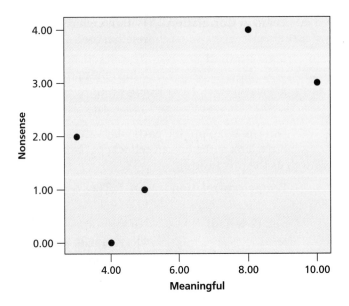

- Move the variable "Nonsense" to the *Y-Axis*.
- Move the variable "Meaningful" to the *X-Axis*.

The resulting scatterplot is shown above:

There does not appear to be any curvilinear pattern to the data.

Interpreting Correlations

If a correlation coefficient is significant, we can describe its strength further. Generally, a correlation of less than $\pm.20$ is weak, a correlation between $\pm.40$ and $\pm.70$ is moderate, and a correlation above $\pm.70$ is strong. Note that with very few data points, a moderate correlation may not be statistically significant. Conversely, when there are many data points, a small correlation coefficient may be statistically significant.

The Coefficient of Determination

To be able to interpret a correlation, we need to go one step further and calculate the **coefficient of determination,** which tells us how much of the variability in one variable (Y) is accounted for by the other variable (X). Calculating the **coefficient of determination** turns the correlation coefficient into a proportion. This is similar to computing Cohen's *d* to estimate effect size; in the case of a correlation, the size of the effect can be thought of as how much variance the variables under investigation have in common.

To determine the coefficient of determination, we square the value of *r* that we obtained. Remember that the correlation between the number of

The **coefficient of determination** tells us how much of the variability in the *Y* variable is accounted for by the *X* variable.

meaningful and nonsense words recalled was + .705. The coefficient of determination is

$$r^2 = .705^2, \text{ or } r^2 = (.705)(.705) = .4970$$

If we were dealing with real data, we would interpret this result as saying that approximately 50% of the variability we observed in the recall of meaningful words is associated with the number of nonsense words recalled.

Note that r^2 will always be a positive number! If you have a negative correlation and you square it, the negative number multiplied by the negative number will yield a positive number.

Restriction of Range

Restriction of range is a potential problem in a correlation, resulting in a distorted correlation coefficient due to truncated data.

The range of the data to be analyzed has an effect on the correlation. A potential problem, called **restriction of range**, can occur if you have a truncated data set (a limited range of data). For example, if you were interested in the correlation between reading ability and math ability, the correlation might look like the following:

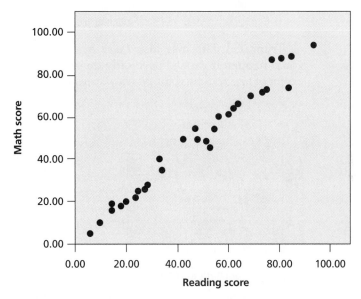

This scatterplot shows a moderate, positive correlation.

If your data set were limited to only those students who scored between 40 and 60 on the reading test when the complete range of scores varied from 0 to 100, your data set would be limited to the data points between 40 and 60, and the scatterplot would change considerably.

The correlation would go from being moderately strong and positive to being weak.

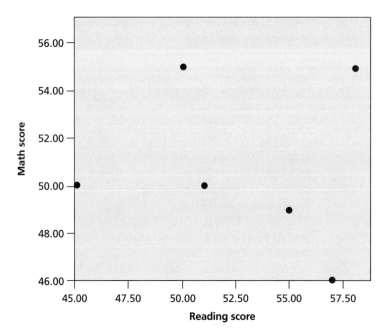

Truncated data may also mask a curvilinear relationship. Look what would happen if we had used only those under age 40 on our original scatterplot of age and need for physical assistance:

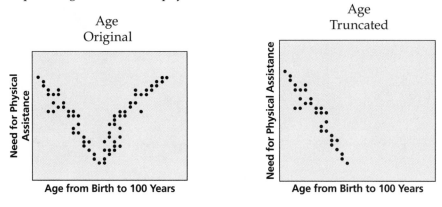

Here, the truncated correlation would appear to show a strong, negative relationship. We would not predict that the relationship between age and assistance rises in later years. This would be a mistake we would not make if we had the full range of data.

Outliers

Scatterplots also help us determine whether we have any outliers in our data. An **outlier** is a data point that is far outside the range of the rest of the data. Outliers may affect the correlation. For example, if we are measuring the height and weight of five-year-old children, we wouldn't expect to obtain a height measurement of 60 inches, nor would we expect to obtain a weight measurement of 25 pounds. Outliers are sometimes the result of measurement error

An **outlier** refers to a data point that is far outside the range of the rest of the data.

and sometimes the result of extreme scores. In either case, they will become obvious in a scatterplot, and then we can decide what to do about them.

Take another look at our scatterplot of age and height without any outliers. For these data, $r = .93$ and the $r^2 = .86$. There is a strong, positive relationship between age and height. With an outlier included in our analysis, $r = .45$ and $r^2 = .20$. We see a change from a very strong relationship to one that is moderate at best. The next two plots illustrate this change:

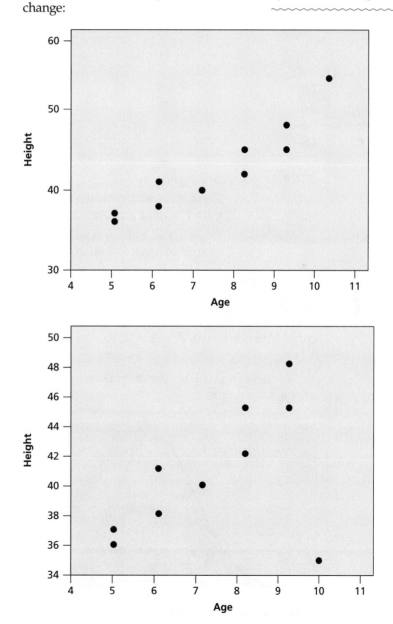

The moral of the story is that we must examine the data for outliers and the potential for a restricted range to make certain that our conclusion about the relationship is appropriate.

Application: Hypothesis Testing with Correlation

Let's now say that you are interested in the relationship between age and the ability to learn a second language. You collect the following data on seven children:

	Age	Score on the (fictional) Second Language Ability Test (SLAT)
Mikaela	10	82
Isabelle	12	80
Ivan	6	97
Benno	8	85
Sulki	15	75
Geno	9	88
Mari	11	81

STEP 1   State the hypotheses.

$H_0: \rho = 0$ There is no relationship between age and the ability to learn a second language, as measured by the SLAT.

$H_1: \rho \neq 0$ There is a significant relationship between age and the ability to learn a second language, as measured by the SLAT.

STEP 2   Determine the critical value of r.

In this example, we will set alpha to .05. In order to determine the critical value of r, we need to compute the degrees of freedom. Remember, for the correlation coefficient, $df = n - 2$, so

$$df = 7 - 2 = 5$$

We now can look up the critical value of r. With 5 df, and $p < .05$, the critical value of r is $\pm.754$, as is shown in the following:

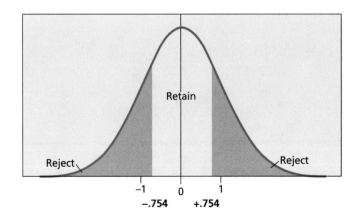

	Age	SLAT_SCORE
1	10	82
2	12	80
3	6	97
4	8	85
5	15	75
6	9	88
7	11	81

FIGURE 3 Sample SPSS data file: Age and SLAT scores.

STEP 3 Compute *r*.

Begin by creating an SPSS file. It should look like that presented in Figure 3.
    Once the data are entered into an SPSS file, you are ready to have SPSS compute *r*.

To perform a correlation test, select

> *Analyze → Correlate → Bivariate.*

- Move the variables "Age" and "SLAT" to the "Variables Box."

Upon completion, your output should look like that shown in Figure 4.

STEP 4 Evaluate the hypothesis.

With 5 *df* and $p < .05$, the critical value of *r* is $\pm.754$. The observed value of *r* is $-.929$. Therefore, we reject the null hypothesis. We found evidence of a significant linear relationship.
    This is a strong negative relationship. If we were dealing with real data, we would conclude that as age increases, the ability to learn a second language decreases.
    Let's look at these data on a scatterplot. We can use SPSS to make the plot.

To have SPSS create a scatterplot, select

> *Graph → Legacy Dialogs → Scatter/Dot → Simple Scatter → Define.*

- Move the variable "SLAT" to the *Y*-Axis.
- Move the variable "Age" to the *X*-Axis.

The plot is as follows:
    How much of the variability in the SLAT can be explained by age? To answer this question, we need to calculate the coefficient of determination, r^2. For this example, r^2 is 86%:

$$r^2 = (-.93)(-.93) = .86$$

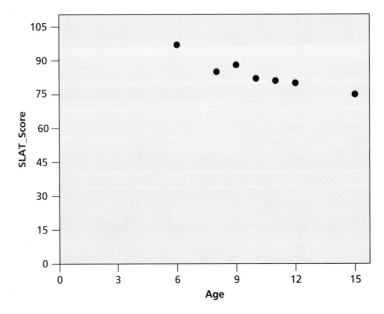

Correlations

		Age	SLAT_Score
age	Pearson Correlation	1	−.929(**)
	Sig. (2-tailed)		.002
	N	7	7
SLAT_Score	Pearson Correlation	−.929(**)	1
	Sig. (2-tailed)	.002	
	N	7	7

** Correlation is significant at the 0.01 level (2-tailed).

FIGURE 4 SPSS output: Correlation between Age and SLAT scores.

Be Here Now

A social worker is assessing the effect of family size on ratings of agency services. Using a scale of 1 (least helpful) to 10 (most helpful), she invites 15 families to rate the services. She notices that the overall ratings tend to increase as the number of children in the families goes up, so she decides to examine this relationship more closely by using correlation. Here are the data:

Family ID	Number of Children	Overall Rating
1	4	8
2	2	6
3	0	6

Be Here Now	Family ID	Number of Children	Overall Rating
	4	1	3
	5	8	9
	6	3	7
	7	6	9
	8	1	2
	9	5	6
	10	1	5
	11	2	6
	12	1	2
	13	2	5
	14	3	5
	15	4	6

Is there a significant relationship between the number of children in a family and the rating for the agency services?

Answer

STEP 1   State the hypotheses.
   $H_0: \rho = 0$ There is no relationship between the number of children and the agency rating.

   $H_1: \rho \neq 0$ There is a significant relationship between the number of children and the agency rating.

STEP 2   Determine the critical value of r.
   At alpha = .05 and $df = n - 2 = 13$, the critical value is .5139.

STEP 3   Compute r.
   The correlation for these data is $r = +.76$.

STEP 4   Evaluate the hypothesis.
   This value of r that we obtained exceeds the critical value of r. Thus, there is a significant relationship between the number of children in a family and the ratings the family gives to a social services agency. Notice that in SPSS, the correlation found is flagged as significant, using a two-tailed test, at an alpha level of .01. You need to calculate the coefficient of determination by hand. Here, $r^2 = .58$. So, just over half of the variability in the overall ratings can be explained by the number of children in the family. The social worker should also look at a scatterplot, to see if the data are linear and to see if there are any outliers.

(*Continued*)

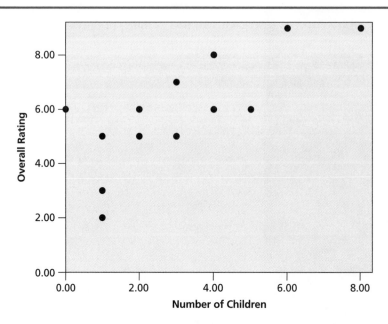

The data appear to be linear and there do not appear to be any outliers.

Reporting Correlation Results in APA Style

In reporting results in APA style, you may use the any of the following formats:

The correlation between family size and rating of agency services was significant, $r(13) = .76$, $p < .01$.

   or

The correlation between age and SLAT scores was significant, $r(7) = -.93$, $p = .01$.

   or

The correlation between recall of nonsense and meaningful words was not significant, $r(5) = .71$, $p = .18$.

   or

The correlation between recall of nonsense and meaningful words was not significant, $r(5) = .71$, *ns*.

Notice that in each example, r is lowercase and in italics and the number of degrees of freedom is shown in parentheses. If you report the probability, use a lowercase p in italics. You may use an equals sign with the actual probability, or you may use the preset alpha level with a less-than sign. Use two decimal places to report the correlation and probability. If you use the abbreviation *ns* for not significant, use italics.

In reporting the coefficient of determination, italicize the r^2 and give a brief explanation. For example, you might say,

The coefficient of determination was $r^2 = .75$, indicating that three-quarters of the variation in blood pressure was explained by the intensity of physical activity.

The Correlation Matrix

When you use SPSS to conduct a test for correlation, the output that the software provides is in the form of a correlation matrix. When you are correlating only two variables, the correlation matrix may not seem necessary. However, if you are interested in the correlations among a number of variables—such as cumulative GPA, SAT score, high school class rank, and number of advanced placement (AP) courses taken—then a correlation matrix provides a concise visual summary of the analyses.

Let's consider an example of a correlation matrix produced by SPSS:

		GPA	SAT	HSRANK	APCOURSE
GPA	Pearson Correlation	1	.938**	−.929**	.959**
	Sig. (2-tailed)		.000	.000	.000
SAT	N	10	10	10	10
	Pearson Correlation	.938**	1	−.771**	.895**
	Sig. (2-tailed)	.000		.009	.000
HSRANK	N	10	10	10	10
	Pearson Correlation	−.929**	−.771**	1	−.901**
	Sig. (2-tailed)	.000	.009		.000
APCOURSE	N	10	10	10	10
	Pearson Correlation	.959**	.895**	−.901**	1
	Sig. (2-tailed)	.000	.000	.000	
	N	10	10	10	10

The same information is shown above and below the diagonal made up of 1s. The correlation coefficients on the diagonal of a correlation matrix will always be 1, a perfect positive correlation (the correlation of a variable with itself).

The correlation matrix allows us to present the results of several analyses in one table. Look at Table 1 to see how the correlation matrix looks in APA format; notice that it is a little different from the SPSS correlation matrix.

The APA-style correlation matrix uses two decimal places in reporting the correlations, and it contains neither the redundant information provided in SPSS nor the coefficients on the diagonal.

Table 1

Intercorrelations between GPA, SAT, High School Rank, and Number of AP Courses ($n = 10$)

Variable	GPA	SAT	HSRank	APCourses
GPA	–	.94**	−.93**	.96**
SAT		–	−.77**	.90**
HSRank			–	−.90**
APCourses				–

Note. ** $p < .01$.

It's Out There...

Klomegah (2007) examined the predictors of academic performance of college students. He found a significant positive correlation between self-efficacy scores and course grades, $r(95) = .32$, $p < .01$. In particular, high levels of self-efficacy tended to occur with high levels of academic performance. The correlation between distractions in the study area and academic performance was not significant, $r(85) = .18$, $p > .05$.

Summary of Correlation

Correlation is used to examine relationships between two variables; it does *not* imply cause and effect. The plus or minus sign tells the direction of the relationship; the number indicates the strength of the relationship. The closer the correlation is to ±1.00, the stronger is the relationship. We calculate the coefficient of determination to establish how much of the variation in **Y** is explained by **X**. We always look at a scatterplot of the variables to determine the direction and shape of the relationship and whether there are any outliers. The range of scores in our data will affect the correlation; therefore, we also watch for potential problems with restrictions of ranges.

Regression

What Is Regression?

Simple regression uses a formula to predict values of *Y* for any given value of *X*.

Multiple regression allows us to relate a group of independent (predictor) variables to a dependent variable.

Regression is an approach that is similar to correlation, except that, in regression, we use one variable to predict another. When we have only two variables, we use what is called **simple regression**. Like correlation, simple regression looks at the relationship between two variables. In simple regression, we compute a formula that allows us to predict values of one variable (*Y*) for any given value of another variable (*X*). **Multiple regression** expands simple regression. In multiple regression, we relate a **group** of independent variables (predictors) to predict a dependent variable. We'll take a look at how simple regression works first, and then we'll use SPSS to see how the ideas can be extended by the use of more than two variables.

Calculating Simple Regression

In simple regression, we want to predict how one variable will behave on the basis of how another variable behaves. As with correlation, we are not implying that one variable *causes* the other to behave in a certain way. With regression, we establish a linear (straight-line) relationship that determines how much we expect Y to change when X changes by one unit. This approach allows us to systematically predict the behavior of one variable from that of the other.

Let's examine simple regression by using the example of shopping for a new car. If you were looking for a new car, aside from choosing the color, you might consider how many miles per gallon (MPG) the car gets, what the car's horsepower is, the number of cylinders the car has, the car's safety rating, and how quickly the car accelerates from zero to 60 miles per hour. Following are data pertaining to 10 cars:

Car Number	MPG	Horsepower	Cylinders	Safety Rating (1 = *low*, 10 = *high*)	Acceleration
1	18	130	8	2	16
2	24	84	6	5	22
3	22	94	6	4	15
4	45	67	4	9	14
5	65	93	4	10	17
6	38	85	4	8	19
7	19	150	6	4	17
8	20	122	8	5	14
9	28	79	4	6	17
10	35	65	4	8	15

For now, let's just look at data for the MPG and for the safety rating. We'll start by making a scatterplot of these data, with MPG on the x-axis and the safety rating on the y-axis (see graph on top of next page).

Notice that a line has been drawn so that it cuts through the center of the data points. Using this scatterplot, with the line through the center of the data, we can now make predictions about the variables. If, for example, we wanted to predict the safety rating for a car that gets 50 MPG, we'd draw a line from 50 on the x-axis, up to the line, and then across to the y-axis (see 2nd graph on next page).

We would predict that a car getting 50 MPG would get a safety rating of around 8.5. You could also make predictions going from the safety rating to MPG. How many MPG would you predict that a car with a safety rating of 5 would get?

It looks like the predicted value would be around 25 MPG.

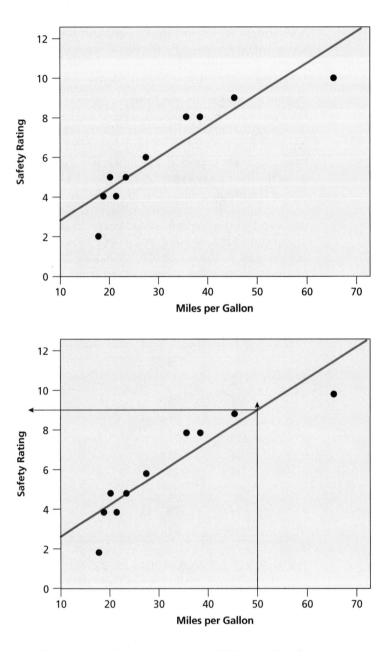

The regression calculation establishes a line for which the squared distance from each data point to the line is as small as it can be. This line is called the **least squares solution**, and it is what permits the best predictions possible. Of all lines that could be drawn, the least squares solution is the best because the distance between any point and the line is as small as possible.

Regression uses a **least squares solution** so that the distance between any point and the line is as small as possible.

Let's examine this concept more closely by using just one data point. We have one car in our data set that gets 45 miles per gallon and has a safety

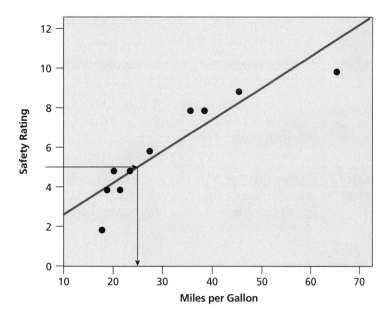

rating of 9. If we return to our graph and draw a line from 45 miles per gallon on the *x*-axis to the diagonal (least squares solution line), then move horizontally to the *y*-axis in order to predict a safety rating, you'd see that the *predicted* safety rating would be 8. The distance between the actual data point and the predicted data point is the error. We want as little error as possible, to get the best possible predictions. That is achieved with the least squares solution.

If we look again at our original scatterplot, we can see that some of the actual points are above the line and some are below:

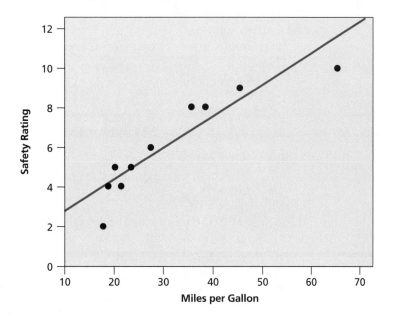

So, some of our error will be positive and some will be negative, depending upon whether the actual data point is above or below the line.

The symbol for an actual data point is **Y**.
The symbol for a predicted data point is **Y'**.

If we took each value of **Y** and subtracted its corresponding value of **Y'**, squared the resulting number, and then summed all the squares, we'd have the total squared error:

$$\Sigma(Y - Y')^2 = \text{Total Squared Error}$$

A **regression line** is the line created by the least squares solution.

Notice that $Y - Y'$ is like a deviation score; therefore, we need to square each number before summing, or we'd get a sum of zero. In regression, the line produced minimizes the total squared error.

The regression equation is

$$Y' = bX + a$$

where

Y' = value predicted for **Y**
b = slope
X = value of **X**
a = Y-intercept (point at which the line crosses the Y-axis).

The formulas for **a** and **b** are found in the Synthesis (Appendix D).

All of this may look familiar to you if you took algebra; it's the formula for a straight line.

The **regression equation** tells us how much change we would predict in **Y** for each one-unit change in **X**. Once we know **b** (the slope) and **a** (the Y-intercept), those two values remain constant. If **b** is positive, then as **X** increases, **Y'** also increases. If **b** is negative, then as **X** increases, **Y'** decreases. This is similar to what happens with a positive correlation and a negative correlation.

The **regression equation** tells us how much change we would predict in *Y* for each one-unit change in *X*.

Using the regression equation, we can plot the regression line with the least squares solution.

For example, if we calculate $b = 2$ and $a = 10$, the regression equation would be

$$Y' = bX + a, \quad \text{or} \quad Y' = 2X + 10$$

If $X = 1$, we have

$$Y' = 2(1) + 10 = 12$$

If $X = 2$, we have

$$Y' = 2(2) + 10 = 14$$

And, if $X = 3$, we have

$$Y' = 2(3) + 10 = 16$$

For this regression equation, each time **X** is increased by 1, we would predict that **Y** will increase by 2.

Reflect

In statistics, regression results in prediction.

<table>
<tr><td>**Be Here Now**</td><td>Use the regression equation $Y = bX + a$.
If $b = 3$ and $a = -2$, what would Y' be
when $X = 6$?
when $X = -2$?

Answers

When $X = 6$, $Y' = 16$.
When $X = -2$, $Y' = -8$.</td></tr>
</table>

So, simple regression uses one variable to predict another. More often, regression uses several variables (predictor or independent variables) to predict one dependent variable. Multiple regression is a useful statistical method that expands on simple regression and allows us to relate a group of variables to a dependent variable.

Let's return to our car data, supposing now that we were most concerned with safety. Which of the other variables in the data set would prove most helpful in explaining the variability in safety rating?

Car Number	MPG	Horsepower	Cylinders	Safety Rating (1 = *low*, 10 = *high*)	Acceleration
1	18	130	8	2	16
2	24	84	6	5	22
3	22	94	6	4	15
4	45	67	4	9	14
5	65	93	4	10	17
6	38	85	4	8	19
7	19	150	6	4	17
8	20	122	8	5	14
9	28	79	4	6	17
10	35	65	4	8	15

To use all of these variables—MPG, horsepower, the number of cylinders, and the rate of acceleration from zero to 60 mph—to predict safety ratings, we need to use multiple regression—and SPSS. The safety rating (SAFETY) becomes the dependent variable in the multiple regression; MPG, HORSEPOWER, CYLINDERS, and ACCELERATION are the independent variables.

Application: Regression and Multiple Regression

➡️ **To conduct a regression analysis, select**

Analyze → Regression → Linear.

- Highlight the dependent variable ("SAFETY"), and use the arrow to move it to the "Dependent Variable" box.
- Highlight the remaining variable(s), and use the arrow to move them to the box for "Independent Variables." SPSS labels the variables you use as predictors—that is, the independent variables.

If you are using multiple regression, you must choose a regression method. This is not necessary if you are using simple regression.

Regression Method

In SPSS, there are several types of regression models from which to choose. The *enter* method, which is the default procedure in SPSS, enters all predictor variables into the regression equation simultaneously. This is the method we will use.

You also have several options regarding the statistics provided via SPSS. One option that is typically of interest is the descriptive statistics. You can get the descriptive statistics if you click on the "Statistics" button.

Statistics → Descriptive Statistics.

Regression Output

The SPSS output is provided in several sections: Descriptives, Correlations, Variables Used, Model Summary, ANOVA model, and Regression Coefficients.

The SPSS output will show the descriptive statistics first, with the mean, standard deviation, and *n* given. This output is presented in Figure 5.

The first row provides the labels for each column. Each row provides the descriptive statistics for each variable.

Descriptive Statistics

	Mean	Std. Deviation	N
SAFETY	6.1000	2.55821	10
MPG	31.4000	14.86383	10
HORSEPOWER	96.9000	28.10476	10
CYLINDERS	5.4000	1.64655	10
ACCELERATION	16.6000	2.45855	10

FIGURE 5 SPSS output for regression analysis: Descriptive statistics.

First Column   Indicates the name of each variable included in the analysis.

Mean   The mean of each variable is provided. For example, the mean safety rating is 6.1.

Std. Deviation   The standard deviation of each variable is given. For example, the standard deviation for the safety rating is 2.55821.

N   The number of observations for each variable is provided (10).

Figure 6 shows the correlation matrix.

Notice that this matrix shows the simple correlations between each pair of variables. By looking at the section of the matrix labeled "Sig. (1-tailed)," you can see which correlations are statistically significant.

First Column   Lists the statistic presented. There are three sections: the correlation, the significance level, and *N* (the number in the sample). On the right-hand side of the first column, the variables included in the

Correlations

		Safety	Mpg	Horse-power	Cylinders	Accele-ration
Pearson Correlation	SAFETY	1.000	.911	−.677	−.828	−.046
	MPG	.911	1.000	−.515	−.715	−.016
	HORSEPOWER	−.677	−.515	1.000	.738	−.060
	CYLINDERS	−.828	−.715	.738	1.000	−.121
	ACCELERATION	−.046	−.016	−.060	−.121	1.000
Sig. (1–tailed)	SAFETY	-	.000	.016	.002	.450
	MPG	.000	-	.064	.010	.482
	HORSEPOWER	.016	.064	-	.007	.434
	CYLINDERS	.002	.010	.007	-	.370
	ACCELERATION	.450	.482	.434	.370	-
N	SAFETY	10	10	10	10	10
	MPG	10	10	10	10	10
	HORSEPOWER	10	10	10	10	10
	CYLINDERS	10	10	10	10	10
	ACCELERATION	10	10	10	10	10

FIGURE 6 SPSS output for regression analysis: Correlation matrix.

regression analysis are listed for each statistic (correlation, significance level, and N).

First Row   Across the first row are the variables included in the regression.

Pearson Correlation   The correlation between each pair of variables is reported. For example, the correlation between SAFETY and MPG is .911. For each pair, remember to look at the labels for the column and row to identify the correlation.

Sig (1-tailed)   The one-tailed level of significance (alpha) is presented for each correlation coefficient.

N   The number of pairs of observations used to compute each correlation is presented. Ten observations were used to generate each correlation.

Figure 7 is the SPSS output that shows the variables used in the regression.

Model   You may select more than one model to run. We selected only one.

Variables Entered   All predictor variables that entered into the regression equation are listed. In our example, these variables are ACCELERATION, MPG, HORSEPOWER, and CYLINDERS.

Variables Removed   Some regression models allow for some variables to be removed from the equation. The enter method does not, so no variables appear in the Variables Removed column.

Method   The method we selected, the default option, is Enter.

Notice that the dependent variable, the variable to be predicted, is identified in footnote b of Figure 7. In our example, it is SAFETY.

Variables Entered/Removed[b]

Model	Variables Entered	Variables Removed	Method
1	ACCELERATION, MPG, HORSEPOWER, CYLINDERS[a]		Enter

[a]All requested variables entered.

[b]Dependent Variable: Safety

FIGURE 7 SPSS output for regression analysis: Variables used.

Figure 8 shows the Model Summary.

The first row of the model summary labels the statistics computed.

Model   There was one model, so we see a 1 under Model.

R   R indicates the multiple correlation coefficient for the predictor (independent) variables and the dependent variable. In this case, **R** = .954.

R Square   R Square is the proportion of the variance in the dependent variable (SAFETY) that is predictable from the independent variables. In our example, **R** Square (*or* R^2) = .911.

Adjusted R Square R^2 is adjusted for the number of variables in the regression compared with the number of participants. Here, Adjusted R^2 is .839. If you are interested in the formula, see the Synthesis (Appendix D).

Standard Error of Estimate   The last box is the Standard Error of the Estimate. This is the average difference between the predicted value and the actual value of the dependent variable. In our example, it is 1.02504.

The next output section is the ANOVA for the model, as shown in Figure 9.

Across the top row of Figure 9 are the labels for the statistics computed. The ANOVA tests whether the independent variables *as a set* show a relationship

Model Summary

Model	R	R Square	Adjusted R Square	Std. Error of the Estimate
1	.954(a)	.911	.839	1.02504

[a]Predictors: (Constant), Acceleration, MPG, Horsepower, Cylinders

FIGURE 8 SPSS output for regression analysis: Model summary.

ANOVA[b]

Model		Sum of Squares	df	Mean Square	F	Sig.
1	Regression	53.646	4	13.412	12.764	–008(a)
	Residual	5.254	5	1.051		
	Total	58.900	9			

[a] Predictors: (Constant), Acceleration, MPG, Horsepower, Cylinders

[b] Dependent Variable: Safety

FIGURE 9 SPSS output for regression analysis: ANOVA summary.

to the dependent variable that is stronger than we would predict to occur by chance.

First Column	The rightmost part of the first column, with no heading, lists the sources of variance. The variance due to the regression model is provided in the first row. The variance due to error that is not accounted for by the regression model is given in the row labeled "Residual." Total describes the variance across all participants and measures.
Sum of Squares	For each source of variance, the sum of squares (SS) is computed. $SS_{regression}$ (regression) is 53.646, while $SS_{residual}$ (error) is 5.254.
df	The *df* column provides the number of degrees of freedom associated with each source of variance. Four degrees of freedom were associated with the regression model, and five degrees of freedom were associated with the residual.
Mean Square	For each source of variance, the mean square (MS) is computed by dividing SS by *df*. $MS_{regression}$ is 13.412. $MS_{residual}$ is 1.051.
F	The *F* ratio, computed by dividing $MS_{regression}$ by $MS_{residual}$ is 12.764.
Sig	Looking at the Sig. box, we see that $p = .008$, which tells us that we have a significant model. Therefore, the independent variables *as a set* show a relationship to the dependent variable that is stronger than we would predict to occur by chance.

As shown in Figure 10, the last output box presents the information that we need to construct a regression equation and achieve a least squares solution.

Shown in Figure 10 are the statistics computed for each predictor variable.

Model	The information provided is for regression model 1, the only one we asked SPSS to compute. The statistics computed are for each predictor variable and the constant.
Unstandardized Coefficients B	The column labeled "Unstandardized Coefficients B" shows the values for the regression equation. These values are the weights assigned to each variable in the regression equation to predict the values of the dependent variable (the safety ratings).

Coefficients(a)

Model		Unstandardized Coefficients		standardized Coefficients		
		B	Std. Error	Beta	t	Sig.
1	(Constant)	7.572	3.761		2.013	.100
	MPG	.110	.033	.639	3.306	.021
	HORSEPOWER	−.014	.018	−.157	−.794	.463
	CYLINDERS	−.410	.384	−.264	−1.068	.334
	ACCELERATION	−.080	.142	−.077	−.564	.597

ᵃ Dependent Variable: Safety

FIGURE 10 SPSS output for regression analysis: Regression coefficients.

	The B associated with MPG is .110.
Unstandardized Coefficients Std. Error	The next column shows the standard errors. We leave their interpretation for a more advanced text.
Standardized Coefficients Beta	The column labeled "Beta" shows a set of coefficients that represent standardized values for the variables, as if they had been converted to z-scores. The Beta weight for MGP is .639.
t	Each coefficient is tested against the null hypothesis that the coefficient equals zero in the population. The t value printed is used in this test. The t testing the Beta weight for MPG is 3.306.
Sig.	The alpha at which each t is significant is printed next. In our example, only MPG is significant, with $p = .021$. This indicates that even after the other variables have been taken into account, in the end MPG is the only variable that accounted for a significant portion of the variance in the safety ratings.

We can use the output to create a regression equation for our car example. For this set of variables, the regression equation would be

$$Y'_{safety} = bX_{MPG} + bX_{Horsepower} + bX_{cylinder} + bX_{acceleration} + \text{constant}$$

Replacing the constant and b's with the numbers from the output, we get

$$Y'_{safety} = (.11)X_{MPG} + (-.01)X_{Horsepower} + (-.41)X_{cylinder}$$
$$+ (-.08)X_{acceleration} + 7.572$$

For car 1 in the data set, we can use the regression equation to obtain a predicted value for SAFETY:

$$Y'_{safety} = (.11)18_{MPG} + (-.01)130_{Horsepower} + (-.41)8_{cylinder}$$
$$+ (-.08)16_{acceleration} + 7.572$$

$$Y'_{safety} = 1.98_{MPG} + (-1.30)_{Horsepower} + (-3.28)_{cylinder}$$
$$+ (-1.28)_{acceleration} + 7.572$$

$$Y' = 3.692$$

The *predicted* SAFETY rating for the first car is $Y' = \mathbf{3.692}$. The *actual* value for the first car for SAFETY is $Y = \mathbf{2.00}$.

So, how should we interpret this regression equation? Well, if these were real data, then, for any car we were considering buying, we would want to look closely at the MPG as a predictor of the safety rating for that car. This is because MPG was the only predictor that accounted for a significant amount of the variance in Safety Rating ($t = 3.31$, $p < .05$).

> **Reflect**
>
> Sometimes it takes a vast amount of information to make a single prediction. Sometimes one piece of information is all that is needed.

Reporting Regression Results in APA Style

In writing the results of a regression analysis in APA format, it is advisable to use a table that shows the unstandardized coefficients, the standard error, and the standardized coefficients, indicating which variables were significant in the *t*-tests. Report the regression method used and the value of n in the title of the table. You can report R Square and Adjusted R Square in a note under the table or in the text. Be certain to give a brief explanation of the table in the text.

From our example, you would write the following:

Table 2 shows the results of the multiple regression analysis using the enter method for predicting the safety rating of cars. MPG was a statistically significant predictor, $p < .01$. $R^2 = .91$; Adjusted $R^2 = .84$.

Table 2
Summary of Multiple Regression Analysis Using the Enter Method for Variables Predicting the Safety Rating of Cars ($n = 10$)

Variable	B	SE B	β
Miles per gallon	.11	.03	.64*
Horsepower	−.10	.02	−.16
Number of cylinders	−.41	.38	−.26
Acceleration	−.08	.14	−.08

Note. $R^2 = .91$; Adjusted $R^2 = .84$.
* $p < .05$

It's Out There...

Poelzer, Zeng, and Simonsson (2007) were interested in developing a model to predict how well pre-service teachers would do on a Teacher Certification Test. As shown in Table 3, a combination of scores from the ACT Composite, Benchmark Teaching Test, and TASP Reading Test accounts for 53 percent of the variance in Teacher Certification Test scores.

TABLE 3

Summary of Multiple Regression Analysis for Variables Predicting Scores on a Teacher Certification Test ($n = 87$)

Variable	B	SE B	β
ACT Composite	.72	.21	.35**
Benchmark Teaching Test	1.04	35	.25**
TASP Reading Test	.14	.05	.29**

NOTE. $R^2 = .53$; Adjusted $R^2 = .51$.
**$p < .01$.

Be Here Now

Recall the example that we used to illustrate correlation: the relationship between age and height. The data were as follows:

Age in Years	Height in Inches
5	36
6	38
7	40
8	42
9	45
6	41
8	45
9	48
10	54
5	37

Taking age as the independent variable and height as the dependent variable, use SPSS to answer the following questions:

1. What is the regression equation?

2. What height would you predict for a three-year-old? A seven-year-old?

3. Is age significant as an independent variable? How do you know?

4. How much of the variability in height is explained by age in this example? How do you know?

(Continued)

Answers

Coefficients^a

Model	Unstandardized Coefficients		standardized Coefficients		
	B	Std. Error	Beta	t	Sig.
1 (Constant)	21.246	2.995		7.094	.000
age	2.925	.400	.933	7.316	.000

^bDependent Variable: SAFETY

1. $Y' = 2.93X + 21.25$
   B weight for age = 2.93   Constant = 21.25
2. When $X = 3$: $Y' = 2.93(3) + 21.25 = 30.04$ inches
   When $X = 7$: $Y' = 2.93(7) + 21.25 = 41.76$ inches
3. Yes, because the *t*-test result for age is significant at $<.001$.
4. Approximately 85%, because the Adjusted R Square is .854.

SUMMARY OF REGRESSION

Simple regression and multiple regression are useful statistical methods for predicting values of one dependent variable from values of one or more independent (predictor) variables. Like correlation, simple regression and multiple regression do not imply causation; rather, they address the relationships among the variables. All predicted values in a regression equation fall on the regression line. Although the actual values of data points will reflect some error, the amount of squared error is kept as small as possible with the least squares solution.

PRACTICE

Scatter Diagrams

1. For each of the following scatterplots, identify the shape of the relationship:

a.

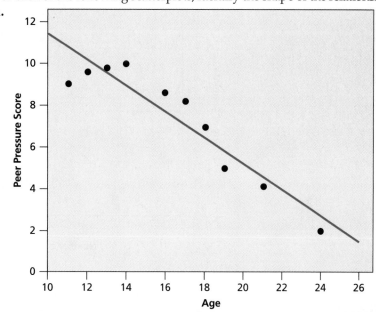

b.

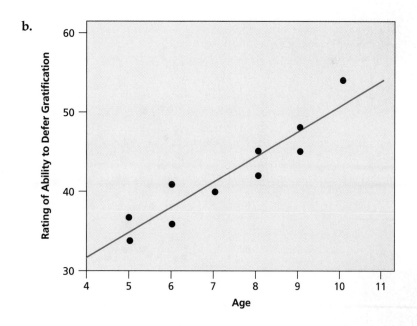

c.

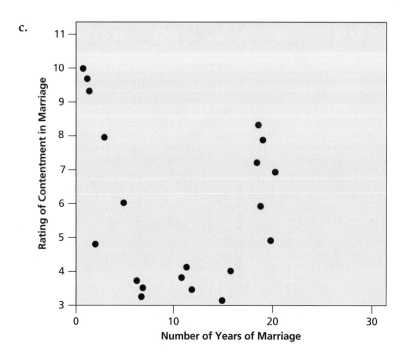

2. Look at the following scatterplot and estimate the strength of the correlation:

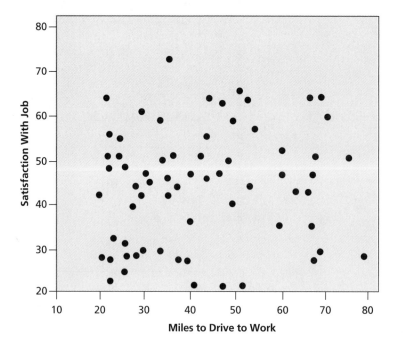

3. Describe a potential problem with the following scatterplot:

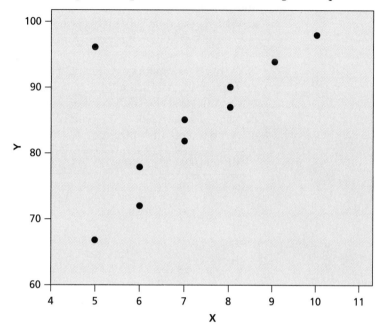

Correlation

4. Using SPSS, enter the data set shown and answer the following questions:
   a. What is the relationship between population density and quality of medical care? Give the correlation coefficient and interpret the relationship in words.
   b. Is this relationship significant? How do you know?
   c. What is the coefficient of determination for these data? How would you interpret the coefficient of determination?
   d. If the researcher collected data only on areas with a low population density, what potential problem would result?
   e. Make a scatterplot of the data. What does the plot show?

Population Density (in thousands)	Rating of Quality of Medical Care (1 = low, 10 = high)
20,500	4
3,700	8
17,800	7
8,500	5
600	9
33,900	3
226	9
10,400	6
41,100	4
1,100	8

5. Using SPSS, enter the data set shown and answer the following questions:
   a. What is the relationship between time spent per day exercising and weight loss? Give the correlation coefficient and interpret the relationship in words.
   b. Is this relationship significant? How do you know?
   c. What is the coefficient of determination for these data? How would you interpret the coefficient of determination?
   d. Make a scatterplot of the data. What does the plot show?
   e. Write the results of this correlation in APA format.

Minutes/Day Exercising	Weight Loss (pounds/week)
45	.50
30	.25
50	.40
75	1.00
90	2.00
20	.33
10	.10
60	1.50

6. Using SPSS, enter the data set shown and answer the following questions:
   a. What is the relationship between years of education and income? Give the correlation coefficient and interpret the relationship in words.
   b. What is the coefficient of determination for these data? How would you interpret the coefficient of determination?
   c. Write the results of this correlation in APA format.

Years of Education	Annual Income (in thousands)
8	20
10	12
12	24
13	15
13	35
16	29
17	45
17	62
18	74
22	55
22	58
24	275

Regression

7. Some people believe that they can predict the weather on the basis of less-than-scientific factors, such as aching joints or the appearance of snails in the garden. Enter the data set shown into SPSS, use amount of rainfall as the dependent variable, and answer the following questions:
   a. What is the regression equation?
   b. Is the model significant? How do you know?
   c. Which of the independent variables are significant? How do you know?
   d. How much of the variation in rainfall is explained by these variables? How do you know?

Rainfall (in mm)	Rating (1 = low, 10 = high), Aching Joints	# Snails in the Garden	Barometric Pressure (mB)
2.04	6	5	1013
2.56	4	6	1017
3.75	7	6	1020
1.38	5	4	1007
.05	6	5	1000
1.07	4	3	1003

8. A researcher is interested in which factors predict young adults' popularity. Using SPSS and the data shown, answer the following questions:
   a. What is the regression equation?
   b. What are the mean and standard deviation for the dependent variable?
   c. Are there any significant correlations between the independent variables and popularity? If so, which ones?
   d. How would you write the results in APA format?

# of Voice mail Messages/Day	# of Invitations to Parties/Month	Monthly Income (in Thousands)	Empathy Scale Score (1 = low, 10 = high)	Popularity Rating (1 = low, 10 = high)
8	11	1.80	9	10
6	8	2.60	8	8
7	6	2.10	7	8
5	5	3.00	7	7
8	7	2.75	6	7
4	2	2.90	5	6
4	2	2.40	4	5
2	1	3.10	4	4

9. A child development specialist is interested in maternal factors that might predict low birth weight. He has collected the following data:

Birth Weight	# Cigarettes Smoked/Day	# Alcoholic Drinks/Week
5.15	10	7
8.25	0	0
7.50	4	2
9.75	0	0
4.40	20	17
4.75	16	14
5.00	10	7
4.60	15	12
9.25	0	0
7.15	2	0

Using SPSS, answer the following questions about these data:
a. What is the regression equation?
b. What birth weight would you predict for the newborn of a mother who smoked seven cigarettes per day and had five alcoholic drinks a week?
c. What birth weight would you predict for the newborn of a mother who did not smoke and had one drink a week?
d. Which of the independent variables are significant? How do you know?

10. Those who play sports often engage in superstitious behavior. But does it work? Using SPSS to analyze the fictitious data shown, determine which, if any, of the superstitious behaviors listed predict batting average. Explain your answer.

Wearing lucky Socks (1 = Yes, 0 = No)	# Times Touches Bat to Ground	# Times Spitting	Batting Average
1	6	3	.133
1	5	4	.225
1	8	5	.150
0	3	1	.333
0	2	0	.375
0	2	1	.500

SOLUTIONS

Scatter Diagrams

1. **a.** negative relationship
    b. positive relationship
    c. curvilinear relationship

2. near zero or no relationship

3. There is an outlier that may distort the correlation.

Correlation

4. **a.** The correlation coefficient is $-.85$. As population density increases, the rating for the quality of medical care decreases.
    b. The correlation is significant at alpha $< .05$. This can be seen on the output line labeled "Sig. (2-tailed)."
    c. The coefficient of determination is $r^2 = .72$. For these data, approximately 72% of the variability in the ratings can be explained by the population density.
    d. restriction of range
    e. The scatterplot shows a moderately strong negative relationship between these variables.

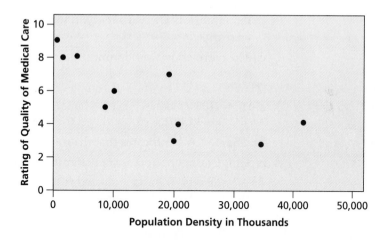

5. **a.** The correlation is .88. As the number of minutes exercised per week increased, so did the weight loss.
    b. The correlation is significant at alpha $< .05$. This can be seen on the output line labeled "Sig. (2-tailed)."
    c. The coefficient of determination is $r^2 = .77$. For these data, approximately 77% of the variability in weight loss can be explained by the amount of time spent exercising.

d. This scatterplot shows the positive relationship between minutes per day spent exercising and weight loss in pounds per week:

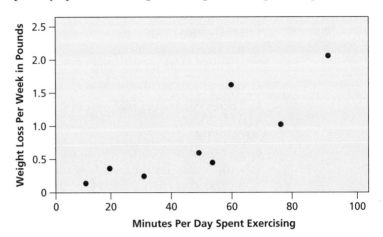

e. The correlation was significant, $r(7) = .88$, $p < .05$.

6. a. The correlation is .68. As years of education increased, so did income.

b. The coefficient of determination is $r^2 = .46$. For these data, approximately 46% of the variability in income can be explained by years of education.

c. The correlation was significant, $r(11) = .68$, $p < .05$.

Regression

7. a. $Y' = .133$ aches $+ (-.307)$ snails $+ .184$ barometric pressure $+ (-183.20)$

b. The ANOVA indicates that the model is significant.

c. Only barometric pressure is significant, as shown by the t-test.

d. R Square is .973 and Adjusted R Square is .933, which indicate that this model explains almost all of the variance in rainfall.

8. a. $Y' = .287$ voicemail $+ (-.137$ invitations$) + (-.775$ income$) + .863$ empathy $+ 2.618$

b. The mean and the standard deviation of the dependent variable popularity are 6.88 and 1.89, respectively.

c. The correlation matrix indicates that all of the variables are significantly correlated with popularity.

d. Table 4 shows the results of the multiple regression analysis using the enter method for variables predicting popularity in young adults. Empathy was the only statistically significant predictor, $p < .05$. $R^2 = .99$; Adjusted $R^2 = .97$.

Table 4

Summary of Multiple Regression Analysis Using the Enter Method for Variables Predicting the Popularity of Young Adults ($n = 8$)

Variable	B	SE B	β
Voicemail Messages	.29	.14	.33
# Monthly Invitations	−.14	.17	−.25
Monthly Income	−.78	.38	−.19
Rating of Empathy	.86	.22	.84*

Note. $R^2 = .99$; Adjusted $R^2 = .97$.
*$p < .05$.

9. **a.** $Y' = -.91$ cigarettes $+ .77$ drinks $+ 9.08$
   b. 6.56 lb
   c. 9.85 lb
   d. Both smoking and drinking are significant, as shown by the *t*-tests.

10. The ANOVA model was not significant, and none of the superstitious behaviors were significant in the regression equation.

-15-

The Chi-Square Test: Hypothesis Tests for Frequencies

FOCUS | Chi-square tests allow us to answer questions about nominal or categorical data. Do the numbers of males and females at the university match the numbers in the general population? Is children's toy selection related to gender? In each of these cases, we look at frequencies. We are simply counting. We will describe two chi-square tests: a test for goodness of fit and a test for independence.

The Chi-Square Statistic

The chi-square statistic revolves around the differences between frequencies we *observe* and frequencies we *expect*. Sometimes, what we expect is based upon information available in the population. For example, if we know that 49% of all adult residents in the state are male, then we would expect 49% of registered voters to be male. Sometimes, however, our expectations are based upon what we would expect by chance. For example, if there are no differences between how boys and girls select toys, then, given a

choice of toys, boys and girls should make their selections by chance. Small differences between what we observe and what we expect could be due to sampling error; after all, life is messy. However, the larger the difference between what we observe and what we expect, the more likely it is that the difference is not due to sampling error.

In the broadest sense, chi-square is

$$\chi^2 = \frac{\text{The sum of the squared differences between the frequencies we observe and those we expect}}{\text{The frequencies we expect}}$$

For each of the categories that we are examining, we square the differences between the frequencies we observe and the frequencies we expect, and then we sum the squared differences. If we didn't square the differences first, they would sum to zero, even when differences exist between what we observe and what we expect.

We are ready to introduce the conceptual formula for chi-square, which is symbolized with the Greek letter χ, squared, or χ^2

$$\chi^2 = \frac{\text{Sum of the squared differences between what we observe and what we expect}}{\text{Expected frequencies}}$$

Expected frequency values are symbolized as f_e.

We use the symbol f_e to refer to **expected frequency**. In one variation of the chi-square test—the chi-square test for goodness of fit—the expected frequencies are determined by our prior knowledge. In another variation of the test—the chi-square test for independence—the expected frequencies are based upon what we might expect by chance.

Observed frequency values are symbolized as f_o and are based on the frequencies we actually observe.

We use the symbol f_o to refer to **observed frequency**. These are the frequencies we actually observe.

Using the preceding notation, we state the formula for χ^2 as

$$\chi^2 = \sum \frac{(f_o - f_e)^2}{f_e}$$

The formula produces larger values of χ^2 for larger differences between observed and expected frequencies. If there are no differences between any observed and expected frequencies, χ^2 will be zero. Because what we observe may differ from what we expect, we use χ^2 to help us decide whether there is a significant difference between our observed and expected frequencies.

> **Reflect**
>
> *Sometimes, what we expect is not what we observe.*

Assumptions of the Chi-Square Test

Several requirements must be met in conducting a chi-square test:

- Each observation must fall into one and only one category.
- Observations must be independent of each other.
- Observations are measured as frequencies.
- The expected frequency for any category cannot be less than 5. Some statisticians suggest that if we cannot meet this requirement, then a mathematical correction called Yates' correction for continuity should be employed. There is, however, some controversy over this suggestion, as well as over other conditions in which the correction should be used (e.g., see Brown, 1988; Hatch & Lazaraton, 1991).

The Chi-Square Test for Goodness of Fit

The **chi-square test for goodness of fit** examines differences between observed frequencies and frequencies we expect based upon prior knowledge.

The American Psychological Association (2002) reported that 75% of undergraduate psychology majors are female. It is possible to look at the number of male and female psychology majors at any given university and compare those data with the national profile of psychology majors. Consider, for example, a class of 40 students currently taking a psychology course. If we observe that there are only six males in the class, does the class depart significantly from the national profile? The **chi-square test for goodness of fit** can examine whether this class "fits" the national profile.

Hypotheses for the Chi-Square Test for Goodness of Fit

Let's begin by formulating a null and an alternative hypothesis:

H_0: The gender composition of the class is not different from the national profile for psychology majors.

    or

$$H_0: f_o = f_e$$

H_1: The gender composition of the class is different from the national profile.

    or

$$H_1: f_o \neq f_e$$

The chi-square test for goodness of fit examines the differences between the frequencies of males and females that we *observe* and the frequencies of males and females that we would *expect* if the class conformed to—or *fit*—the national profile.

Let's consider the frequencies of males and females that we would expect if the null hypothesis were true. On the basis of the national profile, in a class of 40 students we expect 25% of the students in the class to be male; thus, we would expect 10 males [25% of 40 = (.25)(40) = 10]. In a class of 40 students, we expect 75% of the students in the class to be female; thus, we would expect 30 females [75% of 40 = (.75)(40) = 30]. The expected frequencies for the chi-square test for goodness of fit are determined by our knowledge of the population, in this case the national gender trend for psychology majors. So, f_e for males = 10 and f_e for females = 30.

Now let's consider the frequencies of males and females that we *actually* observed. We observed 6 males and 34 females in our example. These are our observed frequencies, or f_o. So, f_o for males = 6 and f_o for females = 34.

As we mentioned, if the null hypothesis were true and life were perfect, we would observe 10 males and 30 females in our class. In that case, the differences between what we observe and what we expect would be zero, and χ^2 would be zero. The question now is: Are the observed frequencies in our example sufficiently different from our expected frequencies to cause us to reject the null hypothesis and conclude that our class differs significantly from (or does not fit) the population gender profile? The chi-square test for goodness of fit will provide the statistic necessary for us to make a decision.

Degrees of Freedom for the Chi-Square Test for Goodness of Fit

The number of degrees of freedom that we calculate for the chi-square test for goodness of fit reflects the number of *categories* that we are comparing, minus 1. We symbolize the number of categories as *c*.

Thus, for the chi-square test,

$$df = c - 1$$

In our example,

$$df = 2 - 1$$
$$df = 1$$

We are ready to find the critical value of χ^2 in the table of chi-square values (Appendix A5).

Because χ^2 cannot be negative (remember, 0 is as small as it can be), we identify a single critical region. Chi-square is a one-tailed test, and if our calculated value of χ^2 falls within this critical region, we will reject the null hypothesis.

In our example, we'll use an alpha level of .05, and with 1 *df*, the critical value of χ^2 is 3.84. Notice that the distribution of χ^2 values begins at 0:

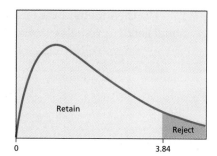

Next, we will compare a calculated value of x^2 with the critical value of x^2. If the calculated value does not exceed the critical value (and thus is not in the critical region), we fail to reject the null hypothesis. If the calculated value falls into the critical region, we reject the null hypothesis and conclude that our class departs significantly from the national gender profile.

Application: Chi Square Test for Goodness of Fit

As before, we'll follow the steps of the hypothesis-testing procedure, and we will use SPSS for our calculation of x^2.

STEP 1   State the null and alternative hypotheses.

H_0: The gender composition of the class is not different from the national profile for psychology majors.

   or

$$H_0: f_o = f_e$$

H_1: The gender composition of the class is different from the national profile.

   or

$$H_1: f_o \neq f_e$$

Next, let's organize the information that we have. Recall that APA (2002) suggests that 75% of psychology majors are female and 25% are male. Thus, in a sample of 40 students, we would expect 10 males and 30 females. In our class, we observed 6 males and 34 females. We can organize the information that we have into a table, or matrix:

	f_o	f_e
Males	6	10
Females	34	30

STEP 2   Choose the alpha level and draw the distribution.

For this example, we use an alpha level of .05. We've calculated $df = c - 1 = 2 - 1 = 1$. We determined that the critical value of $\chi^2 = 3.84$.

To reject the null hypothesis and conclude that the class does not fit the national gender profile, our calculated value of x^2 must fall beyond 3.84:

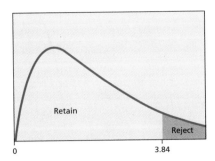

STEP 3   Calculate the appropriate test statistic.

We will use SPSS to calculate x^2.

Data entry into SPSS for the x^2 test uses numeric codes to signify whether a student is male or female. In using SPSS, it is important to remember that each row in the data entry page (the spreadsheet) represents a participant (in this case, a student) and each column represents a variable (in this case, gender). We must click on the tab for "Variable View" and define only one variable: "Gender." In this example, we will use "1" to represent males, and "2" to represent females.

Now we click on the tab labeled "Data View," and we enter either a "1" or a "2" for each student.

Once we have entered all of the data, our spreadsheet consists of 40 entries, with each row presenting the coded information for one student. An example of what the SPSS spreadsheet would look like is presented in Figure 1.

➡ We are now ready for SPSS to calculate the Chi-Square statistic

Choose Analyze → Nonparametric Tests → Chi-Square.

- Move the variable "Gender" over to the "Test Variable List."
- Because the expected frequencies are not equal in our example ($f_e = 10$ for males and $f_e = 30$ for females), we must click on the option "Values" and enter our expected frequency values. Always enter the expected frequency values in the exact same order as the codes that were used in the spreadsheet, so here we will first enter 10 (the f_e for code "1," for males) and then click "Add." Next, we will enter 30 (the f_e for code "2," for females) and click "Add."

Interpretation: Chi-Square Test for Goodness of Fit

Figure 2 presents the SPSS output for the results of the chi-square test for goodness of fit for the gender profile of the psychology class.

Two output sections are printed. One is labeled Gender, the other Test Statistics.

	Gender
1	1
2	1
3	1
4	1
5	1
6	1
7	2
8	2
9	2
10	2
11	2
12	2
13	2
14	2
15	2
16	2
17	2
18	2
19	2
20	2

	Gender
21	2
22	2
23	2
24	2
25	2
26	2
27	2
28	2
29	2
30	2
31	2
32	2
33	2
34	2
35	2
36	2
37	2
38	2
39	2
40	2

FIGURE 1 SPSS spreadsheet for data entry for the x^2 test for goodness of fit.

Gender

In the first output box, the top row labels the calculations made by SPSS. Below and to the left, each row is labeled with the codes used to designate the categories, in this case 1.00 for males, 2.00 for females, and Total for the entire sample.

Observed N	Consists of the observed frequencies (f_o) for males and females. For example, the observed frequency for males was 6.
Expected N	Consists of the expected frequencies (f_e) for males and females. For example, the expected frequency for males was 10.

NPar Tests
Chi-Square Test
Frequencies

Gender

	Observed N	Expected N	Residual
1.00	6	10.0	–4.0
2.00	34	30.0	4.0
Total	40		

Test Statistics

	Gender
Chi-Square[a]	2.133
df	1
Asymp. Sig.	.144

a 0 cells (.0%) have expected frequencies less than 5.
  The minimum expected cell frequency is 10.0.

FIGURE 2 SPSS output for the x^2 test for goodness of fit for the gender profile of the psychology class.

Residual   Consists of the differences between f_o and f_e.

For example, the difference between the observed and expected frequency for males is -4.

Test Statistics
The second output box contains the results of the chi-square test. The top row labels the variable analyzed, in this case "Gender." Below and to the left, each row is labeled with the statistic calculated.

Chi-Square   Indicates the calculated value of $x^2 = 2.133$.

df   Indicates that *df* = 1.

Asymp. Sig.   Indicates the probability of making a Type I error: $p = .144$.

STEP 4   Decision

On the basis of the results, we need to decide whether to reject the null hypothesis.

We can compare our calculated value of x^2 with our critical value of x^2, 3.84. As shown here, our calculated x^2 of 2.133 does not fall into the critical region.

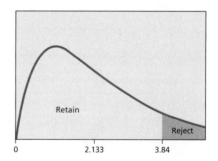

Or, in evaluating our hypothesis, we can use the value of p in the column labeled "Asymp. Sig." Because the p value of .144 is greater than our preset alpha of .05, we fail to reject the null hypothesis.

Either way, in our example we fail to reject the null hypothesis, and we conclude that we do not have evidence to suggest that our class differs from the national gender profile.

Let's consider another example. A psychologist hypothesized that birth order (first-born, middle-born, or last-born) is related to the motivation to achieve, as reflected in one's ambition to attend graduate school. Suppose the psychologist distributes a questionnaire to a group of 60 upper-level undergraduates who attend a workshop on preparing applications to graduate programs. One of the questions on the survey asks whether the student is first-, middle-, or last-born. The psychologist is interested in whether birth-order is related to aspiration to attend graduate school; in other words, are there significant differences in the numbers of first-born, middle-born, and last-born students? We use SPSS to answer this question as follows:

STEP 1   State the null and alternative hypotheses.

H_0: Birth order is not related to ambition to attend graduate school.
    or

$$H_0: f_o = f_e$$

H_1: Birth order is related to ambition to attend graduate school.
    or

$$H_1: f_o \neq f_e$$

If birth order has no relationship to ambition to attend graduate school, we will see equivalent numbers of first-, middle-, and last-born students in the workshop. Thus, we would expect the frequencies of first-, middle-, and last-born students in the workshop to be equal. Because we have 60 students in the workshop, we would expect 20 students in each of the birthorder categories. Our observed frequencies, based upon student responses, are also included in the matrix.

	f_o	f_e
First-Born	31	20
Middle-Born	13	20
Last-Born	16	20

STEP 2 Choose the alpha level and draw the distribution.

For this example, we will use an alpha level of .05. We calculate $df = c - 1 = 3 - 1 = 2$. We find the critical value of $x^2 = 5.99$ (see Appendix A5).

To reject the null hypothesis, the calculated value of x^2 must fall beyond 5.99.

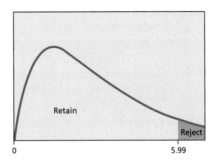

STEP 3 Calculate the appropriate test statistic.

Remember that in order to enter the data for a x^2 test, we use numeric codes to signify whether a student is first-born, middle-born, or last-born. Also, recall that each row in the data entry page (the spreadsheet) is a participant and each column is a variable. We click on the tab for "Variable View," and define only one variable: "BirthOrder." In this example, we will use "1" to represent first-born, "2" to represent middle-born, and "3" to represent last-born.

When we now click on the tab labeled "Data View," we can enter a "1" or a "2" or a "3" for each student.

Once we have entered all of the data, our spreadsheet consists of 60 entries, with each row presenting the information on one participant. A sample of what the SPSS spreadsheet would look like is presented in Figure 3.

We move the variable "BirthOrder" over to the "Test Variable List."

In this example, the expected frequencies are equal ($f_e = 20$), so we make sure that the option "All categories equal" is selected. It is the default value.

Figure 4 presents the SPSS output for the results of the chi-square test for goodness of fit testing differences between first-, middle-, and last-born students' graduate school ambitions.

	BirthOrder
1	1
2	1
3	1
4	1
5	1
6	1
7	1
8	1
9	1
10	1
11	1
12	1
13	1
14	1
15	1
16	1
17	1
18	1
19	1
20	1
21	1
22	1
23	1
24	1
25	1
26	1
27	1
28	1
29	1
30	1

	BirthOrder
31	1
32	2
33	2
34	2
35	2
36	2
37	2
38	2
39	2
40	2
41	2
42	2
43	2
44	2
45	3
46	3
47	3
48	3
49	3
50	3
51	3
52	3
53	3
54	3
55	3
56	3
57	3
58	3
59	3
60	3

FIGURE 3 SPSS spreadsheet for data entry for the χ^2 for goodness of fit.

NPar Tests
Chi-Square Test
Frequencies

BirthOrder

	Observed N	Expected N	Residual
1.00	31	20.0	11.0
2.00	13	20.0	–7.0
3.00	16	20.0	–4.0
Total	60		

Test Statistics

	BirthOrder
Chi-Square[a]	9.300
df	2
Asymp.Sig.	.010

a 0 cells (.0%) have expected frequencies less than 5.
The minimum expected cell frequency is 20.0.

FIGURE 4 SPSS output for the χ^2 test for goodness of fit of birthorder and ambition to attend graduate school.

BirthOrder
In the first output box, the top row labels the calculations made by SPSS. Below and to the left, each row is labeled with the codes used to designate the categories, in this case 1.00 for first-born, 2.00 for middle-born, and 3.00 for last-born.

Observed N	Consists of the observed frequencies (f_o) for first-, middle-, and last-born students. For example, the observed frequency for first-born students is 31.
Expected N	Consists of the expected frequencies (f_e) for first-, middle-, and last-born students. For example, the expected frequency for first-born students is 20.
Residual	Consists of the differences between f_o and f_e. For example, the difference between the observed and expected frequencies of first-born students is 11.

Test Statistics

The second output box contains the results of the chi-square test. The top row labels the variable analyzed, in this case "BirthOrder." Below and to the left, each row is labeled with the statistic calculated.

Chi-Square	Indicates the calculated value of $\chi^2 = 9.30$.
df	Indicates that $df = 2$.
Asymp. Sig.	Indicates the probability of making a Type I error: $p = .01$.

STEP 4 On the basis of the results decide whether to reject the null hypothesis.

We can compare our calculated value of χ^2 with our critical value of χ^2, or we can use the value of p in the column labeled "Asymp. Sig." and reject the null hypothesis if $p < .05$. Using either method for our example, we reject the null hypothesis. On the basis of this sample of students, we conclude that birth order is related to ambition to attend graduate school. In other words, there are significant differences, by birth order, in aspirations to attend graduate school.

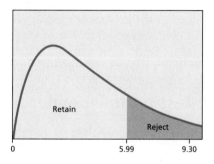

It is important to point out that when we use χ^2 to reject the null hypothesis, we must be prudent in our conclusions. Inferences about a specific birth order cannot be made without additional tests. We *can* conclude that the pattern of our data is different from what would be expected if birth order is unrelated to graduate school ambition to attend graduate school, but we cannot say more than that.

Remember that there are several requirements for conducting the chi-square test. One of them is the assumption that each observation must fall into one and only one category. That requirement is worth revisiting in the context of this example. Consider how we might handle an only child in our example: You *cannot* count an only child in both the first-born and last-born categories.

Another requirement of the chi-square test is that the expected frequency in any given cell cannot be less than 5. In fact, the output in SPSS (in the form of a note at the bottom of the Test Statistics box) reports whether the expected frequency in any cell is less than 5. When the expected frequency in any cell is less than 5, we can apply **Yates' correction** (Yates, 1934; Shavelson, 1996). Yates' Correction is relatively simple to calculate for the chi-square test for goodness of fit: If the observed frequency is greater than the expected frequency in a category, we subtract .5, and if the observed frequency is less than the expected frequency, we add .5. The correction will reduce the final value of x^2 and will "smooth" the data. SPSS does not apply Yates' correction at all when performing the chi-square test for goodness of fit, so the correction must be calculated and applied to x^2 by hand. Of course, we do not need to apply Yates' correction in our example, because the expected frequency is greater than 5 in each of our cells.

Reporting the Results in APA Style

Chi-square test results are reported in a format similar to those of other statistical tests.

For our gender example, we might write:

The gender composition of the class was not different from the APA (2002) gender profile for psychology majors, $\chi^2(1, N = 40) = 2.133$, **ns**. Or,

There was no significant difference between the class and the 2002 gender profile for psychology majors, $\chi^2(1, N = 40) = 2.133$, **ns**.

For our birth-order example, we might write:

On the basis of the sample, ambition to attend graduate school varied by birth order, $\chi^2(1, N = 60) = 9.30$, $p < .05$. Or,

Birth order was found to be related to ambition to attend graduate school in the sample of students canvassed, $\chi^2(1, N = 60) = 9.30$, $p = .01$.

The Chi-Square Test for Independence

The **chi-square test for independence** examines whether two variables are independent of each other in the population.

The chi-square test for goodness of fit allows us to assess whether a set of observed frequencies for a single nominal variable (gender, birth order, importance of parental relationship) fits a set of expected frequencies on the basis of prior information. The **chi-square test for independence** allows us to look at the relationship between *two* nominal variables. For example, we can look at two nominal variables—say, participation in team sports *and* introversion/extroversion—to determine whether they are independent of each other in the population. The chi-square test for independence is also called a **contingency table chi-square** test. In considering whether team sports participation is independent of introversion/extroversion, we would be considering a 2 (participant vs. nonparticipant) by 2 (introverted vs. extroverted) factorial design.

It's Out There...

Hang-up rates among callers put on hold was examined by Munichor and Rafaeli (2007). In particular, the hang-up rates of callers for whom music was played was compared with the hang-up rates of those who were told their estimated wait time. A significant difference was found, $\chi^2(1, N = 123) = 8.44$, $p < .005$. More callers (69.4%) abandoned the call while music was playing, relative to those who were told their place in the queue (35.9%).

Be Here Now

A therapist asks incoming clients to complete a questionnaire about the perceived importance of various family relationships. The questionnaire asks clients who were raised by both parents whether their relationship with their mother or father was the more important. In a sample of 22 clients, 13 indicate that their relationship with their mother was the more important one. The therapist wonders whether clients are equally likely to report their relationship with their mother or father as the more important. Test at $\alpha = .05$.

Answer

H_0 = Clients are equally likely to report their relationship with their mother or father as the more important.

$$H_0: f_o = f_e$$

H_1 = Clients are not equally likely to report their relationship with their mother or father as the more important.

$$H_1: f_o \neq f_e$$

With $c = 2$, our $df = 2 - 1 = 0$, and our critical value of χ^2 is 3.84. We have the following table of frequencies:

	f_o	f_e
Mother	13	11
Father	9	11

The SPSS output is as follows:

Relationship

	Observed N	Expected N	Residual
1.00	13	11.0	2.0
2.00	9	11.0	−2.0
Total	22		

Test Statistics

	Relationship
Chi-Square(a)	.727
df	1
Asymp. Sig	.394

We fail to reject the null hypothesis, because we failed to find evidence that clients report one relationship as more important than the other, $\chi^2(1, N = 22) = .727$, *ns*.

For example, suppose that a school psychologist is interested in whether participation in team sports during high school is related to the introversion or extroversion of the student. She administers a questionnaire and an introversion/extroversion inventory to the entire class of sophomores. She collects two vital pieces of information from all of the students: whether they participate in team sports and whether their introversion/extroversion test results indicate that they are introverted or extroverted.

In the chi-square test for independence, we compare the observed frequencies within categories (introverted participants, extroverted participants, introverted nonparticipants, and extroverted nonparticipants) with what we would expect if participants were assigned to the categories by chance.

Hypotheses for the Chi-Square Test for Independence

Unlike the chi-square test for goodness of fit, in which we have prior information to guide our hypotheses, the chi-square test for independence focuses on the *relationship* between the nominal variables that we are examining. The null hypothesis for the chi-square test for independence proposes that the variables are *independent*, or unrelated. If the variables are unrelated, we would expect participants to be assigned to the categories by chance. The alternative hypothesis suggests that there is a relationship between the variables.

STEP 1   State the null and the alternative hypotheses.

H_0: Participation in team sports and introversion/extroversion are independent of each other in high school sophomores.
    or

$$H_0: f_o = f_e$$

where f_e is the expected value if no relationship exists.

H_1: Participation in team sports and introversion/extroversion are related in high school sophomores.
    or

$$H_1: f_o \neq f_e$$

where f_e is the expected value if no relationship exists.

A **contingency table** presents frequencies for combinations of two nominal variables.

We can organize the data into a matrix called a **contingency table** that we present in Table 1. Notice that the two nominal variables (Team Sports and Personality) are listed in the table, which contains our observed frequencies (f_o) that are based upon the frequency counts in the investigation. SPSS will calculate our expected frequencies (f_e).

TABLE 1

Numbers of Introverted and Extroverted High School Sophomores, by Participation in Team Sports

	Team Sports	
Personality	Participant	Nonparticipant
Introvert	29	76
Extrovert	87	18

Degrees of Freedom for the Chi-Square Test for Independence

In the chi-square test for independence, we examine the relationship between two nominal variables, and our degrees of freedom must reflect this examination. One of the variables, "Team Sports," has two possible categories: Participant and Nonparticipant. The other variable, "Personality," also has two possible categories: Introvert and Extrovert. In order to calculate the number of degrees of freedom for the chi-square test for independence, we need to account for the two categories—or two columns (or c)—of Team Sports participation and two types of Personality presented in two rows (or r). This configuration is easily seen in the contingency table, in which sports participation distinguishes c by columns and personality defines r by rows.

To calculate the number of degrees of freedom for the chi-square test for independence, we use the formula

$$df = (c - 1)(r - 1)$$

In our example,

$$df = (2 - 1)(2 - 1)$$
$$df = (1)(1)$$
$$df = 1$$

STEP 2   Choose the alpha level and draw the distribution.

For this example, we will use an alpha level of .05.

With 1 df, we find that the critical value of χ^2 is 3.84 (Appendix A5), and we can draw the hypothetical distribution of χ^2 values if the null hypothesis were true (see next page).

If the null hypothesis were true, we would expect no differences between our observed frequencies and the frequencies we would expect if students were randomly assigned to the different categories.

Our critical value of χ^2 is 3.84. To reject the null hypothesis, our calculated value of χ^2 must fall beyond 3.84.

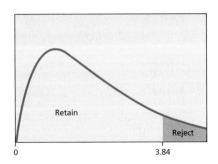

Application: Chi-Square Test for Independence

STEP 3 Calculate the appropriate test statistic.

We will use SPSS to calculate χ^2.

Data entry into SPSS for the chi-square test for independence uses numeric codes for two variables—in our example, participation in sports and personality. We click on the tab for "Variable View," and we define one variable "Sports" and a second variable "Personality." For each variable, we use the codes "1" and "2." For the variable "Sports," we'll use "1" to represent those who participate and "2" for students who do not participate. For the variable "Personality," we'll use "1" to represent introverted students and "2" to represent extroverted students.

When we now click on the tab labeled "Data View," we enter pairs of codes—combinations of the numbers "1" and "2"—to represent the information on the class of high school sophomores. In our example, there were 29 students who participated in sports and who were introverted, so we enter 29 pairs of "1" and "1" (Participant and Introvert), 87 pairs of "1" and "2" (Participant and Extrovert), 76 pairs of "2" and "1" (Nonparticipant and Introvert), and 18 pairs of "2" and "2" (Nonparticipant and Extrovert).

Once we have entered all of the data (and we have a lot to enter!), our spreadsheet consists of 210 rows, with each row containing a pair of codes. A portion of what the SPSS spreadsheet would look like is presented in Figure 5.

➡ **We are now ready for SPSS to calculate the Chi-Square statistic**

Choose Analyze → Descriptive Statistics → Crosstabs.

- Move the column variable, "Sports," to the "Column(s)" box.
- Move the row variable, "Personality," to the "Row(s)" box.
- Click on the *Statistics* button, and select "Chi Square" at the top.
- Click *Continue* in the Statistics dialog box, and then *OK*.

Interpretation: Chi Square Test of Independence

Figure 6 presents the output for this chi-square test for independence.

Three sections of output are presented (Case Processing Summary, Personality*Sports Crosstabulations, and Chi Square Tests).

	Sports	Personality
1	1.00	1.00
2	1.00	1.00
3	1.00	1.00
4	1.00	1.00
5	1.00	1.00
6	1.00	1.00
7	1.00	1.00
8	1.00	1.00
9	1.00	1.00
10	1.00	1.00
11	1.00	1.00
12	1.00	1.00
13	1.00	1.00
14	1.00	1.00
15	1.00	1.00
16	1.00	1.00
17	1.00	1.00
18	1.00	1.00
19	1.00	1.00
20	1.00	1.00
21	1.00	1.00

	Sports	Personality
190	2.00	2.00
191	2.00	2.00
192	2.00	2.00
193	2.00	2.00
194	2.00	2.00
195	2.00	2.00
196	2.00	2.00
197	2.00	2.00
198	2.00	2.00
199	2.00	2.00
200	2.00	2.00
201	2.00	2.00
202	2.00	2.00
203	2.00	2.00
204	2.00	2.00
205	2.00	2.00
206	2.00	2.00
207	2.00	2.00
208	2.00	2.00
209	2.00	2.00
210	2.00	2.00

FIGURE 5 Portion of SPSS spreadsheet for data entry for the χ^2 test for independence.

Case Processing Summary

In the first output box, the top row labels the statistics analyzed for the cases entered into SPSS. The bottom row presents the statistics calculated, namely, N and Percent, for Valid, Missing, and Total cases.

Personality * Sports   Consists of $N = 210$ Valid cases (or 100.0%) of the possible combinations of Personality and Sports. No cases (.0%) were missing. The Total $N = 210$ (100.0%).

Case Processing Summary

	Cases					
	Valid		Missing		Total	
	N	Percent	N	Percent	N	Percent
Personality * Sports	210	100.0%	0	.0%	210	100.0%

Personality * Sports Crosstabulation

Count

		Sports		Total
		1.00	2.00	
Personality	1.00	29	76	105
	2.00	87	18	105
Total		116	94	210

Chi-Square Tests

	Value	df	Asymp. Sig. (2-sided)	Exact Sig. (2-sided)	Exact Sig. (1-sided)
Pearson Chi-Square	64.787(b)	1	.000		
Continuity Corrections(a)	62.572	1	.000		
Likelihood Ratio	68.845	1	.000		
Fisher's Exact Test				.000	.000
Linear-by-Linear Association	64.479	1	.000		
N of Valid Cases	210				

a Computed only for a 2x2 table
b 0 cells (.0%) have expected count less than 5. The minimum expected count is 47.00

FIGURE 6 SPSS output for the χ^2 test for independence of sports participation and introversion/extroversion.

Personality * Sports Crosstabulation
In the second output box, we see SPSS's version of the contingency table, with the word "Count" reminding us that this is the summary of the counts of our observed frequencies. The topmost row is labeled "Sports," with the next row beneath indicating the numeric codes 1.00 (remember, 1 = Partici-pants) and 2.00 (2 = Nonparticipants), and Total observed frequencies across the two categories of sports participation.

Personality This variable is subdivided into each of the two personality categories (1.00 = Introvert and 2.00 = Extrovert). For example, for the Introverts (Personality = 1.00), 29 students were counted in the combined categories of Introvert and Participant, 76 in the combined categories of Introvert and Nonparticipant, and a Total of 105 students across both Sports categories who were introverts.

Total Here we see the totals for each of the two Sports categories across the two Personality categories. For example, there were 116 students who participated in sports (Sports = 1.00).

Chi-Square Tests

The third output box contains the results of several different chi-square tests that SPSS calculates. The top row indicates the statistic presented in each column (Value, for the calculated value of χ^2, *df* for the *df* calculated for each chi-square test, and Asymp. Sig (2-sided), which presents the calculated value of *p*). For our purposes, we will discuss only the first two chi-square test results.

Pearson Chi-Square We use the Pearson chi-square value for our calculated value of chi-square because it gives the basic chi-square statistic. The other choices of chi-square tests in the output box are related to special instances (e.g., whether all the cells are filled). In this example, the calculated value of $\chi^2 = 64.787$ with 1 *df* and *p* = .000.

Continuity Correction Here, SPSS calculates a corrected value of χ^2 because *df* = 1. Notice that the footnote at the bottom of the output box reveals that this correction is calculated only for a 2×2 table. The reason is that, for the chi-square test for independence, only a 2×2 table would produce *df* = 1. Some statisticians suggest that Yates' continuity correction should be used whenever *df* = 1 for the chi-square test; SPSS calculates Yates' continuity correction for us.

The corrected value of $\chi^2 = 62.575$ with *df* = 1 and *p* = .000. Notice that the χ^2 value produced by Yates' correction is slightly smaller than that produced by the Pearson chi-square.

STEP 4 On the basis of the results, decide whether to reject the null hypothesis or fail to reject the null hypothesis.

Whether we compare our calculated value of χ^2 with our critical value of χ^2 or use the value of p in the column labeled "Asymp. Sig. (Two-Sided)," we reject the null hypothesis. We conclude that participation in team sports and introversion/extroversion are related in this sample of high school sophomores.

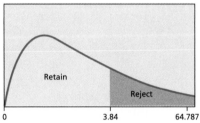

Reporting the Results in APA Style

In APA format, and using our earlier presented Table 1, we might report our findings as follows:

Table 1 presents the frequencies of introverted and extroverted high school sophomores and their participation in team sports. There was a significant relationship between participation in team sports and introversion/extroversion, $\chi^2(1, N = 210) = 62.58, p < .001$, continuity correction applied.

Table 1

Numbers of Introverted and Extroverted High School Sophomores, by Participation in Team Sports

	Team Sports	
Personality	**Participant**	**Nonparticipant**
Introvert	29	76
Extrovert	87	18

Notice that we do not report that $p = .000$. Although it appears that way in the output, the actual alpha level is not 0; rather, it is so small that, when rounded, it effectively becomes zero.

Notice also that we reported the value of chi-square that was corrected for continuity because $df = 1$.

It's Out There...

Corsi, Kwiatkowski, and Booth (2007) examined variables related to intravenous drug users' decision to return for a follow-up interview. Being in case management was found to be related to the decision to return for a follow-up interview: $\chi^2(1, N = 642) = 4.07, p < .05$.

Be Here Now

A principal is interested in determining whether there is a relationship between gender and color preference for a new school uniform. Children in the cafeteria are asked whether they would prefer to wear a red, blue, or green school uniform. Each child's gender is noted. The data are organized in the following contingency table:

	Male	Female
Red	14	10
Blue	15	19
Green	11	6

Are gender and preference for color of school uniforms related?

Answer

H_0 = Gender and preference for school uniforms are independent.
H_0: $f_o = f_e$
H_1 = Gender and preference for school uniforms are related. H_1: $f_o \neq f_e$
With $c = 2$ and $r = 3$, it follows that $df = (2 - 1)(3 - 1) = (1)(2) = 2$, and our critical value of χ^2 is 5.88.

Crosstabs

Case Processing Summary

	Cases					
	Valid		Missing		Total	
	N	Percent	N	Percent	N	Percent
Color * Gender	75	100.0%	0	.0%	75	100.0%

Color* Gender Crosstabulation

Count

		Gender		Total
		1.00	2.00	
Color	1.00	14	10	24
	2.00	15	19	34
	3.00	11	6	17
Total		40	35	75

(Continued)

Chi-Square Tests

	Value	df	Asymp. Sig. (2-sided)
Pearson Chi-Square	2.285(a)	2	.319
Likelihood Ratio	2.300	2	.317
Linear-by-Linear Association	.053	1	.818
N of Valid Case	75		

a 0 cells (.0%) have expected count less than 5. The minimum expected count is 7.93

We fail to reject the null hypothesis. Thus, we conclude that gender and color preference for school uniforms are unrelated, $\chi^2(2, N = 75) = 2.285$, *ns.*

SUMMARY

Chi-square tests allow us to answer questions about nominal data in the form of frequency counts. There are two chi-square tests: a test for goodness of fit and a test for independence. The chi-square test for goodness of fit allows us to assess whether a set of observed frequencies for a single nominal variable fits a set of expected frequencies. The chi-square test for independence allows us to look at the relationship between *two* categorical variables.

Reflect

In the end, there is no fear.

PRACTICE

Chi-Square Test for Goodness of Fit

1. A neuroscientist examines whether rats, like humans, are predisposed to favor either their right or left side. She counts the number of times out of 50 trials a rat chooses to take the left alley or right alley in a maze. The rat chooses to take the right alley 34 times. Does this rat show a significant preference for the right side?

2. A university sets up a "hot line" to provide technical assistance to members of the university community. The director of the computer center wants to know whether an equal number of first-year, second-year,

third-year, and fourth-year students utilize a "help hot line." During a one-week period, the following frequencies are observed:

Class Year	f
First Year	26
Second Year	38
Third Year	20
Fourth Year	31

Does utilization of the hot line vary by class year?

3. Concerned about the healthy development of women's attitudes toward their bodies, a developmental psychologist asks fifth-grade girls whether they believe that they are "underweight," "overweight," or "just right." Out of 30 fifth-grade girls, 4 select "underweight," 23 select "overweight," and 3 indicate that they believe that their weight is "just right." Is there a significant difference in the numbers of girls who identify themselves as underweight, overweight, or of an appropriate weight?

Chi-Square Test for Independence

4. A developmental psychologist is interested in whether gender-stereotypic toys are preferred by each gender. She allows each child tested to select from a doll, a toy truck, a "Slinky®," and a picture book. The following are the observed frequencies:

	Doll	Truck	Slinky®	Book
Male	6	28	15	19
Female	23	8	17	14

Does toy selection vary by gender?

5. A psychologist hypothesizes that the most important characteristic in selecting a mate may vary by socioeconomic status. He distributes a questionnaire at a local shopping mall. The questionnaire contains a question requesting the respondent's annual income, divided into four categories: below $20,000; $20,000–$49,999; $50,000–$79,999; and $80,000 or above. Characteristics that may be important in selecting a mate include education, physical appearance, sexual compatibility, and religious affiliation. Participants are asked to select only one characteristic as the most important in selecting a mate. The following contingency table presents the data:

	Education	Appearance	Sex	Religion
Below $20,000	9	10	16	14
$20,000–$49,999	21	38	17	7
$50,000–$79,999	33	15	11	8
$80,000 or above	20	12	19	6

Do those of different socioeconomic status report different characteristics as the most important in selecting a mate?

6. A geriatric psychologist speculates that elderly men and women report suffering from depression equally often. A survey is administered at a senior-citizen complex, requesting senior citizens to report their gender as well as whether they consider themselves depressed. The following data are collected:

	Depressed	Not Depressed
Male	19	28
Female	33	20

Is the reporting of depression related to the gender of the elderly person?

Presenting Results in APA Format

7. Present the results of Practice Problem 3 in APA format.

8. Present the results of Practice Problem 4 in APA format.

SOLUTIONS

Chi-Square Test for Goodness of Fit

1. H_0 = There is no right–left preference. H_0: $f_o = f_e$
   H_1 = There is a right–left preference. H_1: $f_o \neq f_e$
   With $c = 2$, it follows that $df = 2 - 1 = 1$, and our critical value of x^2 is 3.84 for alpha = .05.

   When entering the data into SPSS, we will code right-alley choice as "1" and left-alley choice as "2."

Preference

	Observed N	Expected N	Residual
1.00	34	25.0	9.0
2.00	16	25.0	–9.0
Total	50		

Test Statistics

	Preference
Chi-Square[a]	6.480
df	1
Asymp. Sig.	.011

a 0 cells (.0%) have expected frequencies less than 5. The minimum expected cell frequency is 25.0.

We reject the null hypothesis. There is a preference,
$\chi^2(1, N = 50) = 6.480, p = .011$.

2. H_0 = Hot line utilization does not vary by class year. $H_0: f_o = f_e$

H_1 = Hot line utilization varies by class year. $H_1: f_o \neq f_e$

With $c = 4$, it follows that $df = 4 - 1 = 3$, and our critical value of χ^2
is 7.81 for alpha $= .05$.

When entering the data into SPSS, we will code first-year students as
"1," second-year students as "2," third-year students as "3," and fourth-
year students as "4."

Class Year

	Observed N	Expected N	Residual
1.00	26	28.8	– 2.8
2.00	38	28.8	9.3
3.00	20	28.8	– 8.8
4.00	31	28.8	2.3
Total	115		

Test Statistics

	Preference
Chi-Square[a]	6.078
df	3
Asymp.Sig	.108

a 0 cells (.0%) have expected frequencies less than 5.
   The minimum expected cell frequency is 28.8.

We fail to reject the null hypothesis. There is no evidence to suggest
that hot line utilization varies by class year, $\chi^2(3, N = 115) = 6.078$, *ns*.

3. H_0 = Girls are equally likely to characterize themselves as underweight,
overweight, or of appropriate weight. $H_0: f_o = f_e$

H_1 = Girls are not equally likely to characterize themselves as underweight, overweight, or of appropriate weight. $H_1: f_o \neq f_e$

With $c = 3$, it follows that $df = 3 - 1 = 2$, and our critical value of χ^2
is 5.99 for alpha $= .05$.

When entering the data into SPSS, we will code responses of
Underweight as "1," Overweight as "2," and Just Right as "3."

BodyImage

	Observed N	Expected N	Residual
1.00	4	10.0	– 6.0
2.00	23	10.0	13.0
3.00	3	10.0	– 7.0
Total	30		

Test Statistics

	BodyImage
Chi-Square[a]	25.400
df	2
Asymp.Sig.	.000

a 0 cells (.0%) have expected frequencies less than 5.
   The minimum expected cell frequency is 10.0.

We reject the null hypothesis. Fifth-grade girls do not characterize themselves as underweight, overweight, or of appropriate weight equally often, $\chi^2(2, N = 30) = 25.40$, $p < .001$.

Chi-Square Test for Independence

4. H_0 = Gender is unrelated to toy selection. $H_0: f_o = f_e$
   H_1 = Gender is related to toy selection. $H_1: f_o \neq f_e$
   With $c = 4$ and $r = 2$, it follows that $df = (4 - 1)(2 - 1) = (3)(1) = 3$, and our critical value of χ^2 is 7.81 at alpha $= .05$.
   When entering the data into SPSS, for toy selection, we will code Doll as "1," Truck as "2," Slinky® as "3," and Book as "4." For gender, we will code Male as "1" and Female as "2."

Case Processing Summary

	Cases					
	Valid		Missing		Total	
	N	Percent	N	Percent	N	Percent
Gender * Toy	130	100.0%	0	.0%	130	100.0%

Gender * Toy Crosstabulation

Count

		Toy				Total
		1.00	2.00	3.00	4.00	
Gender	1.00	6	28	15	19	68
	2.00	23	8	17	14	62
Total		29	36	32	33	130

Chi-Square Tests

	Value	df	Asymp. Sig. (2-sided)
Pearson Chi-Square	21.729(a)	3	.000
Likelihood Ratio	23.010	3	.000
Linear-by-Linear Association	3.026	1	.082
N of Valid Cases	130		

a 0 cells (.0%) have expected count less than 5. The minimum expected count is 13.83.

We reject the null hypothesis. Gender is related to toy selection, $\chi^2(2, N = 130) = 21.73$, $p < .001$.

5. H_o = Mate selection criteria are not related to socioeconomic status. H_0: $f_o = f_e$

H_1 = Mate selection criteria are related to socioeconomic status. H_1: $f_o \neq f_e$

With $c = 4$ and $r = 4$, it follows that $df = (4 - 1)(4 - 1) = (3)(3) = 9$, and our critical value of χ^2 is 16.92 at alpha = .05.

When entering the data into SPSS, for mate selection criteria, we will code Education as "1," Appearance as "2," Sex as "3," and Religion as "4." For income, we will code the lowest income as "1," $20,000–$49,999 as "2," $50,000–$79,999 as "3," and the highest income as "4."

Case Processing Summary

	Cases					
	Valid		Missing		Total	
	N	Percent	N	Percent	N	Percent
Income * Criterion	256	100.0%	0	.0%	256	100.0%

Income * Criterion Crosstabulation

Count

		Criterion				Total
		1.00	2.00	3.00	4.00	
Income	1.00	9	10	16	14	49
	2.00	21	38	17	7	83
	3.00	33	15	11	8	67
	4.00	20	12	19	6	57
Total		83	75	63	35	256

Chi-Square Tests

	Value	df	Asymp. Sig. (2-sided)
Pearson Chi-Square	37.333(a)	9	.000
Likelihood Ratio	34.817	9	.000
Linear-by-Linear Association	6.709	1	.010
N of Valid Cases	256		

a 0 cells (.0%) have expected count less than 5. The minimum expected count is 6.70.

We reject the null hypothesis. Mate selection criteria are related to socioeconomic status, $\chi^2(9, N = 256) = 37.33$, $p < .001$.

6. H_0 = Self-reports of depression are not related to gender in the elderly
H_0: $f_o = f_e$
H_1 = Self-reports of depression are related to gender in the elderly
H_1: $f_o \neq f_e$
With $c = 2$, and $r = 2$, our $df = (2 - 1)(2 - 1) = (1)(1) = 9$, and our critical value of χ^2 is 3.84 at alpha $= .05$.
When entering the data into SPSS, for self-reported depression, we will code Depressed as "1" and Not Depressed as "2." For gender, we will code Male as "1" and Female as "2."

Case Processing Summary

	Cases					
	Valid		Missing		Total	
	N	Percent	N	Percent	N	Percent
Gender * SelfReport	100	100.0%	0	.0%	100	100.0%

Gender * SelfReport Crosstabulation

Count

		SelfReport		Total
		1.00	2.00	
Gender	1.00	19	28	47
	2.00	33	20	53
Total		52	48	100

Chi-Square Tests

	Value	df	Asymp. Sig. (2-sided)	Exact Sig. (2-sided)	Exact Sig. (1-sided)
Pearson Chi-Square	4.760(b)	1	.029		
Continuity Correction(a)	3.925	1	.048		
Likelihood Ratio	4.795	1	.029		
Fisher's Exact Test				.044	.024
Linear-by-Linear Association	4.712	1	.030		
N of Valid Cases	100				

a Computed only for a 2x2 table
b 0 cells (.0%) have expected count less than 5. The minimum expected count is 22.56.

We reject the null hypothesis. Self reports of depression are related to gender in the elderly, $\chi^2(1, N = 100) = 3.925$, $p = .048$.

Notice that we use the corrected value of chi-square because we have 1 *df*.

Presenting Results in APA Format

7. The development of women's attitudes toward their body was examined among 30 fifth-grade girls. The numbers of girls who characterized themselves as underweight (13.33%), overweight (76.67%), or of an appropriate weight (10.0%) were compared. A significant difference was found, $\chi^2(2, N = 30) = 25.40$, $p < .001$.

8. Gender preferences for toys were assessed; the data are presented in Table 2. Among the 68 males, 28 selected a truck; among the 62 females, 23 selected a doll. Gender was found to be significantly related to toy selection, $\chi^2(2, N = 130) = 21.73$, $p < .001$.

Table 2
Toys Selected by Gender

Gender	Toy			
	Doll	Truck	Slinky®	Book
Male	6	28	15	19
Female	23	8	17	14

References

Adams II, T., Graves, M., & Adams, H. (2006). The effectiveness of a university level conceptually-based health-related fitness course on health-related fitness knowledge. *Physical Educator, 63*, 104–112.

Betram, J. (2002). Hypothesis testing as a laboratory exercise: A simple analysis of human walking with a physiological surprise. *Advances in Physiology Education, 26*, 110–119.

Boneau, C. A. (1960). The effects of violations of assumptions underlying the t-test. *Psychological Bulletin, 57*, 49–64.

Box, G. E. (1954). Some theorems of quadratic forms applied in the study of analysis of variance problems. I. Effect of inequality of variance in the one-way classification. *Educational and Psychological Measurement, 25*, 290–302.

Campbell, D. T., & Stanley, J. C. (1963). *Experimental and quasi-experimental designs for research.* Chicago: Rand McNally.

Cattell, R. B. (1965). *The scientific analysis of personality.* Baltimore: Penguin.

Cheung, Y., Brogan, P., Pilla, C., Dillon, M., & Redington, A. (2002). Arterial distensibility in children and teenagers: normal evolution and the effect of childhood vasculitis. *Archives of Disease in Childhood, 87*, 348–352.

Cohen, J. (1988). *Statistical power analysis for the behavioral sciences* (2nd Ed.). Hillsdale, NJ: Lawrence Earlbaum Associates.

Corsi, K., Kwiatkowski, C., & Booth, R. (2007). Treatment entry and predictors among opiate-using injection drug users. *The American Journal of Drug and Alcohol Abuse, 33*, 121–127.

Curhan, J., & Pentland, A. (2007). Thin slices of negotiation: Predicting outcomes from conversational dynamics within the first 5 minutes. *Journal of Applied Psychology, 92*, 802–811.

Freud, S. (1958). Observations on transference-love. In J. Strachey (Ed.), *Standard Edition of the Complete Psychological Works of Sigmund Freud* (Vol. 12, pp. 159–171). London: Hogarth Press.

Fromm, Erich, retrieved July 7, 2007, http://www.brainyquote.com/quotes/authors/e/erich_fromm.html

Glindermann, K. E., Wiegand, D. M., & Geller, E. S. (2007). Celebratory drinking and intoxication: A contextual influence on alcohol consumption. *Environment and Behavior, 39*, 352–366.

Hays, W. L. (2007). *Statistics* (6th Ed.). Belmont, CA: Wadsworth Publishing.

Ironsmith, M., & Eppler, M. A. (2007). Mastery learning benefits low-aptitude students. *Teaching of Psychology, 34*(1), 28–31.

Joplin, J. (1968). Reported in Kristofferson, K., & Foster, F. (1969). Me and Bobby McGee [Recorded by Janis Joplin]. On *Janis Joplin's Greatest Hits.* New York: Columbia (1973).

Kennedy, J. J., & Bush, A. J. (1985). *An introduction to the design and analysis of experiments in behavioral research.* Lanham, MD: University Press of America.

Kerlinger, F. N. (1999). *Foundations of behavioral research* (4th Ed.). Belmont, CA: Wadsworth Publishing.

Kirk, R. E. (2007). *Statistics: An introduction* (5th Ed.). Belmont, CA: Wadsworth Publishing.

Klomegah, R.Y. (2007). Predictors of academic performance of university students: An application of the goal efficacy model. *College Student Journal, 41*, 407–415.

Larson, B. (2007). Adventure camp programs, self-concept, and their effects on behavioral problem adolescents. *Journal of Experimental Education, 29,* 311–330.

Layton, D., Heeley, E., & Shakir, S. (2004). Identification and evaluation of possible signal of exacerbation of colitis during rofecoxib treatment, using prescription-event monitoring data. *Journal of Clinical Pharmacy & Therapeutics, 29,* 171–181.

Leik, R. K. (1997). *Experimental design and the analysis of variance.* Thousand Oaks, CA: Pine Forge Press.

McGrath, A. (Ed.). (2006). *US News and World Report Ultimate College Guide 2007.* Naperville, IL: Sourcebooks, Inc.

Meyer, M.C. (2005). Who wants airbags? *Chance, 18* (2), 3–15.

Miller, D.C. (2000). *Kindergarten Diagnostic Instrument (2nd Ed.; KDI-II) Administration Manual.* Denton, TX: Kindergarten Interventions and Diagnostic Services, Inc.

Munichor, N., & Rafaeli, A. (2007). Numbers or apologies? Customer reactions to waiting time fillers. *Journal of Applied Psychology, 92,* 511–518.

Pedhazur, J. (1997). *Multiple regression in behavioral research* (3rd Ed.). New York: Harcourt Brace.

Poelzer, G., Zeng, L., & Simonsson, M. (2007). Teacher certification tests using linear and logistic regression models to predict success of secondary pre-service teachers. *College Student Journal, 41,* 305–313.

Ragothaman, S., Lavin, A., & Davies, T. (2007). Perceptions of accounting practitioners and educators on e-business curriculum and web security issues. *College Student Journal. 41,* 59–68.

Raina, K. D., Rogers, J. C., & Holm, M. B. (2007). Influence of the environment on activity performance in older women with heart failure. *Disability and Rehabilitation, 29,* 545–557.

Ram Dass. (1971). *Be Here Now.* New York: Crown Publishing.

Rosnow, R. L., & Rosenthal, R. (1996). Computing contrasts, effect sizes, and counternulls on other people's published data: General procedures for research consumers. *Psychological Methods, 1,* 331–340.

Shavelson, R. J. (1996). *Statistical reasoning for the behavioral sciences* (3rd Ed.). New York: Allyn & Bacon.

Shin, S., & Koh, M. (2007). A cross-cultural study of teachers' beliefs and strategies on classroom behavior management in urban American and Korean school systems. Thousand Oaks, CA: SAGE Publications.

Snedecor, G. W., & Cochran, W. G. (1989). *Statistical methods* (8th Ed.). Ames, IA: Iowa State University Press.

Stevens, J. P. (1996). *Applied multivariate statistics for the social sciences* (3rd Ed.). Mahwah, NJ: Lawrence Erlbaum Associates.

Winer, B. J., Brown, D. R., & Michels, K. M. (1991). *Statistical principles in experimental design* (3rd Ed.). New York: McGraw Hill

Witte, R. S., & Witte, J. S. (2006). *Statistics* (8th Ed.). Hoboken, NJ: John Wiley & Sons.

Wise, S. L., Dennison, S. G., & Yang, S. (2006). Taking the time to improve the validity of low-stakes test: The effort-monitoring CBT. *Educational Measurement: Issues and Practice, 25* (2), 21–30.

Tukey, J. W. (1977). *Exploratory data analysis.* Boston: Addison-Wesley.

Yates, F. (1934). Contingency table involving small numbers and the X^2 test. *Journal of the Royal Statistical Society (Supplement), 1,* 217–235.

Appendix A

Statistical Tables

TABLE A.1
The Unit Normal Table (z)*

*Column A lists z-score values. A vertical line drawn through a normal distribution at a z-score location divides the distribution into two sections. Column B identifies the proportion in the larger section, called the *body*. Column C identifies the proportion in the smaller section, called the *tail*.

Note: Because the normal distribution is symmetrical, the proportions for negative z-scores are the same as those for positive z-scores.

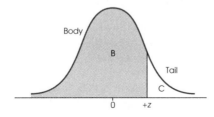 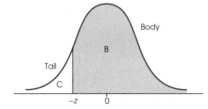

(A) z	(B) Proportion in body	(C) Proportion in tail	(A) z	(B) Proportion in body	(C) Proportion in tail
0.00	.5000	.5000	0.09	.5359	.4641
0.01	.5040	.4960	0.10	.5398	.4602
0.02	.5080	.4920	0.11	.5438	.4562
0.03	.5120	.4880	0.12	.5478	.4522
0.04	.5160	.4840	0.13	.5517	.4483
0.05	.5199	.4801	0.14	.5557	.4443
0.06	.5239	.4761	0.15	.5596	.4404
0.07	.5279	.4721	0.16	.5636	.4364
0.08	.5319	.4681	0.17	.5675	.4325

(Continued)

TABLE A.1

(A) z	(B) Proportion in body	(C) Proportion in tail	(A) z	(B) Proportion in body	(C) Proportion in tail
0.18	.5714	.4286	0.60	.7257	.2743
0.19	.5753	.4247	0.61	.7291	.2709
0.20	.5793	.4207	0.62	.7324	.2676
0.21	.5832	.4168	0.63	.7357	.2643
0.22	.5871	.4129	0.64	.7389	.2611
0.23	.5910	.4090	0.65	.7422	.2578
0.24	.5948	.4052	0.66	.7454	.2546
0.25	.5987	.4013	0.67	.7486	.2514
0.26	.6026	.3974	0.68	.7517	.2483
0.27	.6064	.3936	0.69	.7549	.2451
0.28	.6103	.3897	0.70	.7580	.2420
0.29	.6141	.3859	0.71	.7611	.2389
0.30	.6179	.3821	0.72	.7642	.2358
0.31	.6217	.3783	0.73	.7673	.2327
0.32	.6255	.3745	0.74	.7704	.2296
0.33	.6293	.3707	0.75	.7734	.2266
0.34	.6331	.3669	0.76	.7764	.2236
0.35	.6368	.3632	0.77	.7794	.2206
0.36	.6406	.3594	0.78	.7823	.2177
0.37	.6443	.3557	0.79	.7852	.2148
0.38	.6480	.3520	0.80	.7881	.2119
0.39	.6517	.3483	0.81	.7910	.2090
0.40	.6554	.3446	0.82	.7939	.2061
0.41	.6591	.3409	0.83	.7967	.2033
0.42	.6628	.3372	0.84	.7995	.2005
0.43	.6664	.3336	0.85	.8023	.1977
0.44	.6700	.3300	0.86	.8051	.1949
0.45	.6736	.3264	0.87	.8078	.1922
0.46	.6772	.3228	0.88	.8106	.1894
0.47	.6808	.3192	0.89	.8133	.1867
0.48	.6844	.3156	0.90	.8159	.1841
0.49	.6879	.3121	0.91	.8186	.1814
0.50	.6915	.3085	0.92	.8212	.1788
0.51	.6950	.3050	0.93	.8238	.1762
0.52	.6985	.3015	0.94	.8264	.1736
0.53	.7019	.2981	0.95	.8289	.1711
0.54	.7054	.2946	0.96	.8315	.1685
0.55	.7088	.2912	0.97	.8340	.1660
0.56	.7123	.2877	0.98	.8365	.1635
0.57	.7157	.2843	0.99	.8389	.1611
0.58	.7190	.2810	1.00	.8413	.1587
0.59	.7224	.2776	1.01	.8438	.1562

TABLE A.1

(A) z	(B) Proportion in body	(C) Proportion in tail	(A) z	(B) Proportion in body	(C) Proportion in tail
1.02	.8461	.1539	1.45	.9265	.0735
1.03	.8485	.1515	1.46	.9279	.0721
1.04	.8508	.1492	1.47	.9292	.0708
1.05	.8531	.1469	1.48	.9306	.0694
1.06	.8554	.1446	1.49	.9319	.0681
1.07	.8577	.1423	1.50	.9332	.0668
1.08	.8599	.1401	1.51	.9345	.0655
1.09	.8621	.1379	1.52	.9357	.0643
1.10	.8643	.1357	1.53	.9370	.0630
1.11	.8665	.1335	1.54	.9382	.0618
1.12	.8686	.1314	1.55	.9394	.0606
1.13	.8708	.1292	1.56	.9406	.0594
1.14	.8729	.1271	1.57	.9418	.0582
1.15	.8749	.1251	1.58	.9429	.0571
1.16	.8770	.1230	1.59	.9441	.0559
1.17	.8790	.1210	1.60	.9452	.0548
1.18	.8810	.1190	1.61	.9463	.0537
1.19	.8830	.1170	1.62	.9474	.0526
1.20	.8849	.1151	1.63	.9484	.0516
1.21	.8869	.1131	1.64	.9495	.0505
1.22	.8888	.1112	1.65	.9505	.0495
1.23	.8907	.1093	1.66	.9515	.0485
1.24	.8925	.1075	1.67	.9525	.0475
1.25	.8944	.1056	1.68	.9535	.0465
1.26	.8962	.1038	1.69	.9545	.0455
1.27	.8980	.1020	1.70	.9554	.0446
1.28	.8997	.1003	1.71	.9564	.0436
1.29	.9015	.0985	1.72	.9573	.0427
1.30	.9032	.0968	1.73	.9582	.0418
1.31	.9049	.0951	1.74	.9591	.0409
1.32	.9066	.0934	1.75	.9599	.0401
1.33	.9082	.0918	1.76	.9608	.0392
1.34	.9099	.0901	1.77	.9616	.0384
1.35	.9115	.0885	1.78	.9625	.0375
1.36	.9131	.0869	1.79	.9633	.0367
1.37	.9147	.0853	1.80	.9641	.0359
1.38	.9162	.0838	1.81	.9649	.0351
1.39	.9177	.0823	1.82	.9656	.0344
1.40	.9192	.0808	1.83	.9664	.0336
1.41	.9207	.0793	1.84	.9671	.0329
1.42	.9222	.0778	1.85	.9678	.0322
1.43	.9236	.0764	1.86	.9686	.0314
1.44	.9251	.0749	1.87	.9693	.0307

(Continued)

TABLE A.1

(A) z	(B) Proportion in body	(C) Proportion in tail	(A) z	(B) Proportion in body	(C) Proportion in tail
1.88	.9699	.0301	2.28	.9887	.0113
1.89	.9706	.0294	2.29	.9890	.0110
1.90	.9713	.0287	2.30	.9893	.0107
1.91	.9719	.0281	2.31	.9896	.0104
1.92	.9726	.0274	2.32	.9898	.0102
1.93	.9732	.0268	2.33	.9901	.0099
1.94	.9738	.0262	2.34	.9904	.0096
1.95	.9744	.0256	2.35	.9906	.0094
1.96	.9750	.0250	2.36	.9909	.0091
1.97	.9756	.0244	2.37	.9911	.0089
1.98	.9761	.0239	2.38	.9913	.0087
1.99	.9767	.0233	2.39	.9916	.0084
2.00	.9772	.0228	2.40	.9918	.0082
2.01	.9778	.0222	2.41	.9920	.0080
2.02	.9783	.0217	2.42	.9922	.0078
2.03	.9788	.0212	2.43	.9925	.0075
2.04	.9793	.0207	2.44	.9927	.0073
2.05	.9798	.0202	2.45	.9929	.0071
2.06	.9803	.0197	2.46	.9931	.0069
2.07	.9808	.0192	2.47	.9932	.0068
2.08	.9812	.0188	2.48	.9934	.0066
2.09	.9817	.0183	2.49	.9936	.0064
2.10	.9821	.0179	2.50	.9938	.0062
2.11	.9826	.0174	2.51	.9940	.0060
2.12	.9830	.0170	2.52	.9941	.0059
2.13	.9834	.0166	2.53	.9943	.0057
2.14	.9838	.0162	2.54	.9945	.0055
2.15	.9842	.0158	2.55	.9946	.0054
2.16	.9846	.0154	2.56	.9948	.0052
2.17	.9850	.0150	2.57	.9949	.0051
2.18	.9854	.0146	2.58	.9951	.0049
2.19	.9857	.0143	2.59	.9952	.0048
2.20	.9861	.0139	2.60	.9953	.0047
2.21	.9864	.0136	2.61	.9955	.0045
2.22	.9868	.0132	2.62	.9956	.0044
2.23	.9871	.0129	2.63	.9957	.0043
2.24	.9875	.0125	2.64	.9959	.0041
2.25	.9878	.0122	2.65	.9960	.0040
2.26	.9881	.0119	2.66	.9961	.0039
2.27	.9884	.0116	2.67	.9962	.0038

TABLE A.1

(A) z	(B) Proportion in body	(C) Proportion in tail	(A) z	(B) Proportion in body	(C) Proportion in tail
2.68	.9963	.0037	3.00	.9987	.0013
2.69	.9964	.0036	3.01	.9987	.0013
2.70	.9965	.0035	3.02	.9987	.0013
2.71	.9966	.0034	3.03	.9988	.0012
2.72	.9967	.0033	3.04	.9988	.0012
2.73	.9968	.0032	3.05	.9989	.0011
2.74	.9969	.0031	3.06	.9989	.0011
2.75	.9970	.0030	3.07	.9989	.0011
2.76	.9971	.0029	3.08	.9990	.0010
2.77	.9972	.0028	3.09	.9990	.0010
2.78	.9973	.0027	3.10	.9990	.0010
2.79	.9974	.0026	3.11	.9991	.0009
2.80	.9974	.0026	3.12	.9991	.0009
2.81	.9975	.0025	3.13	.9991	.0009
2.82	.9976	.0024	3.14	.9992	.0008
2.83	.9977	.0023	3.15	.9992	.0008
2.84	.9977	.0023	3.16	.9992	.0008
2.85	.9978	.0022	3.17	.9992	.0008
2.86	.9979	.0021	3.18	.9993	.0007
2.87	.9979	.0021	3.19	.9993	.0007
2.88	.9980	.0020	3.20	.9993	.0007
2.89	.9981	.0019	3.21	.9993	.0007
2.90	.9981	.0019	3.22	.9994	.0006
2.91	.9982	.0018	3.23	.9994	.0006
2.92	.9982	.0018	3.24	.9994	.0006
2.93	.9983	.0017	3.30	.9995	.0005
2.94	.9984	.0016	3.40	.9997	.0003
2.95	.9984	.0016	3.50	.9998	.0002
2.96	.9985	.0015	3.60	.9998	.0002
2.97	.9985	.0015	3.70	.9999	.0001
2.98	.9986	.0014	3.80	.99993	.00007
2.99	.9986	.0014	3.90	.99995	.00005
			4.00	.99997	.00003

From Table IIi of R. A. Fisher and F. Yates, *Statistical Tables for Biological, Agricultural and Medical Research,* 6th ed. London: Longman Group Ltd., 1974 (previously published by Oliver & Boyd Ltd., Edinburgh.) Adapted and reprinted with permission from Addison Wesley Longman Ltd.

TABLE A.2
Table of Critical *t* Values

Table entries are values of *t* corresponding to proportions in one tail or in two tails combined.

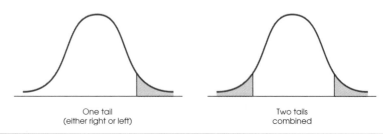

One tail
(either right or left)

Two tails
combined

	Proportion in one tail					
	0.25	0.10	0.05	0.025	0.01	0.005
	Proportion in two tails combined					
df	0.50	0.20	0.10	0.05	0.02	0.01
1	1.000	3.078	6.314	12.706	31.821	63.657
2	0.816	1.886	2.920	4.303	6.965	9.925
3	0.765	1.638	2.353	3.182	4.541	5.841
4	0.741	1.533	2.132	2.776	3.747	4.604
5	0.727	1.476	2.015	2.571	3.365	4.032
6	0.718	1.440	1.943	2.447	3.143	3.707
7	0.711	1.415	1.895	2.365	2.998	3.499
8	0.706	1.397	1.860	2.306	2.896	3.355
9	0.703	1.383	1.833	2.262	2.821	3.250
10	0.700	1.372	1.812	2.228	2.764	3.169
11	0.697	1.363	1.796	2.201	2.718	3.106
12	0.695	1.356	1.782	2.179	2.681	3.055
13	0.694	1.350	1.771	2.160	2.650	3.012
14	0.692	1.345	1.761	2.145	2.624	2.977
15	0.691	1.341	1.753	2.131	2.602	2.947
16	0.690	1.337	1.746	2.120	2.583	2.921
17	0.689	1.333	1.740	2.110	2.567	2.898
18	0.688	1.330	1.734	2.101	2.552	2.878
19	0.688	1.328	1.729	2.093	2.539	2.861
20	0.687	1.325	1.725	2.086	2.528	2.845
21	0.686	1.323	1.721	2.080	2.518	2.831
22	0.686	1.321	1.717	2.074	2.508	2.819
23	0.685	1.319	1.714	2.069	2.500	2.807
24	0.685	1.318	1.711	2.064	2.492	2.797
25	0.684	1.316	1.708	2.060	2.485	2.787
26	0.684	1.315	1.706	2.056	2.479	2.779
27	0.684	1.314	1.703	2.052	2.473	2.771
28	0.683	1.313	1.701	2.048	2.467	2.763
29	0.683	1.311	1.699	2.045	2.462	2.756
30	0.683	1.310	1.697	2.042	2.457	2.750
40	0.681	1.303	1.684	2.021	2.423	2.704
60	0.679	1.296	1.671	2.000	2.390	2.660
120	0.677	1.289	1.658	1.980	2.358	2.617
∞	0.674	1.282	1.645	1.960	2.326	2.576

From Table III of R. A. Fisher and F. Yates, *Statistical Tables for Biological, Agricultural and Medical Research,* 6th ed. London: Longman Group Ltd., 1974 (previously published by Oliver and Boyd Ltd., Edinburgh). Adapted and reprinted with permission of the Addison Wesley Longman Publishing Co.

TABLE A.3
Table of Critical F Values

*Table entries in lightface type are critical values for the .05 level of significance.
Boldface type values are for the .01 level of significance.

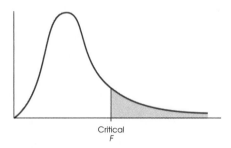

Critical
F

Degrees of freedom: denominator	Degrees of freedom: numerator														
	1	2	3	4	5	6	7	8	9	10	11	12	14	16	20
1	161	200	216	225	230	234	237	239	241	242	243	244	245	246	248
	4052	**4999**	**5403**	**5625**	**5764**	**5859**	**5928**	**5981**	**6022**	**6056**	**6082**	**6106**	**6142**	**6169**	**6208**
2	18.51	19.00	19.16	19.25	19.30	19.33	19.36	19.37	19.38	19.39	19.40	19.41	19.42	19.43	19.44
	98.49	**99.00**	**99.17**	**99.25**	**99.30**	**99.33**	**99.34**	**99.36**	**99.38**	**99.40**	**99.41**	**99.42**	**99.43**	**99.44**	**99.45**
3	10.13	9.55	9.28	9.12	9.01	8.94	8.88	8.84	8.81	8.78	8.76	8.74	8.71	8.69	8.66
	34.12	**30.92**	**29.46**	**28.71**	**28.24**	**27.91**	**27.67**	**27.49**	**27.34**	**27.23**	**27.13**	**27.05**	**26.92**	**26.83**	**26.69**
4	7.71	6.94	6.59	6.39	6.26	6.16	6.09	6.04	6.00	5.96	5.93	5.91	5.87	5.84	5.80
	21.20	**18.00**	**16.69**	**15.98**	**15.52**	**15.21**	**14.98**	**14.80**	**14.66**	**14.54**	**14.45**	**14.37**	**14.24**	**14.15**	**14.02**
5	6.61	5.79	5.41	5.19	5.05	4.95	4.88	4.82	4.78	4.74	4.70	4.68	4.64	4.60	4.56
	16.26	**13.27**	**12.06**	**11.39**	**10.97**	**10.67**	**10.45**	**10.27**	**10.15**	**10.05**	**9.96**	**9.89**	**9.77**	**9.68**	**9.55**
6	5.99	5.14	4.76	4.53	4.39	4.28	4.21	4.15	4.10	4.06	4.03	4.00	3.96	3.92	3.87
	13.74	**10.92**	**9.78**	**9.15**	**8.75**	**8.47**	**8.26**	**8.10**	**7.98**	**7.87**	**7.79**	**7.72**	**7.60**	**7.52**	**7.39**
7	5.59	4.74	4.35	4.12	3.97	3.87	3.79	3.73	3.68	3.63	3.60	3.57	3.52	3.49	3.44
	12.25	**9.55**	**8.45**	**7.85**	**7.46**	**7.19**	**7.00**	**6.84**	**6.71**	**6.62**	**6.54**	**6.47**	**6.35**	**6.27**	**6.15**
8	5.32	4.46	4.07	3.84	3.69	3.58	3.50	3.44	3.39	3.34	3.31	3.28	3.23	3.20	3.15
	11.26	**8.65**	**7.59**	**7.01**	**6.63**	**6.37**	**6.19**	**6.03**	**5.91**	**5.82**	**5.74**	**5.67**	**5.56**	**5.48**	**5.36**
9	5.12	4.26	3.86	3.63	3.48	3.37	3.29	3.23	3.18	3.13	3.10	3.07	3.02	2.98	2.93
	10.56	**8.02**	**6.99**	**6.42**	**6.06**	**5.80**	**5.62**	**5.47**	**5.35**	**5.26**	**5.18**	**5.11**	**5.00**	**4.92**	**4.80**
10	4.96	4.10	3.71	3.48	3.33	3.22	3.14	3.07	3.02	2.97	2.94	2.91	2.86	2.82	2.77
	10.04	**7.56**	**6.55**	**5.99**	**5.64**	**5.39**	**5.21**	**5.06**	**4.95**	**4.85**	**4.78**	**4.71**	**4.60**	**4.52**	**4.41**
11	4.84	3.98	3.59	3.36	3.20	3.09	3.01	2.95	2.90	2.86	2.82	2.79	2.74	2.70	2.65
	9.65	**7.20**	**6.22**	**5.67**	**5.32**	**5.07**	**4.88**	**4.74**	**4.63**	**4.54**	**4.46**	**4.40**	**4.29**	**4.21**	**4.10**
12	4.75	3.88	3.49	3.26	3.11	3.00	2.92	2.85	2.80	2.76	2.72	2.69	2.64	2.60	2.54
	9.33	**6.93**	**5.95**	**5.41**	**5.06**	**4.82**	**4.65**	**4.50**	**4.39**	**4.30**	**4.22**	**4.16**	**4.05**	**3.98**	**3.86**
13	4.67	3.80	3.41	3.18	3.02	2.92	2.84	2.77	2.72	2.67	2.63	2.60	2.55	2.51	2.46
	9.07	**6.70**	**5.74**	**5.20**	**4.86**	**4.62**	**4.44**	**4.30**	**4.19**	**4.10**	**4.02**	**3.96**	**3.85**	**3.78**	**3.67**
14	4.60	3.74	3.34	3.11	2.96	2.85	2.77	2.70	2.65	2.60	2.56	2.53	2.48	2.44	2.39
	8.86	**6.51**	**5.56**	**5.03**	**4.69**	**4.46**	**4.28**	**4.14**	**4.03**	**3.94**	**3.86**	**3.80**	**3.70**	**3.62**	**3.51**
15	4.54	3.68	3.29	3.06	2.90	2.79	2.70	2.64	2.59	2.55	2.51	2.48	2.43	2.39	2.33
	8.68	**6.36**	**5.42**	**4.89**	**4.56**	**4.32**	**4.14**	**4.00**	**3.89**	**3.80**	**3.73**	**3.67**	**3.56**	**3.48**	**3.36**
16	4.49	3.63	3.24	3.01	2.85	2.74	2.66	2.59	2.54	2.49	2.45	2.42	2.37	2.33	2.28
	8.53	**6.23**	**5.29**	**4.77**	**4.44**	**4.20**	**4.03**	**3.89**	**3.78**	**3.69**	**3.61**	**3.55**	**3.45**	**3.37**	**3.25**

(Continued)

TABLE A.3

Degrees of freedom: denominator	Degrees of freedom: numerator														
	1	2	3	4	5	6	7	8	9	10	11	12	14	16	20
17	4.45	3.59	3.20	2.96	2.81	2.70	2.62	2.55	2.50	2.45	2.41	2.38	2.33	2.29	2.23
	8.40	6.11	5.18	4.67	4.34	4.10	3.93	3.79	3.68	3.59	3.52	3.45	3.35	3.27	3.16
18	4.41	3.55	3.16	2.93	2.77	2.66	2.58	2.51	2.46	2.41	2.37	2.34	2.29	2.25	2.19
	8.28	6.01	5.09	4.58	4.25	4.01	3.85	3.71	3.60	3.51	3.44	3.37	3.27	3.19	3.07
19	4.38	3.52	3.13	2.90	2.74	2.63	2.55	2.48	2.43	2.38	2.34	2.31	2.26	2.21	2.15
	8.18	5.93	5.01	4.50	4.17	3.94	3.77	3.63	3.52	3.43	3.36	3.30	3.19	3.12	3.00
20	4.35	3.49	3.10	2.87	2.71	2.60	2.52	2.45	2.40	2.35	2.31	2.28	2.23	2.18	2.12
	8.10	5.85	4.94	4.43	4.10	3.87	3.71	3.56	3.45	3.37	3.30	3.23	3.13	3.05	2.94
21	4.32	3.47	3.07	2.84	2.68	2.57	2.49	2.42	2.37	2.32	2.28	2.25	2.20	2.15	2.09
	8.02	5.78	4.87	4.37	4.04	3.81	3.65	3.51	3.40	3.31	3.24	3.17	3.07	2.99	2.88
22	4.30	3.44	3.05	2.82	2.66	2.55	2.47	2.40	2.35	2.30	2.26	2.23	2.18	2.13	2.07
	7.94	5.72	4.82	4.31	3.99	3.76	3.59	3.45	3.35	3.26	3.18	3.12	3.02	2.94	2.83
23	4.28	3.42	3.03	2.80	2.64	2.53	2.45	2.38	2.32	2.28	2.24	2.20	2.14	2.10	2.04
	7.88	5.66	4.76	4.26	3.94	3.71	3.54	3.41	3.30	3.21	3.14	3.07	2.97	2.89	2.78
24	4.26	3.40	3.01	2.78	2.62	2.51	2.43	2.36	2.30	2.26	2.22	2.18	2.13	2.09	2.02
	7.82	5.61	4.72	4.22	3.90	3.67	3.50	3.36	3.25	3.17	3.09	3.03	2.93	2.85	2.74
25	4.24	3.38	2.99	2.76	2.60	2.49	2.41	2.34	2.28	2.24	2.20	2.16	2.11	2.06	2.00
	7.77	5.57	4.68	4.18	3.86	3.63	3.46	3.32	3.21	3.13	3.05	2.99	2.89	2.81	2.70
26	4.22	3.37	2.98	2.74	2.59	2.47	2.39	2.32	2.27	2.22	2.18	2.15	2.10	2.05	1.99
	7.72	5.53	4.64	4.14	3.82	3.59	3.42	3.29	3.17	3.09	3.02	2.96	2.86	2.77	2.66
27	4.21	3.35	2.96	2.73	2.57	2.46	2.37	2.30	2.25	2.20	2.16	2.13	2.08	2.03	1.97
	7.68	5.49	4.60	4.11	3.79	3.56	3.39	3.26	3.14	3.06	2.98	2.93	2.83	2.74	2.63
28	4.20	3.34	2.95	2.71	2.56	2.44	2.36	2.29	2.24	2.19	2.15	2.12	2.06	2.02	1.96
	7.64	5.45	4.57	4.07	3.76	3.53	3.36	3.23	3.11	3.03	2.95	2.90	2.80	2.71	2.60
29	4.18	3.33	2.93	2.70	2.54	2.43	2.35	2.28	2.22	2.18	2.14	2.10	2.05	2.00	1.94
	7.60	5.42	4.54	4.04	3.73	3.50	3.33	3.20	3.08	3.00	2.92	2.87	2.77	2.68	2.57
30	4.17	3.32	2.92	2.69	2.53	2.42	2.34	2.27	2.21	2.16	2.12	2.09	2.04	1.99	1.93
	7.56	5.39	4.51	4.02	3.70	3.47	3.30	3.17	3.06	2.98	2.90	2.84	2.74	2.66	2.55
32	4.15	3.30	2.90	2.67	2.51	2.40	2.32	2.25	2.19	2.14	2.10	2.07	2.02	1.97	1.91
	7.50	5.34	4.46	3.97	3.66	3.42	3.25	3.12	3.01	2.94	2.86	2.80	2.70	2.62	2.51
34	4.13	3.28	2.88	2.65	2.49	2.38	2.30	2.23	2.17	2.12	2.08	2.05	2.00	1.95	1.89
	7.44	5.29	4.42	3.93	3.61	3.38	3.21	3.08	2.97	2.89	2.82	2.76	2.66	2.58	2.47
36	4.11	3.26	2.86	2.63	2.48	2.36	2.28	2.21	2.15	2.10	2.06	2.03	1.98	1.93	1.87
	7.39	5.25	4.38	3.89	3.58	3.35	3.18	3.04	2.94	2.86	2.78	2.72	2.62	2.54	2.43
38	4.10	3.25	2.85	2.62	2.46	2.35	2.26	2.19	2.14	2.09	2.05	2.02	1.96	1.92	1.85
	7.35	5.21	4.34	3.86	3.54	3.32	3.15	3.02	2.91	2.82	2.75	2.69	2.59	2.51	2.40
40	4.08	3.23	2.84	2.61	2.45	2.34	2.25	2.18	2.12	2.07	2.04	2.00	1.95	1.90	1.84
	7.31	5.18	4.31	3.83	3.51	3.29	3.12	2.99	2.88	2.80	2.73	2.66	2.56	2.49	2.37

TABLE A.3

Degrees of freedom: denominator	Degrees of freedom: numerator														
	1	2	3	4	5	6	7	8	9	10	11	12	14	16	20
42	4.07	3.22	2.83	2.59	2.44	2.32	2.24	2.17	2.11	2.06	2.02	1.99	1.94	1.89	1.82
	7.27	**5.15**	**4.29**	**3.80**	**3.49**	**3.26**	**3.10**	**2.96**	**2.86**	**2.77**	**2.70**	**2.64**	**2.54**	**2.46**	**2.35**
44	4.06	3.21	2.82	2.58	2.43	2.31	2.23	2.16	2.10	2.05	2.01	1.98	1.92	1.88	1.81
	7.24	**5.12**	**4.26**	**3.78**	**3.46**	**3.24**	**3.07**	**2.94**	**2.84**	**2.75**	**2.68**	**2.62**	**2.52**	**2.44**	**2.32**
46	4.05	3.20	2.81	2.57	2.42	2.30	2.22	2.14	2.09	2.04	2.00	1.97	1.91	1.87	1.80
	7.21	**5.10**	**4.24**	**3.76**	**3.44**	**3.22**	**3.05**	**2.92**	**2.82**	**2.73**	**2.66**	**2.60**	**2.50**	**2.42**	**2.30**
48	4.04	3.19	2.80	2.56	2.41	2.30	2.21	2.14	2.08	2.03	1.99	1.96	1.90	1.86	1.79
	7.19	**5.08**	**4.22**	**3.74**	**3.42**	**3.20**	**3.04**	**2.90**	**2.80**	**2.71**	**2.64**	**2.58**	**2.48**	**2.40**	**2.28**
50	4.03	3.18	2.79	2.56	2.40	2.29	2.20	2.13	2.07	2.02	1.98	1.95	1.90	1.85	1.78
	7.17	**5.06**	**4.20**	**3.72**	**3.41**	**3.18**	**3.02**	**2.88**	**2.78**	**2.70**	**2.62**	**2.56**	**2.46**	**2.39**	**2.26**
55	4.02	3.17	2.78	2.54	2.38	2.27	2.18	2.11	2.05	2.00	1.97	1.93	1.88	1.83	1.76
	7.12	**5.01**	**4.16**	**3.68**	**3.37**	**3.15**	**2.98**	**2.85**	**2.75**	**2.66**	**2.59**	**2.53**	**2.43**	**2.35**	**2.23**
60	4.00	3.15	2.76	2.52	2.37	2.25	2.17	2.10	2.04	1.99	1.95	1.92	1.86	1.81	1.75
	7.08	**4.98**	**4.13**	**3.65**	**3.34**	**3.12**	**2.95**	**2.82**	**2.72**	**2.63**	**2.56**	**2.50**	**2.40**	**2.32**	**2.20**
65	3.99	3.14	2.75	2.51	2.36	2.24	2.15	2.08	2.02	1.98	1.94	1.90	1.85	1.80	1.73
	7.04	**4.95**	**4.10**	**3.62**	**3.31**	**3.09**	**2.93**	**2.79**	**2.70**	**2.61**	**2.54**	**2.47**	**2.37**	**2.30**	**2.18**
70	3.98	3.13	2.74	2.50	2.35	2.23	2.14	2.07	2.01	1.97	1.93	1.89	1.84	1.79	1.72
	7.01	**4.92**	**4.08**	**3.60**	**3.29**	**3.07**	**2.91**	**2.77**	**2.67**	**2.59**	**2.51**	**2.45**	**2.35**	**2.28**	**2.15**
80	3.96	3.11	2.72	2.48	2.33	2.21	2.12	2.05	1.99	1.95	1.91	1.88	1.82	1.77	1.70
	6.96	**4.88**	**4.04**	**3.56**	**3.25**	**3.04**	**2.87**	**2.74**	**2.64**	**2.55**	**2.48**	**2.41**	**2.32**	**2.24**	**2.11**
100	3.94	3.09	2.70	2.46	2.30	2.19	2.10	2.03	1.97	1.92	1.88	1.85	1.79	1.75	1.68
	6.90	**4.82**	**3.98**	**3.51**	**3.20**	**2.99**	**2.82**	**2.69**	**2.59**	**2.51**	**2.43**	**2.36**	**2.26**	**2.19**	**2.06**
125	3.92	3.07	2.68	2.44	2.29	2.17	2.08	2.01	1.95	1.90	1.86	1.83	1.77	1.72	1.65
	6.84	**4.78**	**3.94**	**3.47**	**3.17**	**2.95**	**2.79**	**2.65**	**2.56**	**2.47**	**2.40**	**2.33**	**2.23**	**2.15**	**2.03**
150	3.91	3.06	2.67	2.43	2.27	2.16	2.07	2.00	1.94	1.89	1.85	1.82	1.76	1.71	1.64
	6.81	**4.75**	**3.91**	**3.44**	**3.14**	**2.92**	**2.76**	**2.62**	**2.53**	**2.44**	**2.37**	**2.30**	**2.20**	**2.12**	**2.00**
200	3.89	3.04	2.65	2.41	2.26	2.14	2.05	1.98	1.92	1.87	1.83	1.80	1.74	1.69	1.62
	6.76	**4.71**	**3.88**	**3.41**	**3.11**	**2.90**	**2.73**	**2.60**	**2.50**	**2.41**	**2.34**	**2.28**	**2.17**	**2.09**	**1.97**
400	3.86	3.02	2.62	2.39	2.23	2.12	2.03	1.96	1.90	1.85	1.81	1.78	1.72	1.67	1.60
	6.70	**4.66**	**3.83**	**3.36**	**3.06**	**2.85**	**2.69**	**2.55**	**2.46**	**2.37**	**2.29**	**2.23**	**2.12**	**2.04**	**1.92**
1000	3.85	3.00	2.61	2.38	2.22	2.10	2.02	1.95	1.89	1.84	1.80	1.76	1.70	1.65	1.58
	6.66	**4.62**	**3.80**	**3.34**	**3.04**	**2.82**	**2.66**	**2.53**	**2.43**	**2.34**	**2.26**	**2.20**	**2.09**	**2.01**	**1.89**
∞	3.84	2.99	2.60	2.37	2.21	2.09	2.01	1.94	1.88	1.83	1.79	1.75	1.69	1.64	1.57
	6.64	**4.60**	**3.78**	**3.32**	**3.02**	**2.80**	**2.64**	**2.51**	**2.41**	**2.32**	**2.24**	**2.18**	**2.07**	**1.99**	**1.87**

TABLE A.4
Table of Critical *r* Values

*To be significant, the sample correlation, *r*, must be greater than or equal to the critical value in the table.

	Level of significance for one-tailed test			
	.05	.025	.01	.005
	Level of significance for two-tailed test			
$df = n - 2$	.10	.05	.02	.01
1	.988	.997	.9995	.9999
2	.900	.950	.980	.990
3	.805	.878	.934	.959
4	.729	.811	.882	.917
5	.669	.754	.833	.874
6	.622	.707	.789	.834
7	.582	.666	.750	.798
8	.549	.632	.716	.765
9	.521	.602	.685	.735
10	.497	.576	.658	.708
11	.476	.553	.634	.684
12	.458	.532	.612	.661
13	.441	.514	.592	.641
14	.426	.497	.574	.623
15	.412	.482	.558	.606
16	.400	.468	.542	.590
17	.389	.456	.528	.575
18	.378	.444	.516	.561
19	.369	.433	.503	.549
20	.360	.423	.492	.537
21	.352	.413	.482	.526
22	.344	.404	.472	.515
23	.337	.396	.462	.505
24	.330	.388	.453	.496
25	.323	.381	.445	.487
26	.317	.374	.437	.479
27	.311	.367	.430	.471
28	.306	.361	.423	.463
29	.301	.355	.416	.456
30	.296	.349	.409	.449
35	.275	.325	.381	.418
40	.257	.304	.358	.393
45	.243	.288	.338	.372
50	.231	.273	.322	.354
60	.211	.250	.295	.325
70	.195	.232	.274	.302
80	.183	.217	.256	.283
90	.173	.205	.242	.267
100	.164	.195	.230	.254

From Table VI of R. A. Fisher and F. Yates, *Statistical Tables for Biological, Agricultural and Medical Research*, 6th ed. London: Longman Group Ltd., 1974 (previously published by Oliver & Boyd Ltd., Edinburgh). Adapted and reprinted with permission from Addison Wesley Longman Ltd.

TABLE A.5

Critical Values of χ^2

*The table entries are critical values of χ^2.

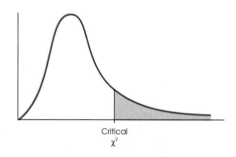

Critical
χ^2

df	Proportion in critical region				
	0.10	**0.05**	**0.025**	**0.01**	**0.005**
1	2.71	3.84	5.02	6.63	7.88
2	4.61	5.99	7.38	9.21	10.60
3	6.25	7.81	9.35	11.34	12.84
4	7.78	9.49	11.14	13.28	14.86
5	9.24	11.07	12.83	15.09	16.75
6	10.64	12.59	14.45	16.81	18.55
7	12.02	14.07	16.01	18.48	20.28
8	13.36	15.51	17.53	20.09	21.96
9	14.68	16.92	19.02	21.67	23.59
10	15.99	18.31	20.48	23.21	25.19
11	17.28	19.68	21.92	24.72	26.76
12	18.55	21.03	23.34	26.22	28.30
13	19.81	22.36	24.74	27.69	29.82
14	21.06	23.68	26.12	29.14	31.32
15	22.31	25.00	27.49	30.58	32.80
16	23.54	26.30	28.85	32.00	34.27
17	24.77	27.59	30.19	33.41	35.72
18	25.99	28.87	31.53	34.81	37.16
19	27.20	30.14	32.85	36.19	38.58
20	28.41	31.41	34.17	37.57	40.00
21	29.62	32.67	35.48	38.93	41.40
22	30.81	33.92	36.78	40.29	42.80
23	32.01	35.17	38.08	41.64	44.18
24	33.20	36.42	39.36	42.98	45.56
25	34.38	37.65	40.65	44.31	46.93
26	35.56	38.89	41.92	45.64	48.29
27	36.74	40.11	43.19	46.96	49.64
28	37.92	41.34	44.46	48.28	50.99
29	39.09	42.56	45.72	49.59	52.34
30	40.26	43.77	46.98	50.89	53.67

(*Continued*)

TABLE A.5

df	Proportion in critical region				
	0.10	0.05	0.025	0.01	0.005
40	51.81	55.76	59.34	63.69	66.77
50	63.17	67.50	71.42	76.15	79.49
60	74.40	79.08	83.30	88.38	91.95
70	85.53	90.53	95.02	100.42	104.22
80	96.58	101.88	106.63	112.33	116.32
90	107.56	113.14	118.14	124.12	128.30
100	118.50	124.34	129.56	135.81	140.17

Appendix B

Glossary of Statistical Symbols

These symbols have been used in this text and represent correct APA format. Note that, except for the null (H_0) and alternative (H_1) hypotheses, italics are used for most alphabetical (Roman) statistical symbols, but not for Greek symbols.

a	in a regression equation, the y-intercept
ANOVA	analysis of variance, symbolized by F for Fisher
b	in a regression equation, the slope
cov	covariance
d	Cohen's measure of effect size
D	difference between two means in a related-samples t-test
$\overline{D}$	average difference between two means in a related-samples t-test
df	degrees of freedom
E_{rc}	expected value for a cell in a chi-square test
f	frequency
f_e	expected frequency
f_o	observed frequency
F	Fisher's F ratio—test statistic associated with ANOVA
H_0	null hypothesis under test
H_1	alternative hypothesis
LSD	Fisher's least significant difference

M	mean (arithmetic average)
Mdn	median
MS	mean square
MS_B	mean square between
MS_W	mean square within
MS_E	mean square error; can be shown as MSE in APA format
n	total number in a sample
N	total number in a population
ns	not significant
p	probability
P	percentage; percentile
r	Pearson product-moment correlation
R	multiple correlation
R^2	multiple correlation squared; measure of strength of relationship, coefficient of determination
s	standard deviation for a sample
s^2	variance for a sample
SD	standard deviation
$S\overline{D}$	standard deviation of the difference between the means
SE	standard error (of measurement)
s_M	standard error of the mean for the sample

s_{M1-M2}	pooled standard error of the mean for two samples	η^2	eta squared; measure of strength of relationship
s_p^2	pooled variance for two samples	ρ	rho; symbol for correlation in the population
SS	sum of squares	Σ	sigma; sum or summation
t	computed value of t-test	σ	standard deviation for the population
x	symbol for the abscissa (X-axis) of a graph	σ^2	variance for the population
y	symbol for the ordinate (Y-axis) of a graph	σ_M	standard error of the mean for the population
z	a standard score	μ	mean for the population
α	alpha; the probability of a Type I error	x^2	squared value of x
β	beta; the probability of a Type II error; standardized multiple regression coefficient	χ^2	computed value of a chi-square test
		y^2	squared value of y
$1-\beta$	power	$'$	(prime) indicates an estimate of a statistic

Appendix C

Glossary

a priori planned comparisons; the researcher may select a somewhat liberal follow-up procedure, such as the **LSD**

alpha (α) the potential level of a Type I Error (incorrectly rejecting a true null hypothesis) that we are willing to accept when conducting a statistical test

alternative hypothesis (H_1) hypothesis that states that there is an effect, a difference, or a relationship between variables being tested

analysis of variance statistical procedure used when mean differences between two or more groups or two or more times of measurement are compared

ANOVA abbreviation for analysis of variance

bar graph a graph that uses separated bars to represent frequencies

between-groups designs designs that compare measures between different groups of participants

box plots plots that graphically present the 50th percentile and that identify the middle 50 percent of the scores and the highest and lowest scores

carry-over effect an effect that may occur if effects of one condition affect the conditions that follow

central limit theorem theorem that makes predictions about the characteristics of a distribution of sample means

chi-square test for goodness of fit (X^2) test that examines differences between observed frequencies and frequencies expected on the basis of prior knowledge

chi-square test for independence (X^2) test that examines whether two nominal variables are independent of each other in the population

coefficient of determination (r^2) statistical quantity that tells us how much of the variability in the Y variable is accounted for by the X variable

Cohen's *d* a measure of effect size, not affected by the size of the sample, that shows the magnitude or the strength of a treatment on a standard scale

contingency table table that presents frequencies for combinations of two nominal variables

correlation measure of the relationship between two variables

correlation coefficient (*r*) a value ranging between −1.00 and +1.00 that describes the linear relationship between two variables

correlational studies studies that examine relationships

counterbalancing varying the presentation of conditions in a repeated-measures design

covariance measure that indicates the degree to which two variables vary together

critical region the area, determined by the alpha level, in which we can reject the null hypothesis

degrees of freedom (*df*) the number of measurements that are free to vary

dependent-samples or related-samples *t*-tests tests that examine the change between two measurements taken over time or on related samples

dependent variables the outcomes that are measured

descriptive, or observational, studies studies that describe observations on a single case or group

descriptive statistics statistics in which the data presented are simplified to better understand them

directional alternative hypothesis hypothesis that specifies the direction of the expected outcome

455

effect size the magnitude of a treatment effect

eta squared (η^2) the proportion of the total score variance accounted for by the between-groups (independent) variable

expected frequency (f_e) expected number to be observed

experimental studies controlled studies in which participants are randomly assigned to levels of the independent variable(s)

factor another way to refer to a variable

factorial designs designs that allow for the simultaneous examination of the impact of two or more variables (or factors) on a dependent variable

fatigue effect effect that occurs when participants' performances become worse as they move through conditions because they become tired

frequency distribution data table or figure that summarizes and organizes data

frequency polygon a graph that uses data points to represent frequencies; the points are connected with lines

grand mean the mean of the means

grouped frequency distribution distribution used to organize data with large ranges of scores; data are reported in ranges or intervals, rather than as individual scores

histogram a graph that uses bars to represent frequencies; the bars touch

homogeneity of variance assumption that indicates that the variances of two or more samples are equivalent

hypothesis a testable statement about a possible outcome of a study

hypothesis testing statistical procedure that uses collected data

to systematically investigate research questions

independence of observations condition in which an observation is not related to any other observations within or between conditions

independent-samples *t*-tests tests that examine differences between two independent groups

independent variables variables that are manipulated or controlled by the researcher

inferential statistics statistics that allow us to generalize from a sample to a population

interaction effect that occurs when the impact of one variable changes at one or more levels of another variable

interquartile range the distance between the 25th percentile and the 75th percentile

interval level of measurement a scale on which the distances between any two units are equal

intervals ranges of values of equal size; taken together, intervals include all of the data in a data set

kurtosis property that describes the height of a distribution

least significant difference (LSD) a type of follow-up test used when the comparisons between means are planned prior to data analysis in a statistical test in which there are more than two levels of the independent variable(s)

least squares solution in regression, the solution that ensures that the distance between any point and a straight line is as small as possible

leptokurtic property of a distribution that is relatively tall and slender

level each condition within a factor

longitudinal study study in which the same participants are repeatedly measured over time

LSD see **least significant difference**

main effect the impact of one variable on the dependent variable

matched-samples *t*-tests tests that examine differences in matched pairs of measurements

mean (M) the arithmetic average

measures of central tendency measures that describe the typical score in a distribution

measures of variability measures that describe the distance between scores within a distribution

median (Mdn) the middlemost score

mesokurtic property of a distribution whose height is moderate

midpoint the score in the middle of a range

mixed-factors design design that combines repeated and independent factors in the same analysis

mode (Mo) the most frequently occurring score

multiple comparisons test used to identify where any significant differences exist when an ANOVA yields a significant value of F

multiple regression statistical procedure that allows us to relate more than one independent (predictor) variable to a dependent variable

negative correlation relationship according to which, as one variable increases, the other variable decreases; a perfect negative correlation would yield an r of -1.00

negatively skewed distribution an asymmetric distribution with relatively few values on the low end of the scale; in this kind of distribution, the mean is less than the median

nominal level of measurement level of measurement in which numbers are used to name or label variables, such as red = 1 and blue = 2

nondirectional alternative hypothesis hypothesis that does not specify the direction of the expected outcome

normal distribution a mathematically defined distribution, based on a hypothetical measurement of the entire population, that permits us to use the unit normal table to determine the critical region for a hypothesis test

null hypothesis (H_0) hypothesis that states that there is no effect, difference, or relationship among the variables tested

observed frequency (f_e) set of values based on the frequencies we actually observe

ordinal level of measurement rank order of a variable, such as movie ratings

outlier data point that is far outside of the range of the rest of the data

parameter mathematical characteristic of a population

partial eta squared (partial η^2) the proportion of the total variability attributable to a factor, taken as if it were the only variable

percentage a proportion multiplied by 100

percentile the score associated with a percentile rank

percentile rank the proportion at or below a particular score in a distribution

pie chart a circular graphic representation of the proportion (or percentage) of each category in a set of data

platykurtic property of a distribution that is relatively short in height

population everyone you are interested in

positive correlation relationship in which, as one variable increases, the other variable

increases; a perfect positive correlation would yield an *r* of +1.00

positively skewed distribution an asymmetric distribution with relatively few values on the high end of the scale; in this kind of distribution, the mean is greater than the median

post hoc comparisons not planned prior to data analysis; the researcher may select a somewhat conservative follow-up procedure, such as the **Scheffé** test

power the probability of correctly rejecting a false null hypothesis

practice effect effect that occurs when participants get better as they move through conditions

proportion (*p*) a decimal expressing the number of the total that share the same value

qualitative data in-depth verbal descriptions of behavior, attitudes, or knowledge

quantitative data numerical descriptions of behavior, attitudes, or knowledge

random sampling sampling in which all members of a population have an equal chance of being selected

range the distance between the highest and lowest scores in a distribution

ratio level of measurement variable that has an absolute zero; allows the researcher to make ratio comparisons

raw score an original value

regression equation equation that tells us how much change we would predict in *Y* for each one-unit change in *X*

regression line the line created by the least squares solution

related-samples or **dependent-samples *t*-tests** tests that examine the change between two measurements taken on the same or a matched sample

restriction of range a potential problem in a correlation resulting from a limited or truncated range in data

rho (ρ) the symbol for the correlation coefficient of a population

sample subset of a population; the group of people observed

sample of convenience sample in which the researcher uses participants that are readily available for study

sampling error the difference between a sample statistic and population parameter

scale scores scores transformed to a new scale to make them more easily interpreted

scales of measurement scales that describe the type of data collected

scatterplot, or **scatter diagram** a graph of the paired coordinates associated with the *X* and *Y* points

Scheffé test a type of *post hoc* comparison of means that is not specifically planned prior to data analysis and in which there are more than two levels of the independent variable(s)

sigma (Σ) or **summation notation** a symbol that specifies that whatever follows it must be added together

significant statistic statistic that indicates that the null hypothesis is rejected

simple distributions distributions that present all raw scores

simple effects effects that examine the impact of one variable at each level of another variable

simple regression statistical procedure that uses a formula to predict values of *Y*, for any given value of *X*

single-sample *t*-test test that compares a single sample mean to a population mean when the population standard deviation is not known

standard deviation (*s* or *SD*) the square root of the average squared distance each score is from the mean

standard error of the mean (*SEM*) an index of variability based on the *population* standard deviation; the standard deviation of a theoretical sampling distribution of means

standard score a score that has been transformed to a common metric (numbering system) with a mean of zero and a standard deviation of 1

statistic mathematical characteristic of a sample

statistical notation set of symbols that help label variables or that specify what operations to perform on data

statistics ways to organize, analyze, and interpret data

stem-and-leaf diagrams diagrams that plot data horizontally in a condensed form

sum of squares (*SS*) the sum of the squared deviations from the mean

summation notation, or **sigma (Σ)** a symbol that specifies that whatever follows it must be added together

symmetrical distribution a distribution in which the mean is equal to the median

Type I error error that occurs when we reject the null hypothesis and it is, in fact, true

Type II error error that occurs when we fail to reject the null hypothesis and it is, in fact, false

variable a characteristic or property that can take on different values

variance (s^2) the average of the squared deviations from the mean

weighted mean an average that takes into consideration the number in each group; the weighted mean will be closest to the mean of the group that contains the most scores

within-groups designs designs in which repeated measurements are taken on the same participants

z score a transformed score in which the mean equals 0 and the standard deviation equals 1

z statistic the distance an observed mean is from μ, expressed in standard units

Synthesis
Choosing the Appropriate
Statistic

This appendix constitutes a guide for choosing which statistic you should use. It is based on the type of data with which you are working and uses a simple checklist approach. The formulas for each statistic are also provided, as a reference and for those who would like to use hand calculations.

Begin by asking yourself what you are trying to test and whether your data are **continuous** or **categorical.** (See the glossary for definitions of these terms.) Then follow down each checklist.

Checklist	Example	Statistic and Chapter	Formula(s)
☐ One continuous dependent variable ☐ One set of scores ☐ I want to compute the average of the set of scores	Calculating the average height of a basketball team	Mean Chapter 3	Population $\quad \mu = \dfrac{\sum X}{N}$ Sample $\qquad M = \dfrac{\sum X}{n}$
☐ One continuous dependent variable ☐ Two or more sets of scores with different numbers of participants for each group ☐ I want to compute the average across both sets of scores	Comparing the overall average final-exam grade across four sections of introductory statistics	Weighted Mean Chapter 3	$M_{weighted} = \dfrac{(M_A)n_A + (M_B)n_B + (M...)n...}{n_A + n_B + n...}$
☐ One continuous dependent variable ☐ One set of scores ☐ I want to compute the midpoint of the set of scores	Determining the midpoint of salaries for a local business	Median Chapter 3	Put the scores in order of magnitude (e.g., low to high). With an odd number of scores, the score precisely in the middle is the median. With an even number of scores, the average of the two middle scores is the median.
☐ One categorical dependent variable ☐ One set of frequencies ☐ I want to determine the most frequent response	Determining the favorite flavor of ice cream for a group of 10-year-old children	Mode Chapter 3	Determine the most frequently occurring value.
☐ One continuous dependent variable ☐ One distribution of scores ☐ I want to know the range of the scores in the distribution	Determining the number of scores in a distribution of SAT verbal scores	Range Chapter 4	Take the highest value and subtract the lowest value.

Checklist	Example	Statistic and Chapter	Formula(s)
☐ One continuous dependent variable ☐ One distribution of scores ☐ I want to know the distance between the 25th percentile and the 75th percentile of the distribution	Determining the middle 50% of a distribution of SAT math scores	Interquartile Range Chapter 4	Take the value associated with the 75th percentile and subtract the value associated with the 25th percentile.
☐ One continuous dependent variable ☐ One distribution of scores ☐ I want to know the average squared distance between each score and the mean	Determining how much exam scores differ from the class average	Variance Chapter 4	Population $\sigma^2 = \dfrac{\sum(X-\mu)^2}{N}$ or $\sigma^2 = (\sigma)(\sigma)$ Sample $s^2 = \dfrac{\sum(X-M)^2}{n-1}$ or $s^2 = (s)(s)$
☐ One continuous dependent variable ☐ One distribution of scores ☐ I want to know the average distance between each score and the mean, in standardized form	Comparing IQ scores for a group of individuals with the known distribution of IQ scores in the population	Standard Deviation Chapter 4	Population $\sigma = \sqrt{\dfrac{\sum(X-\mu)^2}{N}}$ or $\sigma = \sqrt{\sigma^2}$ Sample $s = \sqrt{\dfrac{\sum(X-M)^2}{n-1}}$ or $s = \sqrt{s^2}$
☐ One continuous dependent variable ☐ One sample representing a population of scores ☐ The standard deviation of the population is known ☐ I want to compare my sample mean with the mean of the population or a hypothetical mean value	Comparing the current mean GPA at a university with the mean GPA of that university 25 years ago	z statistic Chapter 7	$z = \dfrac{M-\mu}{\sigma_M}$, where $\sigma_M = \dfrac{\sigma}{\sqrt{n}}$

Checklist	Example	Statistic and Chapter	Formula(s)
☐ One continuous dependent variable ☐ One sample ☐ The standard deviation of the population is not known ☐ I want to compare my sample mean with the mean of the population or a hypothetical mean value	Comparing the mean test anxiety score for a sample of students who have been taught relaxation techniques with the known mean anxiety score for all students	Single-sample t-test Chapter 8	$t = \dfrac{M - \mu}{s_M}$ where $s_M = \dfrac{s}{\sqrt{n}}$ $df = n - 1$
☐ One categorical dependent variable ☐ One sample ☐ I want to compare frequencies from my sample with those of the population or a hypothetical distribution	Comparing the number of males and females in the class with what is expected in the population	Goodness-of-fit chi-square (χ^2) Chapter 15	$\chi^2 = \sum \dfrac{(f_o - f_e)^2}{f_e}$ $df = c - 1$
☐ One continuous dependent variable ☐ One independent variable with two levels (two independent groups); participants are assigned to only one level ☐ Participants are measured once ☐ I want to compare the means of the two groups to find out whether they are significantly different	Comparing the mean heart rate of a group of people who exercise with the mean heart rate of a group of people who do not exercise	Independent-samples t-test Chapter 9	$t = \dfrac{M_1 - M_2}{s_{M_1 - M_2}}$ where $s_{M_1 - M_2} = \sqrt{\dfrac{s_p^2}{n_1} + \dfrac{s_p^2}{n_2}}$ and $s_p^2 = \dfrac{SS_1 + SS_2}{df_1 + df_2}$ $df = n_1 + n_2 - 2$ or $df = (n_1 - 1) + (n_2 - 1)$
☐ One continuous dependent variable ☐ One sample ☐ Participants are measured twice	Comparing pretest scores with posttest scores after a new instructional program is introduced	Related-samples (paired-samples) t-test Chapter 12	$t = \dfrac{\overline{D}}{s_{\overline{D}}}$ where $s_{\overline{D}} = \dfrac{s_D}{\sqrt{n}}$ $df = n - 1$

(Continued)

Checklist	Example	Statistic and Chapter	Formula(s)
☐ I want to compare any change in the means from the first measurement to the second measurement			$\bar{D} = \dfrac{\sum D}{n}$
☐ One continuous dependent variable ☐ Two samples that have been matched on one or more factors ☐ Each participant is measured once ☐ I want to compare the means of the matched samples	Comparing posttest scores of students in Classroom A with posttest scores of students in Classroom B, where the students have been matched on gender and IQ	Related-samples (paired-samples) t-test; in this situation, this is called a matched-samples t-test Chapter 12	$t = \dfrac{\bar{D}}{s_{\bar{D}}}$ where $s_{\bar{D}} = \dfrac{s_D}{\sqrt{n}}$ $df = n - 1 \quad \bar{D} = \dfrac{\sum D}{n}$ (note that this formula is the same as that for the related-samples test)
☐ One continuous dependent variable ☐ One independent variable having two or more levels (two or more independent groups); participants are assigned to only one of the levels ☐ Participants are measured once ☐ I want to compare the means of the groups to find out whether they are significantly different	Comparing mean ratings for four different types of apples; each participant tastes only one type of apple	One-way analysis of variance (ANOVA) Chapter 10 See ANOVA Table below	$F = \dfrac{MS_B}{MS_W}$ See ANOVA Table below

ANOVA Table

Source of Variation	df	Sums of Squares	Mean Squares
Between	number of groups − 1	$n\sum_j^a (M._j - \bar{\bar{M}})^2$	SS_B / df_B
Within	n − number of groups	$\sum_i^n \sum_j^a (X_{ij} - M._j)^2$	SS_W / df_W
Total	n − 1	$\sum_i^n \sum_j^a (X_{ij} - \bar{\bar{M}})^2$	

. = a cross/summed across
j = number of groups
n = number in sample
i = each individual person
$\bar{\bar{M}}$ = grand mean

Checklist	Example	Statistic and Chapter	Formula(s)
□ One continuous dependent variable □ Two or more levels of the independent variable □ Participants take part in all levels of the independent variable □ I want to compare the change in the means across each time measurements are taken	Comparing mean scores of participants in a program for reducing anxiety at the beginning, middle, and end of the program	Repeated-measures analysis of variance (ANOVA) Chapter 13	See ANOVA Table below
□ One continuous dependent variable □ Two (or more) independent variables □ Participants take part in only one level (group) for each independent variable	Comparing male and female dogs under age 1 with those over age 3 on the mean amount of time spent with a new type of chew toy	Factorial analysis of variance (ANOVA) Chapter 11	See ANOVA Table below

ANOVA Table

Source of Variation	df	Sums of Squares	Mean Squares	F
Repeated Measures (a)		$\dfrac{\sum_j \left(\sum_i x_{ij}\right)^2}{n} - \overline{\overline{M}}$	SS_a / df_a	MS_a / MS_e
Participants (S)		$\dfrac{\sum_i \left(\sum_j x_{ij}\right)^2}{a} - \overline{\overline{M}}$	SS_s / df_s	
Error		$SS_{TOTAL} - SS_a - SS_s$	SS_e / SS_s	
Total		$\sum_i \sum_j x_{ij}^2 - \overline{\overline{M}}$		

i = each individual
e = error
j = group
$\overline{\overline{M}}$ = grand mean

(Continued)

Checklist	Example	Statistic and Chapter	Formula(s)
☐ I want to determine whether the means of the groups by variable are significantly different		ANOVA Table	

ANOVA Table:

Source of Variation	df	Sums of Squares	Mean Squares	F
Main Effect A(α)	α groups − 1	$n\sum_j^a (M.j. - \overline{\overline{M}})^2$	SS_α/df_α	MS_α/MS_e
Main Effect B(β)	β groups − 1	$nj\sum_k^b (M..k - \overline{\overline{M}})^2$	SS_β/df_β	MS_β/MS_e
Interaction ($\alpha\beta$)	$(\alpha - 1)(\beta - 1)$	$n\sum_j^a \sum_k^b (M.jk - M.j. - M.k. + \overline{\overline{M}})^2$	$SS_{\alpha\beta}/df_{\alpha\beta}$	$MS_{\alpha\beta}/MS_e$
Error (within groups)	$\alpha\beta(n - 1)$	$\sum_i^n \sum_j^a \sum_k^b (X_{ijk} - M.jk)^2$	SS_e/df_e	
Total	$N - 1$	$\sum_i^n \sum_j^a \sum_k^b (X_{ijk} - \overline{\overline{M}})^2$		

i = individual
j = group A
k = group B
$\overline{\overline{M}}$ = grand mean
. = summed across
e = error

Checklist	Example	Statistic and Chapter	Formula(s)
☐ Two categorical dependent variables ☐ Two (or more) independent variables ☐ Participants take part in only one group (level) for each variable ☐ I want to compare whether the frequencies within the two groups are significantly different	Comparing frequencies of responses to a poll to determine whether a recreation center for teenagers should be built. Taxpayers could respond either "yes" or "no" to this question. The data were broken down into families with teenagers and families without teenagers.	Chi-square test of independence (χ^2) Chapter 15	$\chi^2 = \dfrac{(f_0 - f_e)^2}{f_e}$ $df = (c - 1)(r - 1)$ expected frequencies in a cell are calculated by $E_{rc} = \dfrac{f_r f_c}{N}$ where r = row C = column N = total number in sample rc = cell

Checklist	Example	Statistic and Chapter	Formula(s)
☐ Two dependent variables; at least one should be continuous ☐ One sample ☐ One value for each variable ☐ I want to determine whether there is a significant relationship between the two variables	Comparing the relationship between age and reaction time	Correlation Chapter 14	$$r = \frac{cov}{S_x S_y}$$ where $$cov = \frac{\sum(x - M_x)(y - M_y)}{n-1} =$$ $$cov = \frac{\sum xy}{n-1}$$
☐ One continuous dependent variable ☐ One independent variable ☐ I want to determine the degree to which the independent variable predicts the dependent variable	Determining whether visual acuity is a significant predictor of reading speed	Regression (Simple Regression) Chapter 14	$y' = bX + a$ where $$b = r\left(\frac{s_y}{s_x}\right)$$ $$a = M_y - bM_x$$
☐ One continuous dependent variable ☐ More than one independent variable ☐ I want to determine the degree to which the independent variables predict the dependent variable	Determining whether smoking, drinking, exercise, and/or eating yogurt are significant predictors of longevity	Multiple Regression Chapter 14	$y' = a + b_1 X_1 + b_2 X_2 + \cdots + b_j X_j$ for two variables: $$b_1 = \frac{\left(\sum x_2^2\right)\left(\sum x_{1y}\right) - \left(\sum x_1 x_2\right)\left(\sum x_{2y}\right)}{\left(\sum x_1^2\right)\left(\sum x_2^2\right) - \left(\sum x_1 x_2\right)^2}$$ $$b_2 = \frac{\left(\sum x_1^2\right)\left(\sum x_{2y}\right) - \left(\sum x_1 x_2\right)\left(\sum x_{1y}\right)}{\left(\sum x_1^2\right)\left(\sum x_2^2\right) - \left(\sum x_1 x_2\right)^2}$$ $$a = M_y - b_1 M_{x_1} - b_2 M_{x_2}$$

Index